THERMOTROPIC LIQUID CRYSTALS

Thermotropic Liquid Crystals

Recent Advances

Edited by

AYYALUSAMY RAMAMOORTHY

University of Michigan
Ann Arbor, Michigan, USA

 Springer

A C.I.P. Catalogue record for this book is available from the Library of Congress.

ISBN-10 1-4020-5327-4 (HB)
ISBN-10 1-4020-5354-1 (e-book)
ISBN-13 978-1-4020-5327-6 (HB)
ISBN-13 978-1-4020-5354-2 (e-book)

Published by Springer,
P.O. Box 17, 3300 AA Dordrecht, The Netherlands.

www.springer.com

Printed on acid-free paper

TABLE OF CONTENTS

PREFACE

Liquid crystalline materials are omnipresent in daily life. A broad spectrum of powerful applications of these exotic materials has created new avenues in academic and industrial research. Some of the common applications include display devices, temperature and pressure sensors, light valves and biosensors. Yet, there is considerable current interest in the design and development of novel liquid crystalline compounds with various functional properties. In addition, there is a significant interest in the characterization of these compounds at atomistic-level resolution using a variety of modern experimental, theoretical and computational approaches, which would aid the easy creation of high quality functional molecules. The mesogenic properties of liquid crystalline molecules are fascinating to spectroscopists and have been well utilized in the development of a variety of physical techniques including Nuclear Magnetic Resonance spectroscopy. Needless to mention that the increasing number of research teams, reports and meetings related to this interdisciplinary field is an indication of the wealth and remaining challenges of this rapidly growing field.

This book does not intend to cover the whole field of thermotropic liquid crystalline (TLC) materials as it is extremely difficult to cover within a single book. Instead it presents a collection of Chapters written by experts on various exciting topics in the field. Properties of recently developed TLCs (such as banana-type, thiophene-based, and columnar TLCs), phase biaxiality, and novel polymeric TLCs are discussed in detail. Solid-state NMR studies to obtain atomistic-level structural and geometrical information of TLCs are presented. Synthesis of liquid crystalline conjugated polymers, fast switching of nematic materials by an electric field, and photoconducting discotic systems are also presented.

It is my considerable pleasure to offer my thanks to all the authors for their wonderful contributions and the publishers for the help in developing the book. I thank my family for their help in bringing out this book. I also would like to thank my colleague and friend, Dr. Narasimhaswamy (Central Leather Research Institute, Chennai, India), who introduced me to this exciting field of research that has lead to the development of this book.

I sincerely hope researchers in both academia and industries will find the book to be useful for their research.

Ann Arbor,   *Ayyalusamy Ramamoorthy*
Michigan, USA   The University of Michigan
October 17, 2006

CHAPTER 1

MESOPHASE BEHAVIOUR AT THE BORDERLINE BETWEEN CALAMITIC AND "BANANA-SHAPED" MESOGENS

GERHARD PELZL AND WOLFGANG WEISSFLOG[1]

Institut für Physikalische Chemie, Martin-Luther-Universität Halle-Wittenberg,
Mühlpforte 1, 06108 Halle (Saale), Germany
[1] *E-mail: wolfgang.weissflog@chemie.uni-halle.de*

1.   INTRODUCTION

Organic compounds with pronounced shape anisotropy (rod-like or disc-like) are able to form thermotropic mesophases. Rod-like (calamitic) mesogens can exhibit nematic as well as different smectic phases characterized by a layered structure. Disc-shape mesogens preferably form columnar phases of different type characterized by a 2D superstructure. In rare cases they can also form nematic phases. Nematic phases may be characterized as the simplest liquid crystalline phase which is distinguished from the isotropic liquid only by the long-range orientational order. It is possible to describe the spontaneous parallel orientation of the molecules in nematic mesophases only by the anisotropy of repulsive forces coupled with the isotropic part of the dispersion interaction [1, 2]. It does not exclude that in dependence on the molecular structure – especially for the formation of smectic layer structures or columnar structures – other intermolecular interactions can play an important role (e.g. polar forces, charge-transfer complexes, hydrogen bonds). But in addition, the interaction between chemically incompatible moieties [3] as well as special steric interactions can clearly influence the structural features of mesophases. Special steric interactions may occur, if the molecular shape strongly deviates from the classical rod-like (or disc-like) shape. Elongated molecules with bulky branches at the ends of the molecules, the so-called polycatenar compounds are well-known examples. These compounds are able to form not only nematic or smectic phases but also mesophases with 2D superstructure (columnar phases) or 3D superstructure (cubic, tetragonal, rhombohedral phases) [4, 5]. Another impressive example of

1

A. Ramamoorthy (ed.), Thermotropic Liquid Crystals, 1–58.
© 2007 *Springer.*

a special steric interaction is molecules with a pronounced bent shape which are the topic of this chapter. Such substances were already synthesized by Vorländer et al. 1932 [6] and later by Akutagawa et al. [7], but it were Niori et al. [8] who discovered the polar properties of smectic phases formed by bent molecules. It was shown that due their bent shape the molecules can preferably be packed in bent direction giving rise to a long-range correlation of the lateral dipole moments, that means to a macroscopic polarization in the smectic layers.

In the following the most important mesophases formed by bent molecules shall be briefly characterized. In the SmAP phase (P means polar) the molecules are arranged perpendicular to the layers like in SmA phases but they possess a polar packing (in bent direction) within the layer plane. Depending on the direction of the polar axes in adjacent layers the SmAP phase can behave ferroelectric (polar axes parallel) or antiferroelectric (polar axes antiparallel) which is indicated by the subscripts F or A ($SmAP_F$, $SmAP_A$).

In most cases the polar packed bent molecules are tilted with respect to the layer normal. In analogy to the SmAP phase this phase is designated as SmCP which can either be ferroelectric ($SmCP_F$) or antiferroelectric ($SmCP_A$). But in addition, in the case of SmCP phases we have also to distinguish – with respect to the tilt direction – between synclinic or anticlinic interlayer correlation which is indicated in the phase symbol by the subscripts S or A after C. That means, four structural types of the SmCP phase are possible – SmC_SP_A, SmC_AP_A, SmC_SP_F, SmC_AP_F – which are schematically presented in Figure 1-1 [9]. Another aspect is of fundamental interest. The combination of director tilt and polar order leads to the chirality of smectic layers although the constituent molecules are achiral [9, 10]. As shown in Figure 1-1 two equivalent layer structures with antiparallel polar axes exist for a given tilt direction but these structures are mirror images from each other. In Figure 1-1 the opposite chirality is depicted by red or blue molecule symbols. In the so-called "racemic" states (SmC_SP_A, SmC_AP_F) the chirality alternates from layer to layer. In the "homochiral" states (SmC_AP_A, SmC_SP_F) the layer chirality

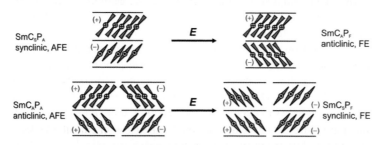

Figure 1-1. The molecular arrangement considering tilt and polarity in adjacent layers of bent-core molecules. The smectic layers are perpendicular to the drawing plane. The different molecule symbols ($\times$), ($\bullet$) correspond to opposite bent direction perpendicular to the drawing plane, i.e. to opposite polar axes. Red and blue molecule symbols indicate layers of different chirality. The arrows indicate the field-induced transition of the antiferroelectric into the ferroelectric state (AFE = antiferroelectric; FE = ferroelectric)

is uniform within a macroscopic domain. In order to escape from the macroscopic polarization, SmAP as well as SmCP phases have mostly an antiferroelectric ground state (alternating polarization in adjacent layers) which can be switched into corresponding ferroelectric states (indicated by arrows in Figure 1-1). This switching (also the switching between the opposite ferroelectric states) usually takes place by the collective rotation of the molecules around the tilt cone like in SmC* phase of calamitic compounds. It can be seen from Figure 1-1 that during the switching the chirality of the layers (red and blue, resp.) does not change.

Another way to avoid bulk polarization in layered structures of bent molecules is to break the layers into 2D modulated structures which can be regarded as columnar structures where the layer fragments represent the columns. In many cases the columnar phases of bent molecules have a rectangular cell (formerly this Col_r phase was also designed as B_1), but there are also examples of an oblique lattice (Col_{ob}) [11–16]. It seems that the columns consisting of layer fragments are polar ones but the macroscopic polarization is avoided by an antiparallel alignment of the polar axes.

It should be mentioned that further phases are formed by bent-core mesogens, which, however, have no relation to the subject of this chapter. Concerning the structure of the B_3, B_4, B_5, B_6 and B_7 phases, we refer therefore to three reviews on banana-shaped liquid crystals [12, 16, 17]

The assignment of the mesophases is based on different experimental techniques. Phase transitions are usually detected by calorimetry but also by polarizing optical microscopy. Textural features obtained by polarizing microscopy give first hints about the mesophase type. The most important method for the structural characterization is X-ray diffraction measurements. If possible these measurements should be performed on oriented samples. The polar properties are investigated by current response measurements, mostly using the triangular wave voltage method [18]. Information about the molecular dynamics can be obtained by dielectric measurements [19–21]. NMR spectroscopy allows the determination of the orientational order parameter by ^1H splitting or ^{13}C anisotropic shift measurements. Using the experimental order parameter the conformation of the bent molecules in the liquid crystalline state may be estimated by ^{13}C NMR [22–25], ^{19}F NMR [26] or ^2H NMR [26–28]. The knowledge of the conformation (especially of the bending angle) is important to understand basic relationships between molecular structure and the mesophase behaviour.

Since the discovery of the polar properties in smectic phases of bent-core mesogens in 1996 [8] chemists have synthesized a large variety of banana-shaped compounds with quite different structure. In most cases the chemical structure corresponds to the molecular design sketched in Figure 1-2.

The molecules consist of at least five rings connected by the linking groups X, Y. The two legs are attached to the central part, which introduces the bending angle in the whole molecule. Mostly the bending angle of about 120° results from the 1,3-phenylene disubstitution, as shown in Figure 1-2. The central phenyl ring can be replaced by naphthalene or by six- or five-membered heterocyclic

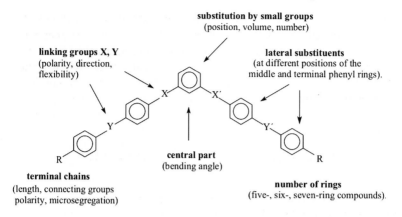

Figure 1-2. Molecular design of bent-core mesogens

rings (e.g. pyridine, oxazole, oxadiazole). But also a short odd-numbered spacer consisting of one (-O-; -S-; -CH$_2$-; -CO-) or three single units (e.g. -CH$_2$OOC-) can be used to introduce the bending angle. The terminal rings are substituted by long hydrocarbon chains in most cases. Aliphatic chains containing perfluorinated moieties or siloxane containing fragments increase the effect of microsegregation in the resulting mesophases. Polar as well as non-polar groups can be attached at the different lateral positions of the phenyl rings which is a successful way to create new phases and phase sequences in banana-shaped liquid crystals. It will be shown that the influence of lateral substituents is of great importance with respect to the subject of this chapter. More detailed information about the relationship between the chemical structure of bent-core mesogens and their mesophase behaviour has been summarized in some reviews [12, 16, 17, 29, 30].

In principle the molecular design to prepare banana-shaped liquid crystals sketched in Figure 1-2 works, but until now some essential questions could not be answered in a satisfying way. An important question is, for example: What are the structural preconditions of bent-core molecules to form definite phases with polar structure or polar structural fragments ("banana phases"). There are bent molecules which form only the classical mesophases of calamitic mesogens, that means nematic or smectic phases. On the other hand, there are bent-core compounds which are able to form nematic and/or smectic phases as well as "banana phases". This indicates that there is a more continuous borderline between bent-core and calamitic mesogens.

In order to get more insight into this borderline we present at first representative examples where "banana phases" as well as nematic and/or conventional smectic phases are observed at the same compound in quite different polymorphism variants. In cases where information about the detailed conformation of the molecules in the mesophases and its change with temperature are available, we will discuss the molecular origin of such complex phase behaviour. Furthermore, in a brief section we will show that also in binary systems between bent-core and suitable calamitic mixing components and in dimers consisting of a calamitic and a bent-core mesogenic unit phase sequences with "banana phases" and nematic and smectic phases can be realized. An important part is devoted to structural features of nematic or smectic phases of bent compounds which can be manifested in unusual properties, for example unusual electro-optical behaviour.

2.   VARIANTS OF POLYMORPHISM WITH SmCP$_A$, CONVENTIONAL SMECTIC (SmA, SmC) AND NEMATIC PHASES

2.1   Dimorphism SmA-SmCP$_A$ in Bent-core Mesogens with Perfluorinated Terminal Chains

The first examples of bent-core mesogens which form a polar SmCP$_A$ phase as well as a SmA phase were reported in 2000 [31]. These five-ring compounds possess a CH$_3$OOC-substituent in 5-position of the central core and perfluorinated terminal chains (Table 1-1).

The high-temperature smectic phase exhibits a homeotropic or a fan-shaped texture which is typical for a SmA phase. In contrast, the low-temperature phase forms a schlieren or a grainy texture, respectively, and shows a polar electro-optical response since the texture of the switched state depends on the polarity of the applied field. Current response measurements give evidence for an antiferroelectric ground state which can be switched into corresponding ferroelectric states. On the base of these experimental findings the low-temperature phase could be identified

Table 1-1. Transition temperatures ($^\circ$C) and enthalpy changes (kJ mol^{-1}; in square brackets below the temperatures) of the compounds **1a-1b**

No	n	Cr		SmCP$_A$		SmA		I
1a	9	•	160 [20.6]	•	165 [6.7]	•	250 [1.8]	•
1b	11	•	175 [19.0]	•	182 [8.8]	•	265 [1.9]	•

Cr: crystalline; SmCP$_A$: antiferroelectric polar smectic C phase; SmA: smectic A phase; I: isotropic liquid; The numbers between the phase symbols designate the phase transition temperatures in $^\circ$C, the numbers below which are the transition enthalpies in kJ mol^{-1} (in square brackets).

A

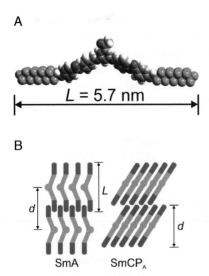

B

Figure 1-3. a) Molecular model of compound **1a**; b) Structure model of the SmA and SmCP$_A$ phase of compounds **1** (copyright (2000) from Liq. Cryst. by L. Kovalenko et al., ref. 31. Reprinted by permission of Tayler & Francis, LLC.http://www.taylerandfrancis.com/)

as an antiferroelectric SmCP phase. It was found that the layer spacing d (47 Å for compound **1a**) is clearly smaller than the molecular length L (57 Å for compound **1a**), but surprisingly d is nearly the same in the SmA and the tilted SmCP$_A$ phase. It is known from NMR measurements and X-ray studies on the SmA phase of analogous three-ring compounds that the difference between L and d is in the order of magnitude of the length of a terminal chain (similar to compounds **1a** or **1b**). Therefore it can be assumed that in the SmA phase the perfluorinated chains are more or less interdigitated (Figure 1-3). At the transition into the SmCP$_A$ phase the interdigitation disappears and the molecules are tilted now. That means, both smectic phases possess a different packing of the molecules in the layers. This clear structural change is also indicated by the relatively high transition enthalpies (see Table 1-1).

2.2   Polymorphism SmA-SmCP$_A$, SmA-SmC-SmCP$_A$ and N-SmA-SmC-SmCP$_A$ in Cyano-substituted Bent-core Compounds

Five-ring bent-core compounds derived from 4-cyanoresorcinol represent the second substance class where phase sequences with SmCP$_A$ phases and non-polar phases (N, SmA, SmC) have been observed [32].

It is seen from Table 1-2 that the dodecyl compound **2a** exhibits a dimorphism SmA-SmCP$_A$ like compounds **1a** and **1b**. The other compounds show trimorphism

Table 1-2. Transition temperatures ($°$C) and enthalpy changes [kJmol^{-1}] of the compounds **2a-2f**

No	R	Cr	SmCP$_A$	SmC	SmA	N	I
2a	C$_{12}$H$_{25}$	• 80 [13.2]	• 124 [1.3]		• 164 [6.8]	–	•
2b	C$_{14}$H$_{29}$	• 90 [19.5]	• 128 [1.8]	• 142 [–]	• 166 [7.6]	–	•
2c	C$_6$H$_{13}$O	• 122 [32.3]	• 103 [0.2]	• 133 [1.8]	• 156 [0.6]	• 165 [0.9]	•
2d	C$_8$H$_{17}$O	• 97 [37.4]	• 142 [1.9]	• 146 [2.1]	• 175 [4.8]	–	•
2e	C$_9$H$_{19}$O	• 62 [45.9]	• 134 [3.5]	• 138.5 [2.7]	• 180 [5.8]	–	•
2f	C$_{12}$H$_{25}$O	• 65 [14.7]	• 122 [4.8]	• 141 [–]	• 188 [7.2]	–	•

Cr: crystalline phase; I: isotropic phase; SmA, SmC: smectic A or smectic C phase, respectively; N: nematic phase; Parentheses indicate monotropic transitions.

SmA-SmC-SmCP$_A$ or tetramorphism N-SmA-SmC-SmCP$_A$ (compound **2c**). For the homologues **2c-2e** the phase transition SmA-SmC is of first order, the corresponding transition enthalpies vary between 1.8 and 2.7 kJmol^{-1}. The different smectic phases could be identified by polarizing microscopy, by electro-optical measurements and by X-ray investigations on oriented samples. The tilt angle of the SmC and SmCP$_A$ phase is found to be relatively low and does not exceed 15°. Figure 1-4 shows the temperature dependence of the orientational order parameter S for the nematic, SmA, SmC, and SmCP$_A$ phases of compounds **2c** determined by ^{13}C-NMR [32]. The bending angle – the opening angle between the two wings of the molecules – has been determined in the liquid-crystalline phases of compounds **2b** and **2c** by NMR studies. In the non-polar SmA and SmC phases the molecules possess a bent shape with a bending angle of about 140° which decreases to 135° in the SmCP$_A$ phase [32]. It can be assumed that in the SmA phase the bent molecules can freely rotate around their long-axes giving rise to the uniaxial structure. In the SmC phase the molecules are weakly tilted ($\approx$15°) but according to dielectric measurements [33] ferroelectric clusters in the short-range order region exist. At a definite temperature the bending angle obviously reaches a critical value where the rotational barrier becomes strong and the polar packing with a long-range correlation of the lateral dipoles is favoured.

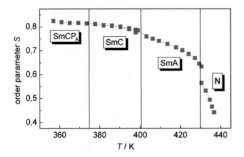

Figure 1-4. Temperature dependence of the orientational order parameter S in the nematic, SmA, SmC, and SmCP$_A$ phase of compound **2c**

Table 1-3. Transition temperatures (°C) of the compounds **3a-3d**

No.	n	A	B	Cr		SmCP$_A$		SmC		SmA		I
3a [34]	12	F	H	•	75	•	129	•	133	•	182	•
3b [35]	12	Cl	H	•	87	•	112	–		•	168	•
3c [35]	8	CH$_3$	H	•	110	(•	76	•	97)	•	117	•
3d [34]	8	H	CH$_3$	•	90	(•	79	–		•	107	•

In contrast to the compounds of series **1** the structural differences between the polar and non-polar smectic phases are less pronounced which is also indicated by the relatively low transition enthalpy SmA–SmCP$_A$ or SmA–mCP$_A$. The main reason for the transition from the polar into the non-polar phases seems to be the clear decrease of the bending angle with decreasing temperature.

Starting from the parent series **2**, the attachment of lateral substituents (F, Cl, CH$_3$) at the outer rings can result in similar phase sequences, see for example compounds **3a–3d** in Table 1-3.

2.3   Five-ring 4,6-Dichlororesorcinol Derivatives with Fluorine or Chlorine Substituents at the Outer Rings

Also the following homologous bent five-ring compounds derived from 4,6-dichlororesorcinol show a similar mesophase behaviour which is summarized in Table 1-4 [22, 36].

Table 1-4. Transition temperatures (° C) and enthalpy changes [kJ mol^{-1}] of the compounds **4a-4f**

No.	n	Cr		SmCP$_A$		SmC		SmA		N		I
4a	3	•	179 [75.9]	–		–		–		(•	(160) [1.3]	•
4b	8	•	127 [49.0]	(•	(95) [–]	–		•	130 [–]^a	•	130.5 [–]^a	•
4c	9	•	113 [44.4]	(•	(100) [0.5]	–		•	133 [4.2]	–		•
4d	10	•	108 [50.5]	(•	(103) [0.5]	–		•	137 [5.1]	–		•
4e	11	•	107 [52.8]	(•	(102) [0.3]	•	116 [–]^b	•	139 [5.7]	–		•
4f	12	•	103 [57.1]	(•	(100) [0.2]	•	125 [–]^b	•	139 [6.1]	–		•

^{a}) The calorimetric peaks of SmA-N and N-iso transition could not be resolved. The sum of these transition enthalpies: ΔH (SmA-N) + ΔH (N-iso) = 2.0 kJ mol^{-1}.
^{b}) The transition is not observable by DSC.

Three distinct variants of polymorphism occur in dependence on the length of the terminal chains: N–SmA–SmCP$_A$ for compound **4b**, SmA–SmCP$_A$ for compounds **4c** and **4d** and SmA–SmC–SmCP$_A$ for compounds **4e–4f** whereas the homologue with the shortest terminal chains **4a** forms a monotropic nematic phase, only. It follows from X-ray patterns of oriented samples and from electro-optical measurements that the tilt of the molecules in the SmC and SmCP$_A$ phase is not larger than 8°. Similar to the homologues of series **2** the layer spacing *d* is nearly the same in all smectic phases and clearly smaller than the molecular length L. Considering the very low tilt angle the results of X-ray measurements can be interpreted by a partial interdigitation of the terminal chains not only in the SmA phase but also in the tilted smectic phases SmC and SmCP. Electro-optical investigations give evidence for an antiferroelectric ground state of the SmCP phase; the switching polarization shows a strong temperature dependence which is unusual for SmCP$_A$ phases (Figure 1-5). On the other hand, the dielectric strength strongly increases on approaching the phase transition SmC → SmCP$_A$ (Figure 1-6a) or SmA → SmCP$_A$ (Figure 1-6b) which is characteristic for a transition from a paraelectric state (SmA, SmC) to a polar (ferro- or antiferroelectric) state. As shown in Figure 1-7 also in these compounds the bending angle ($\alpha = 180 - 2\varepsilon$; see formula in Table 1-4) decreases with decreasing temperature, e.g. it is 162° in the SmA phase and 155° to

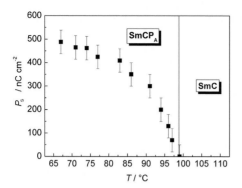

Figure 1-5. Temperature dependence of the switching polarization P_S in the SmCP$_A$ phase of compound **4f** (reprinted with permission from ref. 22, copyright, 2004, The PCCP Owner Societies)

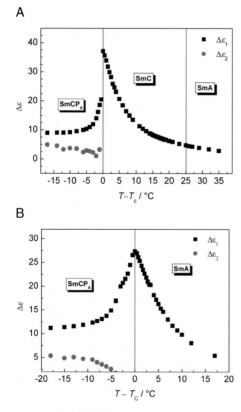

Figure 1-6. a) Dielectric strength in the SmA, SmC, and SmCP$_A$ phase of compound **4f** plotted versus $T - T_C (T_C = $ SmC $-$ SmCP$_A$ transition temperature); b) Dielectric strength in the SmA and SmCP$_A$ phase of compound **4d** plotted against $T - T_C (T_C = $ SmA $-$ SmCP$_A$ transition temperature) (reprinted with permission from ref. 22, copyright, 2004, The PCCP Owner Societies)

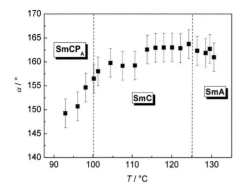

Figure 1-7. Temperature dependence of the bending angle α in the smectic phases of compound **4f** (reprinted with permission from ref. 22, copyright, 2004, The PCCP Owner Societies)

149° in the SmCP$_A$ phase. That means, similar to the 4-cyanosubstituted compounds of series **2** and **3** the direct transition from a non-polar (SmA, SmC) into a polar smectic phase (SmCP$_A$) is related to the continuous decrease of the bending angle.

The only structural difference between series **4** and the next series **5** are the lateral substituents at the outer rings: here fluorine atoms are replaced by chlorine atoms, see Table 1-5 [37].

Table 1-5. Transition temperatures (° C) and enthalpy changes [kJ mol^{-1}] of the compounds **5a–5g**

No	n	Cr		SmCP$_A$		SmC		SmA		N		I
5a	5	•	128 [29.7]	–		–		–		(•	114) [0.8]	•
5b	6	•	114 [39.6]	–		–		–		(•	113) [0.7]	•
5c	7	•	115 [25.9]	–		(•	68) [1.0]	–		•	111) [0.9]	•
5d	8	•	117 [65.9]	(•	71 [0.5]	•	91 [-]^a	•	104 [0.2]	•	109) [1.3]	•
5e	9	•	101 [35.0]	•	75 [0.5]	•	105 [-]^a	•	112 [3.2]	–		•
5f	10	•	103 [35.1]	(•	74) [0.4]	•	112 [-]^a	•	118 [4.7]	–		•
5g	12	•	106 [35.8]	(•	73) [0.4]	•	121 [-]^a	•	122 [5.1]	–		•

^{a}This transition is not indicated by a calorimetric signal.

A

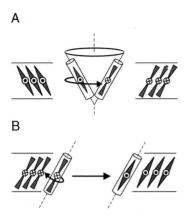

B

Figure 1-8. The possible mechanism of polar switching in the SmCP phases; a) Switching by collective rotation of the molecules around the tilt cone; b) Switching by collective rotation of the molecules around their long axes. Mechanism b) is accompanied by an inversion of layer chirality

Depending on the length of the terminal chains the homologues form a nematic phase, nonpolar smectic phases (SmA, SmC) and the polar smectic phase $SmCP_A$. It follows from X-ray measurements that the layer spacing d of the SmA phase is clearly smaller than the molecular length and it decreases only slightly on cooling into the tilted smectic phases which have a tilt angle not higher than 15°. Since the difference $L - d$ is in the order of magnitude of the length of the corresponding terminal aliphatic chain an intercalated structure of the smectic phases – as also found for the members of series **2** and **4** can be assumed. A significant difference to series **4** is the switching behaviour. The switching mechanism of the $SmCP_A$ phase clearly depends on the temperature. At lower temperatures the switching takes place in the usual way by rotation of the director around the tilt cone. At higher temperatures the switching is based on the collective rotation of the molecules around the long axes which corresponds to the field-induced inversion of the layer chirality (Figure 1-8). Until now, such switching mechanism has been reported for $SmCP_A$ phases of a few bent-core compounds, only[13, 38–41].

2.4 Polymorphism of the Non-polar SmC and/or the Nematic Phase with Several SmCP Phases

The homologues of the next series have a quite similar structure compared with series **2**, only the outer linking groups are ester groups instead azomethine groups [42]:

The common feature is that also in this series non-polar phases (N, SmC) as well as polar phases ($SmCP_A$) can occur at the same substance (Table 1-6).

Table 1-6. Transition temperatures (° C) and enthalpy changes [kJ mol^{-1}] of the compounds **6a-6g**

No	n		Cr	Col	SmCP″$_A$	SmCP′$_A$	SmCP$_A$	SmC	N	I
6a	6	•	105 [44.6]	–	–	–	–	–	• 140 [0.8]	•
6b	7	•	96 [37.7]	(• 68) [4.8]	–	–	–	–	• 132 [0.7]	•
6c	8	•	99 [52.0]	–	–	–	–	–	• 132 [0.8]	•
6d	9	•	92 [47.7]	–	–	–	–	–	• 128 [0.7]	•
6e	10	•	99 [66.1]	–	–	(• 66 [2.2]	• 77) [2.0]	–	• 128 [1.1]	•
6f	12	•	103 [65.7]	–	(• 68 [0.4]	• 75 [0.5]	• 94) [0.2]	• 109 [0.5]	• 129 [1.3]	•
6g	16	•	103 [65.7]	–	(• 74 [0.4]	• 79 [0.4]	• 99) [–]^a	• 133.5 [6.3]	–	•

^{a} This transition could not be detected by calorimetry.

But there are some significant differences. With exception of the hexadecyloxy compound all homologues listed in Table 1-6 exhibit a nematic phase but a SmA phase does not occur. The decyloxy homologue **6e** forms two antiferro-electric SmCP$_A$ phases whereas for the longer chained homologues the non-polar SmC and three SmCP$_A$ phases could be distinguished by calorimetry although the transition enthalpies are very low. On the other side, also the textural changes are very small and X-ray investigations showed no significant differences between the SmCP$_A$ phases. The phase sequences of the homologues **6e**, **6f** and **6g** were only observed in this series. Similar to the series **2** the terminal chains of the smectic phases are partially intercalated, but in contrast to series **2** the tilt angle of the smectic phases is clearly enhanced (between 30...35°). Another remarkable difference is the switching mechanism of the SmCP$_A$ phases. The field-induced transition from the antiferroelectric ground state into the switched ferroelectric states preferably takes place through a collective rotation of the molecules around their long axes. Such switching mechanism is accompanied by an inversion of the layer chirality by reversal of the field polarity (see series **5**). It should be noted that also the nematic phase shows some unexpected properties which are obviously related to the bent shape of the molecules and which will be discussed in section 6.1.

2.5 Liquid Crystal Tetramorphism with a SmA Phase and a Reentrant SmC$_S$P$_A$ Phase

In long-chained members of a new series of five-ring bent-core compounds which contain a chlorine substituent in 4-position of the central phenyl ring and only ester linking groups between the rings, an unusual polymorphism variant occurs: SmA–SmC$_S$P$_A$–Col$_{ob}$–SmC$_S$P$_A$, see Table 1-7 [43].

This phase sequence is elucidated by X-ray diffraction measurements and detailed electro-optical investigations. The SmC$_S$P$_A$ phases are structurally identical and differ only in the mechanism of polar switching. The switching of the high-temperature SmC$_S$P$_A$ phase is based on the collective rotation of the molecules around their long axes whereas the switching of the reentrant SmC$_S$P$_A$ phase takes place though the rotation of molecular long axes around the tilt cone. The Col$_{ob}$P$_A$ phase can be transformed to the SmC$_S$P$_F$ phase by the application of an electric field. The non-polar SmA phase shows a reversible field-induced transition into the SmC$_S$P$_F$ phase in a limited temperature range which will be discussed in section 7.2.

It is remarkable to notice that the short chain homologues **7a** and **7b** exhibit only a SmCP$_A$ phase, that means, lengthening the terminal chains (compounds **7c-e**) can induce a conventional smectic A phase in this series. That is in contrast to the relationships reported on banana-shaped liquid crystals up to now.

It should be noted that also in analogous compounds, however with 4,6-dichloro substitution, phase sequences with "banana phases" and nematic and conventional smectic phases occur, e.g. SmCP$_A$–SmA–N(n = 8); SmCP$_A$–SmC–SmA (n = 12) and SmCP$_A$–SmC (n = 16) [44].

Table 1-7. Transition temperatures (° C) and enthalpy changes [kJ mol^{-1}] of the compounds **7a-7e**

No	n	Cr		SmC$_S$P$_A$		Col$_{ob}$P$_A$		SmC$_S$P$_A$		SmA		I
7a	8	•	113.0 [20.7]	•	145.0 [16.0]	−		−		−		•
7b	12	•	110.0 [50.0]	•	147.0 [18.2]	−		−		−		•
7c	14	•	105.0 [60.3]	•	134.0^a [−]	•	138.0 [0.8]	•	139.5 [2.2]	•	144.0 [8.8]	•
7d	16	•	108.0 [74.9]	•	130.0^a [−]	•	135.0 [1.5]	•	138.5 [2.7]	•	145.5 [8.6]	•
7e	18	•	107.0 [71.9]	•	123.0^a [−]	•	128.0 [1.4]	•	134.0 [3.1]	•	142.5 [8.7]	•

^{a} The enthalpy for this transition could not be determined.

2.6 Benzoyl Derivatives of Secondary Cyclic Amines

There are non-linear compounds where the bend of the molecule is not achieved by 1,3-substitution of the central aromatic ring. Weissflog et al. presented a new class of bent compounds in which the central fragment consists of a benzoyl derivative of a secondary cyclic amine. In this case a carbonyl group links the phenyl ring of one leg with the nitrogen of the piperidine or piperazine ring which is part of the second leg [45, 46]. There are some longer-chained five-ring compounds which form $SmCP_A$ (in one case a Col_r phase), only. But the majority of the studied six-ring compounds **8** exhibit polymorphism variants with $SmCP_A$ or $SmAP_A$ as well as conventional smectic phases.

Compounds **8** X, X´: -N=CH-; -CH=N-
Y, Y´: -COO-; -OOC-
A = H, F, Br, CN, CH3
B = H, CH3

Because the clearing temperatures of the six-ring compounds are rather high, structural investigations and physical measurements are very difficult or impossible, e.g. in some cases $SmCP_A$ and $SmAP_A$ phases could not be distinguished since X-ray studies on oriented samples could not be performed. Depending on the linking groups, on lateral substituents and on the length of the terminal chains following phase sequences with decreasing temperature can occur: $SmA-SmAP_A/SmCP_A$; $SmA-SmAP_A/SmCP_A-SmA'$, $SmA-SmA''-SmCP_A$; $SmC-SmCP_A-SmCP_A'$. The structural differences between SmA and SmA' or SmA'' resp., or between $SmCP_A$- and $SmCP_A'$ are not yet clear [46].

3. POLYMORPHISM WITH POLAR BIAXIAL SmA PHASES

3.1 $SmAP_A$ Phases Formed by Terminal Non-polar Bent-core Mesogens

The existence of a polar SmA phase formed by bent-core mesogens has been predicted by Brand et al. [47]. For the octyloxy derivative of the following series the $SmAP_A$ phase was experimentally proved for the first time [48].

Table 1-8. Transition temperatures (° C) and enthalpy changes [kJ mol^{-1}] of the compounds **9a-9e**

$H_{2n+1}C_nO$ ... CN ... OC_nH_{2n+1}

No	n	Cr		SmAP$_A$		SmA		I
9a [48]	8	•	73 [74.6]	•	145 [1.2]	•	180.1 [16.7]	•
9b [49]	9	•	78 [30.8]	•	143 [0.4]	•	180 [8.1]	•
9c [49]	10	•	81 [33.7]	•	140 [0.7]	•	183 [8.2]	•
9d [49]	11	•	72 [20.5]	•	137 [0.6]	•	184 [8.1]	•
9e [49]	12	•	75 [18.2]	•	133 [0.7]	•	182 [8.1]	•

As seen from Table 1-8 the SmAP$_A$ phase occurs on cooling the SmA phase. If the SmA phase exhibits a fan-shaped texture this texture does not markedly change at the transition to the SmAP$_A$ phase (Figure 1-9b). Only some irregular stripes parallel to the smectic layers appear on further cooling.

However, if the SmAP$_A$ phase is formed from the homeotropically oriented SmA phase it adopts a strongly fluctuating schlieren texture indicating the biaxial nature of this phase, as shown in Figure 1-9a. The orthogonal alignment of the molecules could be clearly proved by X-ray measurements on an oriented sample. The polar SmA phase shows a current response characteristic for an antiferro-electric ground state. The switching polarization was found to be between 800 and 1000 nCcm^{-2} [48, 49].

If the lateral fluorine substituents at the outer rings in the compounds **9** are exchanged by chlorine or bromine also SmAP$_A$ phases occur in the phase sequence SmA-SmAP$_A$-SmCP$_A$, see compounds **10** in Table 1-9 [35].

It was found by X-ray diffraction measurements that the layer spacing is nearly the same in all smectic phases. Furthermore the tilt angle of the SmCP$_A$ does not exceed 10°. The strong temperature dependence of the spontaneous polarization is unusual (Figure 1-10).

In comparison to series **9** and **10**, the lateral substituents at the outer rings are missing in the following compounds **11** and the azomethine linking groups are replaced by ester groups, see Table 1-10 [42]. Also in these compounds a SmAP$_A$ phase occurs in the sequence SmA–SmAP$_A$. The switching polarization (640 and 320 nCcm^{-2} for compounds **11a** and **11b**, respectively) is clearly lower than that of the series **9**.

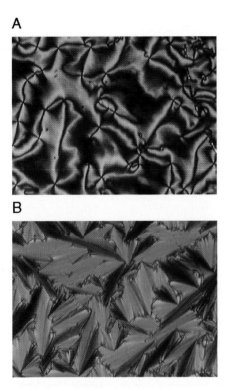

Figure 1-9. a) Schlieren texture of the SmAP$_A$ phase of compound **9a**; b) Fan-shaped texture of the SmAP$_A$ phase of compound **9a**

Table 1-9. Transition temperatures (° C) of the compounds **10a-10c**

No	n	A	Cr		SmCP$_A$		SmAP$_A$		SmA		I
10a	6	Cl	•	92	•	80	•	120	•	142	•
10b	8	Cl	•	91	•	105	•	117	•	156	•
10c	8	Br	•	89	•	90	•	96	•	146	•

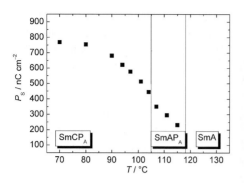

Figure 1-10. Temperature dependence of the switching polarization P_S in the SmAP$_A$ and SmCP$_A$ phase of compound **10b** (reprinted with permission from ref. 35, copyright, 2005, The Royal Society of Chemistry)

Table 1-10. Transition temperatures ($^\circ$C) and enthalpy changes [kJmol^{-1}] of the compounds **11a** and **11b**

No	n	Cr		SmAP$_A$		SmA		I
11a	8	•	113	•	144	•	187	•
			[32.4]		[1.2]		[8.4]	
11b	12	•	106	•	112	•	185	•
			[16.7]		[0.9]		[9.8]	

3.2   SmAP$_A$ Phases with a Partial Bilayer Structure (SmA$_d$P$_A$) Formed by Terminal Polar Bent-core Mesogens

Recently Shreenivasa Murthy et al. [50] have been reported non-symmetric bent-core compounds which contain a nitro group at one end of the molecule

Compounds **12**

The long-chain members of the series **12** ($n = 14, 16, 18$) form two SmA phases the layer spacing of which is significantly larger than the molecular length. This evidently implies a partial bilayer structure of the smectic layers which is also observed in terminal polar calamitic compounds. In contrast to the high-temperature smectic phase (SmA_d) the low-temperature smectic phase is biaxial and shows an antiferroelectric switching; it can be designated as SmA_dP_A.

The same phase sequence SmA_d-SmA_dP_A was found in structurally similar bent-core compounds **13** [51]. In Figure 1-11 the relationship between the length of the terminal chain and the occurrence of this phase sequence is shown for the series **13-I**.

Compounds **13**   **13-I**   A, B = H
                     13-II A = F; B = H (n = 6...18)
                     13-III A = H; B = F (n = 6...18)

These compounds were synthesized already in 2002 but the polar properties of the low-temperature phase were discovered later [52]. In the terminal polar six-ring bent-core compounds **14** also nematic phases occur.

Compounds **14**   **14-I**   A, B = H
                     14-II A = F; B = H
                     14-III A = H; B = F

Depending on the length of the terminal chains and the pattern of lateral substitution apart from the dimorphism SmA_d-SmA_dP_A new phase sequences N–SmA_dP_A and N–SmA_d–SmA_dP_A could be detected [52].

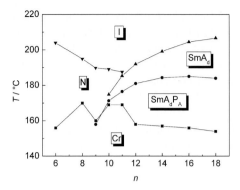

Figure 1-11. Transition temperatures of series **13-I** in dependence on the length of the terminal hydrocarbon chains (reprinted with permission from ref. 52, copyright, 2004, The Royal Society of Chemistry)

4.   DIRECT TRANSITIONS FROM THE NEMATIC TO POLAR SMECTIC OR COLUMNAR PHASES

4.1   Direct Transition from the Nematic Phase into the $SmCP_A$ Phase

There are a few bent-core compounds where the polar $SmCP_A$ phase (or another "banana phase") is directly transformed into a nematic phase. It is plausible that such transitions are not so frequently observed than transitions to non-polar smectic phases since the structural differences are more pronounced. The first examples of a dimorphism $SmCP_A$-N were reported in 2001 by Weissflog et al. [34] and Amaranatha Reddy et al. [53]. The bent compound described in ref. [34] is derived from 4-chlororesorcinol and possesses fluorine substituents at the outer rings.

15   Cr 71 $SmCP_A$ 99 N 103 I

The only special feature of the nematic phase is the occurrence of cybotactic smectic groups in which the molecules are tilted by 35°.

In ref. [53] three homologous bent-core compounds **16** derived from 2,7-dihydroxynaphthalene are presented which also show the direct transition $SmCP_A$–N, see Table 1-11.

Table 1-11. Transition temperatures (° C) and enthalpy changes [kJmol^{-1}] of the compounds **16a –16c**

No	n	Cr		SmCP$_A$		N		I
16a	12	•	119.5 [75.5]	(•	116) [13.2]	•	127 [0.49]	•
16b	13	•	117.5 [59.3]	•	120 [13.9]	•	125 [0.43]	•
16c	14	•	107 [58.6]	•	123 [14.1]	•	124 [0.34]	•

It is understandable that the transition enthalpies SmCP$_A$–N are clearly higher than those of transitions between the SmCP$_A$ and non-polar smectic phases.

Recently the same dimorphism has also been detected in an asymmetric bent-core compound where the bend of the molecules is not achieved by 1,3-substitution of a central aromatic ring but by a carbonyl group which is linked to one of the nitrogen atoms of a central piperazine ring [45, 54], as shown in Tables 1-12 and 1-13.

Table 1-12. Transition temperatures (° C) and enthalpy changes [kJ mol^{-1}] of the compounds **17a-17c**

No	n	Cr		SmCP$_A$		SmA		N		I
17a	6	•	187 [28.3]	–		(•	178) [0.2]	•	201 [0.2]	•
17b	8	•	177 [30.6]	•	198 [0.5]	–		•	201 [0.6]	•
17c	12	•	164 [47.4]	•	215 [0.5]	–		–		•

Table 1-13. Transition temperatures (°C) and enthalpy changes [kJ mol^{-1}] of the compounds **18a-18f**

No	n	Cr		SmCP$_A$		Col		N$_X$		N		I
18a	4	•	201 [34.4]	–		–		(•^{a}	193) [–]	•	212 [0.5]	•
18b	5	•	187 [46.7]	–		–		(•	172) [0.35]	•	192 [0.4]	•
18c	6	•	176 [41.5]	–		(•	157 [5.2]	•	169) [0.3]	•	188 [0.5]	•
18d	7	•	169 [23.2]	•		–		–		–	186 [0.4]	•
18e	8	•	177 [15.4]	•	180 [12.8]^b	–		–		•	180.5	•
18f	9	•	167 [15.4]	•		–		–		–	185 [15.2]	•

^{a} The N$_X$ phase crystallizes immediately after its formation.
^{b} This value corresponds to the sum of the SmCP$_A$–N and N–I transitions.

As demonstrated for the three homologues **17a-17c**, the phase type changes from non-polar conventional phases to the polar "banana phase" SmCP$_A$ with increasing chain length. Only the octyloxy derivative **17b** exhibits the dimorphism N-SmCP$_A$ [54]. On cooling the isotropic liquid of compound **17c** the SmCP$_A$ phase forms a non-birefringent texture which exhibits randomly distributed domains of opposite handedness. Such texture could also be induced above the clearing temperature applying a sufficiently high electric field (see section 7.1) [54].

The homologues **18** with a chain length larger than octyloxy form a SmCP$_A$ phase, only. For homologues with shorter terminal chains (butyloxy to hexyloxy) two different nematic phases could be detected which will be discussed in section 6.1. For the homologue with a middle length of the terminal chains (compound **18e**) the phase sequence SmCP$_A$–N could be detected where the nematic phase exists in a small existence range, see Table 1-13 [45].

A transition between a polar SmCP-like phase and a nematic phase is also observed in the following derivative of 4-chlororesorcinol **19b** (Table 1-14) [55].

The tilt angle of the SmCP phase of compound **19b** is rather high (45°). It follows from ^{13}C NMR measurements that the bending angle of the molecules in the nematic as well as in the smectic phase is about 140°. In the fan-shaped texture of the smectic phase domains of opposite handedness spontaneously arise

Table 1-14. Transition temperatures ($^{\circ}$C) and enthalpy changes [kJmol^{-1}] of the compounds **19a-19b**

No	n	Cr		SmCP		N		I
19a[56]	12	•	98 [38.7]	(•X	80 [9.1]	•	95) [0.7]	•
19b[55]	16	•	99 [34.8]	(•	88 [3.5]	•	92) [1.2]	•

on cooling the nematic phase. The tilted smectic phase shows no noticeable polar electro-optical response above 80° C. But below 80° C the fan-shaped texture can be transformed into an optically isotropic state by application of a sufficiently high electric field. This optically isotropic texture displays randomly distributed domains of opposite handedness. The most remarkable finding is that the chiral domains can be reversibly switched into a state of opposite handedness depending on the field polarity. This switching is bistable and accompanied by a ferroelectric current response (switching polarization: 100 nC cm^{-2}) [55].

Also the short-chained homologues of series **19** (nonyloxy to dodecyloxy) show an interesting transition between a nematic phase and a "banana phase", see for example the dodecyloxy homologue **19a** [56, 57]. The highly viscous "banana phase", designated by the code letter X, is optically isotropic and spontaneously forms domains of opposite handedness. By application of a high electric field (40 Vμm^{-1}) a ferroelectric switching could be proved ($P_S \approx 500$ nC cm^{-2}) [58]. On the base of X-ray diffraction measurements and electro-optical studies a structure model is proposed according to which the isotropic phase consists of weakly inter-connected orthoconic racemic smectic granules with random directions [58].

It should be noted that the nematic phase of these compounds shows some unusual properties which are obviously related to the bent shape of the molecules. On the one side it spontaneously forms large domains of opposite handedness; on the other side, the application of an electric field leads to unusual textures which have never been observed in nematic liquids formed by calamitic molecules. A detailed description of this behaviour will be given in section 6.1.

Gesekus et al. [59] reported a compound **20** which possesses a sugar moiety as a chiral bent core. This compound forms a cholesteric (N*) phase and in addition a smectic phase which shows an antiferroelectric switching with a relatively low switching polarization (30...37 nC cm^{-2}). It is not yet clear if this switchable smectic phase is a "banana-phase" or a chiral SmC* phase.

20 Cr 134 SmX* 140 N 180 I

4.2   Direct Transition from the Nematic Phase into the SmA$_d$P$_A$ Phase

There are only few examples exhibiting a direct transition from the polar SmAP phase to the nematic phase. Such compounds are homologues of the series **14**, presented in connection with polar biaxial smectic A phases (section 3.2). It is remarkable that in all three series **14-I**, **14-II**, and **14-III** the compounds with terminal octyl- and/or nonyl-chains show this direct transition from the SmA$_d$P$_A$ phase into the nematic phase (Table 1-15) [52].

4.3   Variants of Polymorphism with Columnar and Nematic Phases
or Conventional Smectic Phases

Already in 1999 a homologous series of five-ring bent-core mesogens was presented where the short-chained members form only a nematic phase whereas the longer homologues show the phase sequence Col$_{ob}$–N, Col$_{ob}$–SmC–N, and Col$_{ob}$–SmC,

Table 1-15. Transition temperatures (° C) and enthalpy changes [kJ mol^{-1}] of selected derivatives of the compounds **14**

No	n	A	B	Cr		SmA$_d$P$_A$		N		I
14-I-8	8	H	H	•	170.0 [50.2]	(•	151.5) [–]^a	•	195.0 [0.45]	•
14-I-9	9	H	H	•	160.0 [39.2]	(•	158.0) [0.48]	•	189.8 [0.44]	•
14-II-9	9	F	H	•	160.0 [52.3]	(•	152.5) [0.5]	•	183 [0.4]	•
14-III-9	9	H	F	•	154.0 [35.9]	(•	149.0) [0.23]	•	183.1 [0.35]	•

^{a} The enthalpy could not be determined as the sample crystallizes immediately.

Table 1-16. Transition temperatures (°C) and enthalpy changes [kJ mol^{-1}] of the compounds **21a-21g**

No	R	Cr		Col$_{ob}$		SmC		N		I
21a	$C_6H_{13}O$	•	127 [46.2]	•		–		•	165 [1.8]	•
21b	$C_7H_{15}O$	•	115 [35.9]	•		–		•	153 [1.3]	•
21c	$C_8H_{17}O$	•	126 [45.7]	•		–		•	148 [0.9]	•
21d	$C_9H_{19}O$	•	106 [13.9]	(•	88) [1.6]	–		•	143 [0.5]	•
21e	$C_{10}H_{21}O$	•	111 [43.8]	(•	97) [–]	–		•	140 [1.5]	•
21f	$C_{12}H_{25}O$	•	111 [47.2]	•	113 [–]	•	121 [0.9]	•	137 [1.9]	•
21g	$C_{14}H_{29}$	•	94 [33.5]	•	100 [–]	•	117 [6.1]	–		•

respectively, see Table 1-16 [24]. The Col$_{ob}$ phase, in the original paper assigned as undulated SmC phase, was identified on the base of X-ray measurements on oriented samples. By re-investigation it was found that the Col$_{ob}$ phase shows a polar switching which points to an antiferroelectric ground state. The polarization of the tetradecyl derivative **21g** adopts a value of 450 nC cm^{-2} at 60° C. Interestingly, a temperature dependence of the switching mechanism could be detected. At higher temperature the polar switching takes place by the collective rotation of the molecules around their long axes, whereas at lower temperatures (<60° C) the switching is based on the director rotation around the tilt cone (see Figure 1-8).

In the same year, a bent-shaped seven-ring compound **22** which exhibits the dimorphism B$_1$–N (B$_1$ means Col$_r$) was reported by Shen et al. [60].

22   Cr 149 Col$_r$ 197 N 217 I

In 2000 Amaranatha Reddy et al. [61] presented three series of bent-core compounds with a central naphthalene ring. The homologues show different phase sequences depending on the lateral substituents and the length of the terminal chains. It is seen from Table 1-17 that seven compounds exhibit the dimorphism Col_r–N. For two homologues the sequence Col–B_6–N was observed for the first time. The B_6 phase is an intercalated SmC phase formed by bent molecules [12].

Figure 1-12 shows the transition temperatures of the full homologous series including the compounds **23a** and **23b**. Such curves of the B_1–I (that means Col_r–I) and $SmCP_A$–I transition temperatures not overlapping to each other have been relatively often found for bent-core mesogens.

In 2004 the Bangalore group [63] presented three further structurally similar series **24-I** to **24-III** with a central naphthalene ring, however cinnamate fragments

Table 1-17. Transition temperatures (°C) and enthalpy changes [kJ mol^{-1}] of the compounds **23a-i** [61, 62]

No	R	A	B	E	Cr		Col_r	B_6	N		I
23a	C_2H_5	H	F	H	•	188.5 [49.7]	(• 137.5 [–]^a	–	•	170.0) [0.13]	•
23b	C_3H_7	H	F	H	•	171.0 [52.7]	(• 155.0 [10.7]	–	•	174.5 [0.20]	•
23c	C_5H_{11}	F	H	H	•	154.5 [44.9]	(• 145.5 [–]	–	•	150.5) [0.26]	•
23d	C_4H_9O	H	H	H	•	197.5 [39.3]	• 210.0 [14.1]	–	•	213.0 [0.5]	•
23e	C_3H_7O	H	F	H	•	162.0 [29.9]	• 176.0 [0.5]	• 191.5 [9.6]	•	204.5 [0.52]	•
23f	C_4H_9O	H	F	H	•	142.0 [40.3]	• 192.0 [0.2]	• 196.5 [12.6]	•	197.5 [0.32]	•
23g	$C_5H_{11}O$	F	H	H	•	182.0 [51.4]	(•$^{b)}$ 156.5) [10.8]	–	•	169.0 [0.23]	•
23h	$C_6H_{13}O$	F	H	H	•	156.0 [42.9]	•$^{b)}$ 159.0 [10.8]	–	•	163.0 [0.28]	•
23i	C_4H_9O	H	H	F	•	182.0 [47.0]	(• 169.5) [11.0]	–	•	194.0 [0.41]	•

^{a} The enthalpy could not be determined.
^{b} A second Col phase is listed for compounds **23g** and **23h** [62].

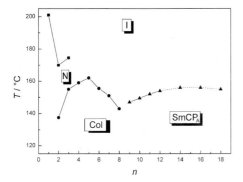

Figure 1-12. Transition temperatures in dependence on the length of the terminal alkyloxy chains of the full series including the compounds **23a** and **23b** (copyright (2000) from Liq. Cryst. by R. Amaranatha Reddy et al., ref. 61. Reprinted by permission of Tayler & Francis, LLC.http://www. taylerandfrancis.com/)

in the legs. The members of these series show different phase sequences depending on the lateral substitution and the length of the terminal chains.

Compounds **24-I** to **24-III**

series **24-I** $E = H$; $A = H$; $B = F$ $n = 5, 6$ Col_r-B_6-N
                                              $n = 7,8$ Col_r-N
                                              $n = 9...14$ Col_{ob}-N

series **24-II** $E = CH_3$; $A = H$; $B = F$ $n = 4...9$ Col_r-N
                                              $n = 11...14$ $SmCP_A$-N

series **24-III** $E = CH_3$; $A = F$; $B = H$ $n = 12$ Col_r-N
                                              $n = 13,14$ $SmCP_A$-N

It should be noted that in a further bent-core naphthalene series (compounds **25**) with different wings at each side the sequence Col_r–N was found for the short-chain members ($n = 2...8$) and $SmCP_A$–N for the long-chain members ($n = 9, 10$), as shown in Figure 1-13 [64].

Recently, a dimorphism Col_r–N has been reported by Niori et al. [65] for a six-ring bent-core mesogen **26** with a lateral methoxy group at the central core.

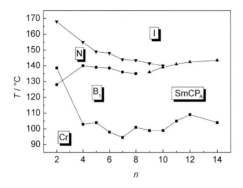

Figure 1-13. Transition temperatures of series **25** in dependence on the chain length (copyright (2004) from Liq. Cryst. by H.N. Shreenivasa Murthy et al., ref. 64. Reprinted by permission of Tayler & Francis, LLC.http://www.taylerandfrancis.com/)

25

26 I 201 N 162 Col$_r$ 56 Cr

An interesting mesophase behaviour has also been observed in a series of 4-bromoresorcinol derivatives **27** [66]. As usual, the homologues with longer terminal chains form the polar SmCP$_A$ phase and the short-chain members form a columnar phase with a hexagonal lattice. At higher temperatures (i.e. above the Col$_h$ phase) an additional mesophase has been found which shows a fan-shaped texture typical for an SmA or SmC phase but in the X-ray pattern sharp layer reflections do not occur. Therefore this phase was assigned as a nematic phase (designated as

N_Y) which is reminiscent of the N_X phase of some members of series **18** described in section 4.1 [45].

27	n = 4, 5	Col$_h$-N$_Y$
	n = 6...8	SmX
	n > 8	SmCP$_A$

There are also cases in which the Col$_r$ banana phase occurs together with conventional smectic phases. One example is the five-ring mesogen **28** with –CH=CH- linking groups between the outer rings and bearing lateral substituents, this compound shows a dimorphism Col$_r$–SmC [13].

28   Cr 142 Col$_r$ 150 SmC 151 I

A very interesting mesophase behaviour is reported for an asymmetric six-ring bent-core compound **29** in which a chiral group is appended to one of the terminal positions [67].

29   Cr 114.5 Col$_{ob}$P$_A$* 119.0 SmC$_\gamma$* 141.0 SmC* 147.3 SmX 148.0 I

The low-temperature phase could be identified as an antiferroelectric oblique columnar phase. Within the layer fragments of the columnar phase the molecules are tilted by 22°. Above the temperature range of the columnar phase a ferrielectric SmC$_\gamma$* and a ferroelectric SmC* phase appear. The high-temperature phase is probably a SmC$_\alpha$* phase. The spontaneous polarization in the smectic phases is rather low: 25 nC cm^{-2} in the SmC* phase and 40 nC cm^{-2} in the SmC$_\gamma$* phase. At the transition into the Col$_{ob}$P$_A$* phase the polarization value increases drastically to about 250 nCcm^{-2}. The authors explain the special phase sequence by an increase of the rotational barrier of the aromatic parts with decreasing temperature. In order to escape from bulk polarization the organization of the layer fragments in the columnar phase is antiferroelectric. In contrast, in the conventional chiral smectic C phases (SmC$_\gamma$*, SmC*) the macroscopic polarization can be compensated by a helical structure. It is interesting to note that the racemate of the compound possesses an analogous phase sequence. The only difference is that the smectic phases show no polar switching [67].

5.   BENT-CORE AND CALAMITIC MESOGENS – MIXED WITH EACH OTHER OR COVALENTLY LINKED TO EACH OTHER

5.1   Transitions Between "Banana Phases" and Nematic and/or Conventional Smectic Phases in Binary Mixtures

It is shown in this chapter that depending on structural features bent-core compounds cannot only form typical "banana phases" (SmCP, SmAP, Col) but also nematic or conventional smectic phases. Soon after the discovery of such compounds in 2000 [31, 33] the question arose if such polymorphism variants can also be realized in binary systems where one mixing component is a bent-core compound and the other one is a calamitic compound. The first preliminary studies in this direction were not so successful. In some cases the phase region of the "banana phase" was clearly separated by isotropic regions from the regions of the nematic or smectic phases. In most cases the heterogeneous regions between the incompatible phases were strongly broadened so that complete transitions between these phases could not be observed. But in the intermediate time it could be shown that by suitable selection of the mixing components it is indeed possible to realize such polymorphism variants. This will be illustrated on the base of three representative binary systems where the bent-core compound **30** (mixing component **A**) forms a SmCP$_A$ phase at relatively low temperatures [68]:

30 (Mixing component **A**)   Cr 76 SmCP$_A$ 130 I

In the first binary system **A/B** [69] the calamitic compound **31** is a thiadiazole derivative [70] designated as mixing compound **B** which forms a SmC phase in a wide temperature range.

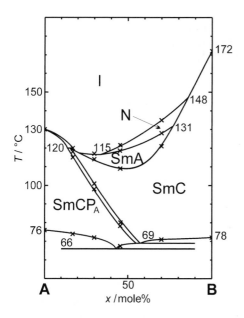

31 (Mixing component **B**) Cr 78 SmC 172 I

It is seen from the isobaric phase diagram in Figure 1-14 that at middle concentrations an intermediate nematic and SmA phase appear. Depending on the concentration different phase sequences can be observed: $SmCP_A$–SmC (78–80 mole% **A**); $SmCP_A$–SmC–SmA (68–78 mole% **A**); $SmCP_A$–SmC–SmA–N (45–65 mole% **A**). The mesophases have been distinguished by their characteristic textures (Figure 1-15), by X-ray and by the electro-optical behaviour [69].

Figure 1-14. Isobaric composition – temperature phase diagram for the binary system with components **A** and **B** (copyright (2002) from Liq. Cryst. by M.W. Schröder et al., ref. 69. Reprinted by permission of Tayler & Francis, LLC.http://www.taylerandfrancis.com/)

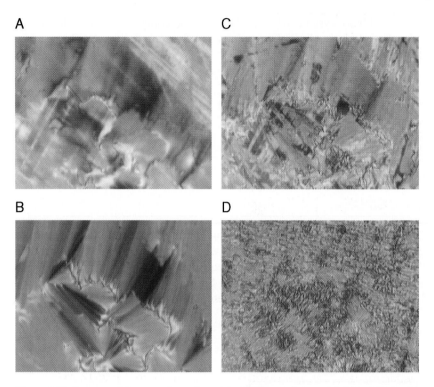

Figure 1-15. Textures of the mesophases formed in the binary mixture with 54 mole% **A**; a) nematic phase (119° C); b) SmA phase (115° C); c) SmC phase (85° C); d) SmCP$_A$ phase (79° C) (copyright (2002) from Liq. Cryst. by M.W. Schröder et al., ref. 69. Reprinted by permission of Tayler & Francis, LLC.http://www.taylerandfrancis.com/)

In the second binary system **A/C** the calamitic compound **32** (mixing component **C**) [71] is also a thiadiazole derivative which forms a nematic phase in a wide temperature range:

H_5C_2O — N–N / S — C_4H_9

32 (Mixing component **C**)   Cr 95 N 205 I

The isobaric phase diagram **A/C** is shown in Figure 1-16. In mixtures between 41 and 65 mole% **A** there is a direct transition from the SmCP$_A$ phase into the nematic phase (texture see Figure 1-17a). At concentrations <40 mole% **A** the SmCP$_A$ phase disappears and instead of this a columnar phase appears (texture see Figure 1-17b) which has a rectangular 2D cell where the molecules are tilted by

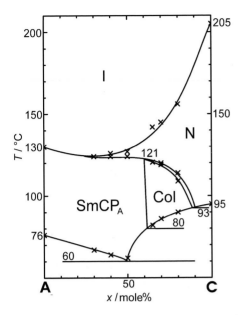

Figure 1-16. Isobaric composition – temperature phase diagram for the binary mixture system with components **A** and **C**

≈30° within the layer fragments of the columnar structure. Up to 12 mole% **A** a phase transition Col$_r$ to N can be observed [72].

In the third example **A/D** the mixing component of the bent-core mesogen **A** is trinitrofluorenone **33** which does not form a mesophase.

33 (Mixing component **D**)   Cr 175 I

It is seen from the phase diagram (Figure 1-18) that between 40 and 85 mole% **A** a SmA phase is induced which is obviously the result of electron-donor-acceptor interactions. On cooling the SmA phase between 85 and 75 mole% **A** the transition into the SmCP$_A$ phase occurs. At lower concentrations another transition takes place where the fan-shaped texture of the SmA phase remains unchanged while the homeotropic texture adopts a schlieren texture which indicates a biaxial structure. X-ray studies on oriented samples give evidence that not only in the SmA but also in the low-temperature phase the molecules are arranged perpendicular to the layer

A

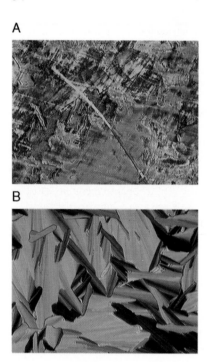

B

Figure 1-17. Mixture **A/C** containing 29 mole% component **A** a) Texture of the nematic phase b) texture of the Col$_r$ phase

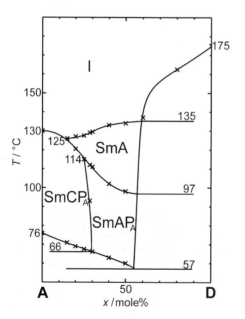

Figure 1-18. Isobaric composition – temperature phase diagram for the binary mixture **A** and **D** (reprinted with permission from ref. 73, copyright, 2002, The Royal Society of Chemistry)

planes. Since the biaxial low-temperature phase shows an antiferroelectric current response it can be assigned as $SmAP_A$ phase. In a limited concentration range (70.5–72 mole% **A**) the phase sequence $SmA-SmAP_A-SmCP_A$ could be detected before this sequence was found for pure compounds (textures see Figure 1-19) [73].

In a binary system **E/F** of a bent-core compound **34** (mixing component **E**) with a calamitic compound **35** (mixing component **F**) three additional mesophases are induced in the mixed phase region. For compositions between 15 and 63 mole% of the calamitic compound the columnar B_1 phase, for mixtures between 63 and 87 mole% of component **F** the intercalated B_6 phase is observed. In the concentration range between 87 and 95.5 mole% compound **F** a new biaxial SmA_b phase appears which is the low-temperature phase with respect to the uniaxial SmA phase. According to X-ray studies both SmA phases possess a bilayer structure and were designated as SmA_2 and SmA_{2b}, respectively [74, 75].

34 (Mixing component **E**)   Cr 114 $SmCP_A$ 126.5 I

35 (Mixing component **F**)   Cr 98.5 SmA 109.5 N 125 I

5.2 Further Binary Systems with Bent-core Mixing Components

Concerning mixing of bent-core compounds with calamitic mesogens or non-mesogenic compounds there are several other interesting aspects.

Gorecka et al. [76] showed that it is possible to transform a ferroelectric SmC* phase to an antiferroelectric SmC_A* phase by doping the SmC* phase with a bent-core compound. If the tilt angle of the SmC* phase is ≈45° the addition of a bent-core dopand leads to the formation of a SmC_A* phase in which the tilt angle of 45° is maintained. Such a SmC_A* phase is optically isotropic when the layers are perpendicular to the substrates. Application of an electric field leads to the transition into the birefringent bright ferroelectric state [77].

A

B

C

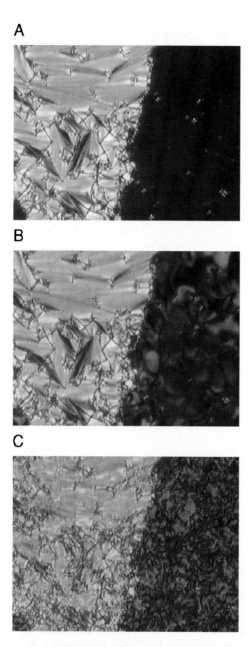

Figure 1-19. Textures of the smectic phases for a sample with 71.5 mole% **A;** a) fan-shaped texture and homeotropic tecture of the SmA phase (115° C); b) fan-shaped texture and fluctuating schlieren texture of the SmAP$_A$ phase (96° C); c) fan-shaped and schlieren texture of the SmCP$_A$ phase (75° C) (reprinted with permission from ref. 73, copyright, 2002, The Royal Society of Chemistry)

In this context it should be noted that by addition of an achiral bent-core mesogen to a calamitic cholesteric compound the twisting power of the cholesteric phase can be significantly enhanced [78].

An interesting doping effect is reported for the B_4 phase. This is a solid-like "banana phase" which spontaneously exhibits domains of opposite handedness. By addition of an achiral compound (4-n-pentyl-4'-cyanobiphenyl) the B_4 phase is stabilized up to ~95 wt% of the calamitic compound. Surprisingly, the size of the chiral domains is significantly enhanced which was interpreted as an intrinsic property of the B_4 phase when it arises directly on cooling the isotropic liquid [79].

Other studies have shown that the polar properties of a $SmCP_A$ host phase can be maintained and tuned by the addition of non-polar solvents like xylene [80] or hexadecane [81]. For example, by mixing up to 30 wt% (77 mole%) of xylene the "banana phase" B_7 of a bent-core compound remains stable. The role of the non-polar solvent is to soften the polar structure by keeping the bent molecules further apart. This leads to a decrease of the clearing temperature and enables polar switching at room temperature [80].

In the binary system **A/G** of the bent-core compound **30** (mixing component **A**) and an amphiphilic glycolipid (dodecyl-β-D-glucopyranoside) **36** (mixing component **G**) the $SmCP_A$ phase remains stable up to 60 wt% glycolipid.

36 (Mixing component **G**)   Cr 80 SmA 142 I

Interestingly, the spontaneous polarization is unaffected even when it is diluted up to 60% lipid. On the other hand, at higher concentrations of the lipid **G** a SmA phase exists. It shows at lower temperature a linear electro-optical switching (up to 98% lipid) which is not observable in the SmA phase of the pure lipid. This switching is reminiscent of the electroclinic switching of chiral SmA* phases [82].

5.3   Dimers Consisting of a Bent-core and a Calamitic Mesogenic Unit

A comparison between binary systems discussed before and corresponding twins containing a covalent connection between the both mesogenic moieties is of topical interest. There are two papers about dimers consisting of a bent-core and a calamitic mesogenic unit [83, 84]. For the compounds **37**, listed in Table 1-18 different columnar phases are proved. The existence of columnar phases is related to the bent-core unit, because this phase type was never observed for dimers built up from two simple calamitic mesogenic units.

It follows from Table 1-18 that the compounds **37a** and **37b** exhibit the dimorphism Col–N. The nematic phases of these compounds possess an unusual behavior

Table 1-18. Transition temperatures of the compounds **37a-c**

No	X	m	p	q	n	Cr		Col		N		I
37a	–	11	1	1	8	•	142.5	•a	154.5	•	158.0	•
37b	–	6	0	1	6	•	148.5	(•b	125.5	•	145.5)	•
37c	C=O	6	1	0	6	•	149.0	(•	141.5)	–		•

a) Col_{ob}.
b) Col_r.

on applying an electric field, which is clearly different from that known for nematic phases formed by calamitic compounds. For example, a field-induced formation of focal-conic textures was observed. Such behavior is only reported for nematic phases of some bent-core compounds, and will be discussed more in detail in section 6.1 [84].

It should be mentioned that the mesophases of a dimer consisting of a five-ring bent-core moiety connected with 4-cyanobiphenyl show also properties of interest: the nematic as well as the SmA phase were described to be biaxial phases [83].

6.   UNUSUAL PROPERTIES OF CONVENTIONAL NEMATIC AND SMECTIC PHASES FORMED BY BENT-CORE MESOGENS

6.1   Nematic Phases of Bent-core Compounds

It was shown in earlier sections that the short-chained homologues of a series of bent-core compounds can form a nematic phase while the longer-chained members exhibit "banana phases" or "banana phases" as well as nematic and/or non-polar smectic phases, for example series **4, 5, 6, 18**. It was found that the nematic phase of bent-core compounds (but also the SmA or SmC phases) show unusual properties which are obviously related to the bent shape of the molecules.

In some cases (for example in several compounds of the series **18** [45] and **19** and analogous bromosubstituted compounds [57]) the nematic phase spontaneously forms domains of opposite handedness which can be distinguished by rotating one polarizer clockwise or anticlockwise from the crossed position (Figure 1-20). These domains are not fixed and can be modified by mechanical stress or by change of the temperature [45, 56, 57].

Later on Niori et al. [85] found a similar behaviour in the nematic phase of the following bent-core compound **38** bearing four chloro atoms in lateral positions. A pitch of the chiral domains in the order of $\approx 13.6\,\mu m$ was reported.

38 Cr 71 N 143 I

The formation of regions of opposite handedness is obviously the result of bend-twist deformations which are promoted by the bent shape of the molecules [86]. The occurence of spontaneous twisted regions in the nematic phase is also seen in computer simulations of bent molecules [87].

By the way, the short-chained members of series **18** [45] form a nematic low-temperature phase (designated as N_X, see Table 1-13) which exhibits the characteristic textures of a SmA phase but an X-ray pattern identical with that of the nematic high-temperature phase. On the base of structural and dielectric measurements a structure model of the N_X phase was proposed where the bent

A

B

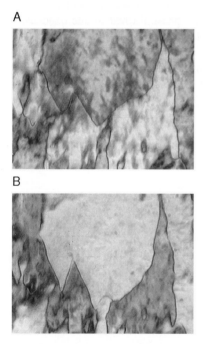

Figure 1-20. Chiral domains in the texture of the nematic phase of 4-bromo-1,3-phenylene bis[4-(4-n-nonyloxybenzoyloxy)benzoate [57] observed by rotating one polarizer by 10° a) clockwise and b) anti-clockwise from the crossed polarizer position indicating domains of opposite handedness (copyright (2004) from Liq. Cryst. by W. Weissflog et al., ref. 57. Reprinted by permission of Tayler & Francis, LLC.http://www.taylerandfrancis.com/)

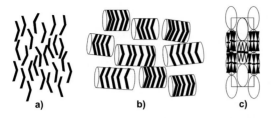

Figure 1-21. Models for the N (a), N_X (b) and Col (c) phases of compounds **18** (reprinted with permission from ref. 45, copyright, 2003, The Royal Society of Chemistry)

molecules are packed in bundles of non-defined length. Between the bundles only short range order exists and in adjacent bundles the bent-direction is antiparallel to avoid bulk polarization (Figure 1-21). On cooling the lateral distances between the bundles are fixed and the N_X passes over into the 2D structure of a Col_r phase, i.e. the N_X phase can be regarded as a precursor of the Col_r phase.

Recently it was shown that the nematic phase of the following 2,5-diphenyl-1,3,4-oxadiazole derivatives **39** (Table 1-19) is a biaxial one which is unambigiously proved by polarizing microscopy, conoscopy and ^{2}H NMR [88], later also by X-ray investigations and ab-initio calculations which explicitly include the bent molecular shape [89]. By the way, also the biaxial nematic phase formed by the oxadiazole **39a** spontaneously forms domains of opposite handedness [90].

There are experimental hints that also in the asymmetric 1,3,4-oxadiazole based liquid crystals **40** containing a naphthalene unit and bearing different substituents in position R the nematic phase probably exhibits a biaxial nature [91].

40   R = CH$_3$, OCH$_3$

Table 1-19. Transition temperatures (° C) and enthalpy changes [kJ mol^{-1}] of the compounds **39a-39b**

No	R		Cr		SmY		SmX		SmC		N_b^+		I
39a	C$_7$H$_{15}$	•	148	•	166	•	173	–		•	222	•	
39b	C$_{12}$H$_{25}$O	•	104	•	148	•	184	•	193	•	204	•	

N_b^+ : biaxial nematic phase.

The nematic phases of compounds **19** [56, 57] as well as of compounds **6** [42] show an unusual electro-optical response. If a d.c. or a low-frequency a.c. field is applied to a planar oriented nematic phase initially a domain pattern with equidistant stripes parallel to the original director direction arises (Figure 1-22a–b). With increasing voltage the long wave nematic director fluctuations become

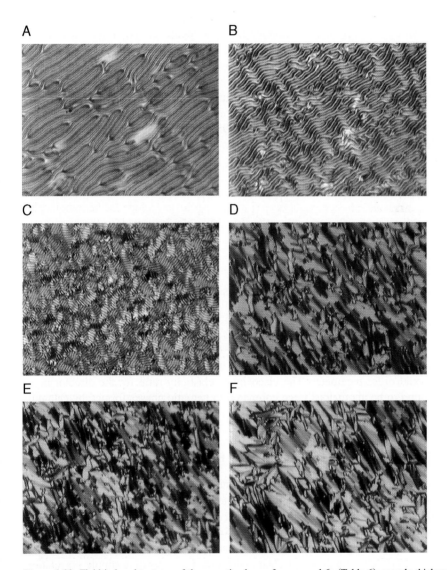

Figure 1-22. Field-induced textures of the nematic phase of compound **6c** (Table 6); sample thickness: 6 μm, temperature 121° C) a) 4 V; b) 6 V; c) 10 V; d) 17 V; e) 26 V; f) 32 V (copyright (2005) from Liq. Cryst. by L. Kovalenko et al., ref. 42. Reprinted by permission of Tayler & Francis, LLC.http://www.taylerandfrancis.com/)

weaker and a myeline texture with fine equidistant stripes appears (Figure 1-22c–d) which is transformed into a fan-like textur with further increasing voltage (Figure 1-22e–f) This field-induced fan-like texture is reminiscent of a smectic or cholesteric phase. If the electric field is removed the texture of the planar oriented nematic phase reappears. Until now the mechanisms of this effect is not yet clear.

It is interesting to notice that for bent-core mesogens the existence of a nematic phase with polar order is not forbidden from the theoretical point of view [92, 93]. The occurence of a spontaneous polarization in a nematic phase formed by an 1,2,4-oxadiazol derivative was reported by Torgova et al., but until now an unambigious proof is missing [94].

The bent shape of the molecules obviously influences the dynamic properties of the molecules in the mesophases. For example, in the nematic phase of the following 4-chlororesorcinol derivative **41** [95] the dielectric fluctuation was studied by dynamic light scattering.

41   Cr 106 (SmC 70) N 124 I

Polarization selection of the scattering cross-section reveals separate modes attributed to uniaxial director fluctuations and lower temperature fluctuations of the biaxial order parameter. The viscosity to elasticity ratio for the director mode was found to be anomalous large compared to nematic phases formed by calamitic molecules [95].

6.2   Non-polar SmA and SmC Phases Formed by Bent-core Compounds

For compounds with $SmCP_A$–SmA or $SmCP_A$–SmC–SmA polymorphism also the non-polar phases (SmA, SmC) can show some peculiar structural features. As shown for members of the series **4** [22] the dielectric permittivity strongly increases on approaching the transition SmA to $SmCP_A$ or SmC to $SmCP_A$, respectively (see Figure 1-6). The same result was reported for compounds of the series **2** [33]. This finding is a clear indication for a positive dipole correlation, that means, for the existence of polar clusters in the short-range order region already in the non-polar smectic phase which significantly grow with decreasing temperature.

In the case of the symmetric 1,3,4-oxadiazole bent-compound **42** which forms a SmC and SmA phase, the SmA$_b$ phase was found to be biaxial which is mainly the consequence of the special molecular shape [96].

42   Cr 220 SmC 235 SmA$_b$ 295 I

It seems that the bent shape of the molecules can change the mechanism of the Freedericksz transition in SmC phases. There are experimental hints that the dielectric reorientation in the SmC phase of compound **5f** (Table 1-5) can also take place by collective rotation of molecules around the long axes driven by the coupling of the field with the effective dielectric anisotropy $\varepsilon_2 - \varepsilon_1$ [37].

7.   FIELD-INDUCED TRANSITIONS FROM A NON-POLAR PHASE TO A POLAR PHASE

7.1   Field-induced Enhancement of the Transition Temperature SmCP$_A$–isotropic

There are some bent-core compounds with SmCP$_A$ phases in which immediately above the clearing temperature the isotropic phase can be transformed into the SmCP$_A$ phase by application of a sufficiently high electric field. This interesting effect is most pronounced for the following bent-core compound **43** which possesses a fluorine substituent in 5-position of the central phenyl ring.

43   Cr 117 (SmCP$_A$ 111) I

This compound forms a monotropic SmCP$_A$ phase with a tilt angle of ≈45° and a switching polarization of 870 nC cm^{-2} [97]. Above the clearing temperature (111°C) an electric field generates nuclei of the SmCP$_F$ phase within the isotropic liquid which coalesce to a grainy texture upon further increase of the field strength.

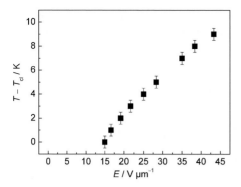

Figure 1-23. The temperature difference $\Delta T = T - T_{cl}$ as function of the threshold voltage where the SmCP$_F$ phase is nucleated within the isotropic liquid (T = the actual temperature; T_{cl} = clearing temperature (compound **43**) (reprinted with permission from ref. 97, copyright, 2003, Wiley-VCH)

When the field is switched off the isotropic phase immediately reappears. The threshold voltage at which the smectic phase nucleates increases with increasing $\Delta T = T - T_{cl}$ where T_{cl} is the clearing temperature. With other words, ΔT increases with increasing voltage (Figure 1-23). It is seen that for a field of $43 \, V \, \mu m^{-1}$ the clearing temperature can be enhanced by 9 K with respect to the clearing temperature of the field-free situation.

In the mean time the same effect was observed for several other bent-core compounds derived from different classes of substances. The enhancement of the clearing temperature varies between 2.5 and 5K. For example, the transition temperature SmCP – I of the six-ring N-benzoylpiperazine derivative **17c** (Table 1-12) could be increased up to 5 K ($40 \, V \, \mu m^{-1}$) [54]. For the laterally substituted azomethine compound **44** $\Delta T = 3 \, K$ ($30 \, V \, \mu m^{-1}$) was measured [35]. In the case of compounds **17c** [54] and **44** [35] the field-induced SmCP$_F$ phase is weakly birefringent and exhibits domains of opposite handedness.

44   Cr 83 SmCP$_A$ 105 I

Up to now there are no systematic investigations concerning the relationship between the chemical structure and the increase of the existence range of the polar phase by applying an electric field. Comparing the data listed in Table 1-20 it seems that the directions of the carboxylic groups and related to these the resulting

Table 1-20. Increase of the SmCP–isotropic transition temperatures on applying an electric field in dependence on the chemical structure of five-ring compounds containing only carboxylic connecting groups

No	n	X	Y	A	$\Delta T = T - T_{cl}$	
45 [39]	10-16	COO	OOC	H	4.0 K	$(40\,V\mu m^{-1})$
46 [98]	12	COO	COO	H	2.5 K	$(40\,V\mu m^{-1})$
47 [41]	12	OOC	COO	H	1.5 K	$(35\,V\mu m^{-1})$
43 [97]	12	COO	OOC	F	9.0 K	$(43\,V\mu m^{-1})$
48 [44]	16	OOC	COO	F	2.5 K	$(35\,V\mu m^{-1})$

dipole moment are of importance. The inversion of one or two of the carboxylic linking groups (see compounds **45**, **46** and **47**) results in a partial compensation of the dipole moments and therefore in lower ΔT values. The same tendency can be observed for the 5-fluorosubstituted derivatives, compare compounds **43** and **48**. Figure 1-24 shows the nucleation of the SmCP$_F$ phase for compound **48** 2K above the clearing temperature.

On the base of the general thermodynamic equilibrium condition a simple formula for the enhancement of the clearing temperature in dependence on the applied field E can be derived [99]. For temperatures near T_{cl}, approximately the following

Figure 1-24. Nucleation of the SmCP$_F$ phase of compound **48** 2 K above the transition temperature SmCP$_A$ - isotropic (applied field 25 V/μm) (reprinted with permission from ref. 44, copyright, 2006, The Royal Society)

equation is valid for the difference between the free energy per unit mass Δg of the liquid-crystalline state and the isotropic liquid state:

$$\Delta g = \frac{-\Delta h_S}{T_{cl}} \Delta T + \varepsilon_0 \left(\varepsilon_2 - \varepsilon_1\right) \frac{(E - E_0)^2}{\rho} + \frac{P_S(E - E_0)}{\rho}$$

Δh_S: specific transition enthalpy
ε_2-ε_1: difference of the dielectric permittivity of the SmCP$_A$ and the isotropic liquid phase
E: field strength
E_0: threshold field strength
P_S: spontaneous polarization
ρ: density
$\Delta T = T - T_{cl}$ (T_{cl} = clearing temperature)
From the equilibrium condition $\Delta g = 0$ follows:

$$\Delta T = T - T_{cl} = \frac{T_{cl}}{\Delta h_S \rho} \left\{\varepsilon_0 \left(\varepsilon_2 - \varepsilon_1\right)(E - E_0)^2 + P_S(E - E_0)\right\}.$$

Using the experimental values of compound **43** (T_{cl}: 384 K; Δh_S : 23.5 J g^{-1}; P_S : 870 nC cm^{-2}, $\rho \approx 1$ g cm^{-3}; $\varepsilon_2 - \varepsilon_1 \approx 25$) an enhancement of ≈ 7 K can be estimated in good agreement with the experimental value.

7.2   Field – induced Enhancement of the Transition Temperature SmCP$_A$ - SmA

As shown in section 2.5, the long-chained members of the series **7** exhibit liquid crystalline tetramorphism SmA–SmC$_S$P$_A$–Col$_{ob}$–SmC$_S$P$_A$ (Table 1-7). The nonpolar SmA phase shows an interesting electro-optical behaviour. Above the transition temperature SmC$_S$P$_A$–SmA (i.e. in the existence region of the SmA phase) the fan-shaped texture of the SmA phase transforms into another texture by application of an electric field. The texture of the switched state corresponds to the texture of the field-induced SmC$_S$P$_F$ state. If the field is switched off the fan-shaped texture of the SmA phase reappears. Since the polar current response above the transition temperature SmCP$_A$–SmA is identical with that obtained below this temperature, the switching above this temperature can be related to a field-induced transition from the nonpolar SmA phase to the polar SmC$_S$P$_F$ phase. For an electric field of 35 V μm^{-1} an enhancement of the transition temperature by about 2K could be detected [43].

7.3   Field-induced Enhancement of the Transition Temperature Crystalline-isotropic

The third example of bent-core compounds which show a field-induced transition of a non-polar into a polar phase are long-chained members of the following series **49**:

Table 1-21. Transition temperatures ($^{\circ}$C) and enthalpy changes [kJmol^{-1}] of the compounds **49a-49c**

No	n	Cr I		Cr II		I
49a	14	•	139 [22.0]	•	162 [46.4]	•
49b	16	•	135 [19.0]	•	165 [58.0]	•
49c	18	•	134 [25.7]	•	157 [57.2]	•

These compounds do not form a mesophase see Table 1-21 [44]. The high-temperature solid modification Cr II shows an unexpected behaviour. By the application of a sufficiently high electric field ($\approx 15\,\mathrm{V\,\mu m^{-1}}$) the birefringence (colour) of the texture is clearly changed while the texture of the switched state depends on the polarity of the field. During this switching one current peak could be recorded per half period of a triangular voltage. At first sight this indicates a ferroelectric ground state. The switching polarization was found to be between 600–700 nC cm^{-2} for the three homologues studied. Since the switching is not really bistable (as expected for ferroelectrics) an antiferroelectric ground state can be assumed. The crystalline nature of this switchable phase is clearly proved by X-ray studies on partially oriented samples; also the high transition enthalpies are an indication for a solid phase. The mechanism of the switching is far from being understood. But it is known that in some crystalline modifications of organic solids formed by elongated molecules collective motions of molecules or parts of molecules are, in principle, possible [100–102].

With regard to the topic of this section it is remarkable that also the transition temperature of this switchable solid modification into the isotropic liquid can be enhanced by at least 2K by an electric field of 25 V μm$^-$. Above the melting temperature nuclei of this unusual solid modification can be induced within the isotropic liquid which grow and coalesce to a beautiful optical texture with further increasing field. Also in this case the field-induced solid phase disappears when the field is removed.

The common feature of all these field-induced effects (sections 7.1–7.3) is that a non-polar phase (isotropic; SmA) could be transformed to a polar phase which is the low-temperature phase with respect to the non-polar one. This means an enhancement of the transition temperature of the polar to the non-polar phase. The plausible assumption to explain this behaviour is that already in the non-polar phase polar clusters exist in the short-range order region. These polar clusters are obviously aligned by the external field giving rise to a phase with macroscopic polarization. This assumption is supported by results of dielectric measurements which were performed above the clearing temperature of an $SmCP_A$ phase. It was found that the dielectric permittivity shows a significant increase on approaching the phase transition from the isotropic to the $SmCP_A$ phase which points to a pronounced positive dipole correlation already in the isotropic liquid [103]. Also for compounds showing a polymorphism SmA-$SmCP_A$ a strong increase of the dielectric permittivity has been observed already in the SmA phase on approaching the transition SmA → $SmCP_A$ [22, 32].

8.   DISCUSSION

The experimental results presented in this chapter demonstrate that the occurrence of "banana-phases" on the one side and the occurrence of nematic or conventional smectic phases on the other hand mainly depends on the size and conformation of the bent core but also on the structure and the length of the terminal chains. The size of the bent core is determined by the number of rings; the conformation is characterized by the bending angle and also by the average torsion angles between the rings. Not only the conformation of the bent core, also the polarity and the distribution of the electronic density (actually also the flexibility of the molecules) are mainly determined by the structure and direction of the linking groups as well as the number, volume, position and the dipole moment of lateral substituents. In this respect it should be taken into account that the lateral substituents are also able to influence the packing of the bent molecules. The role of the structure and length of the terminal chains should not be underestimated because they strongly influence the segregation of the aromatic bent core and the aliphatic parts which is important for the organization of smectic layer structures or columnar structures.

The importance of the *bending angle* can be illustrated by a simple example, by the comparison of a bent-core reference compound without substituents at the central core with corresponding 4-chloro- as well as 4,6-dichloro-substituted compounds, see Table 1-22.

Longer chained members of the series **50-I** and **50-II** form a SmCP phase only, but for the monochloro-substituted compounds the clearing temperatures are clearly reduced [23]. In contrast, the 4,6-dichlororesorcinol derivatives with shorter chains form only a nematic phase whereas the longer chained homologues show the interesting phase sequence Col_{ob}–N, Col_{ob}–SmC–N, or Col_{ob}–SmC [24]. The Col_{ob} phase is really a "banana-phase" which possesses an antiferroelectric structure and shows a polar switching.

It follows from NMR investigations that the bending angle clearly increases in the sequence **50-I, 50-II, 21**. It is 120° in the series **50-I** [25], ≈140° in series **50-II** [23] and ≈160° in series **21** [24]. It is interesting to note that also in the solid state of compound **21-8** a bending angle of 155° could be measured by X-ray structure analysis [24]. These results are a clear indication that lateral substituents in the position A and B change the bending angle. It can be assumed that the

Table 1-22. Influence of chlorine atoms attached to the positions A and/or B on the bending angle

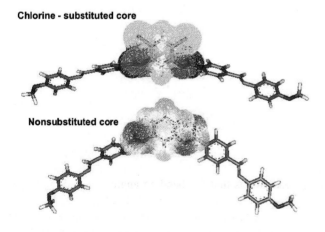

	Compounds	**50-I**	A = H;	B = H	[25]
		50-II	A = H;	B = Cl	[23]
		21 (Table 16)	A = Cl;	B = Cl	[24]

No	RO	A	B	Clearing temperature		Bending angle
50-I-8	$C_8H_{17}O$	H	H	SmCP–I	174° C	$\approx 120°$
50-II-8	$C_8H_{17}O$	H	Cl	SmCP–I	133° C	$\approx 140°$
21-8	$C_8H_{17}O$	Cl	Cl	N–I	148° C	$\approx 160°$

repulsion between the polar substituents and the adjacent electronegatively charged ester groups leads to a torque exercised on the wing of the bent molecule and in this way to an increase of the bending angle. This relationship is schematically shown in Figure 1-25 [22], it is plausible that the effect is stronger in the case of the dichloro-substituted compounds.

In addition, another effect of the bulky substituents must be taken into account. Such substituents can influence the lateral interaction of the bent molecules which makes the packing softer and enables a partial interdigitation of the terminal chains as experimentally proved. This weakening of the lateral interaction can also explain

Chlorine - substituted core

Nonsubstituted core

Figure 1-25. Charge distribution in an energy-minimized conformation of dichloro-substituted and non-substituted bent-core mesogens (reprinted with permission from ref. 22, copyright, 2004, The PCCP Owner Societies)

the clear depression of the clearing temperature of the octyloxy compounds **50-II-8** and **21-8** ($T_{cl} \approx 140°$ C) in comparison to the non-substituted compound **50-I-8** ($T_{cl} \approx 170°$ C) (Table 1-22).

The importance of the bending angle on the mesophase behaviour of bent-core mesogens was also theoretically investigated by Monte Carlo simulations of a minimal excluded volume model of the bent molecules which were described by hard spherocylinder dimers. In this theoretical analysis two geometric parameters (the length-to-breath-ratio, the bending angle) and a thermodynamic parameter (reduced density) are taken into account. The obtained phase diagram as function of the density and the bending angle predicts that the nematic phase becomes thermodynamically unstable for a bending angle <135°. Furthermore, a transition from the non-polar SmA into the polar SmAP$_A$ phase is predicted for an opening angle near 167° [104]. Considering the roughness of the used molecular model that is a satisfying qualitative interpretation of the experimental results.

Of course, a sufficiently high bending angle is not the only important parameter which promotes the formation of nematic or conventional smectic phases. In the case of series **2** (Table 1-2) derived from 4-cyanoresorcinol the bending angle is in the same order of magnitude ($\approx 140°$) as for corresponding 4-chlororesorcinol derivatives **50-II** [23]. But whereas the members of series **50-II** exhibit SmCP$_A$ phases only, the analogous cyano compounds **2** form in addition nematic, SmA and SmC phases depending on the chain length.

The bending angle can also be influenced by the chemical structure of the central ring. Mieczkovski et al. reported a very interesting case where the insertion of a heteroatom into the position Z of the central aromatic ring of compounds **51** drastically changes the phase behaviour [105].

Compounds **51**   X = NO$_2$, J, Cl;   Y = NO$_2$, J, Cl
Z = CH, N

If Z = CH, i.e. if the central core is a phenyl ring, only the non-polar SmA phase occurs independently of the substituents X or Y. If Z = N, i.e. if the central core contains a pyridine ring typical "banana phases" (Col$_r$, B$_7$) have been observed. The authors assume that the pyridyl ring makes the angle between the legs of the molecule smaller than phenyl rings. This assumption is suggested by molecular simulations which showed that exchanging the phenyl by a pyridyl ring leads to a decreased bending angle by about 7–10°. In contrast to all the compounds discussed before, in the derivatives **51** both legs are linked by connecting groups having three single units to the central ring, therefore the conformation and flexibility of the molecules is clearly changed.

It is remarkable that a minor change of the molecular structure, e.g. the reversed *direction of linking groups*, can drastically change the mesophase behaviour. For example, the Bordeaux group reported isomeric homologous series which are only distinguished by the direction of the carboxylic groups at the central core. The members of series **52** form nematic and SmC phases [106]

Compounds **52**   n = 8...10:   N
                    n = 11...13:   N, SmC

If the direction of the ester group is reversed (see compounds **53**) typical "banana phases" occur: $SmCP_A$, Col_r (B_1), and B_6 [107].

Compounds **53**   n = 6, 7: Col_r, B_6
                    n = 8...10: Col_r
                    n = 11...16: $SmCP_A$

By the way, if in the molecules of series **52** a fluorine atom is attached in the neighbourhood of the alkyloxy chain of both terminal rings polymorphism variants with a SmC and three additional polar tilted smectic phases are obtained [107]. This is a convincing example that substituents at the outer rings can also promote the formation of polar "banana phases" although the average intermolecular distance is increased.

There are examples which show that also the *size of the bent core* (the number of rings) can play a role. The most six- and seven-ring compounds exhibit „banana-phases" comparable to the five-ring mesogens. There are examples, however, where „banana-phases" are reported for the five-fing compounds whereas related six-ring derivatives show phases typical for calamitic molecules. In some cases a combination of both phase types exist which is the topic of this chapter.

Murthy et al. [108] reported derivatives of the 3-hydroxybenzoic acid, only different in the number of rings. All members of the homologous series **54** (hexyloxy to octadecyloxy in both terminal positions) forms SmCP and/or Col phases, exclusively. Increasing the number of rings by one results in compounds **55** having an SmA–SmC–SmX polymorphism.

Compounds **54**   e. g. (p = 1; n = 16; m = 16)   Cr 114 SmC$_S$P$_F$ 138 I
Compounds **55**   e. g. (p = 2; n = 16; m = 12)   Cr 141 SmX 183 SmC 197 SmA 200 I

Another example is known from benzoyl derivatives of heterocyclic amines, already presented in sections 2.6 and 4.1. Many of the six-ring compounds **8**, **17** and **18** exhibit variants of polymorphism with conventional smectic as well as „banana-phases". In contrast, for all comparable five-ring compounds derived from piperazine or 4-hydroxypiperidine, only columnar or SmCP$_A$ phases were found [46].

The interpretation of this relationship is not so clear, however, the high clearing temperatures and therefore a higher flexibility of the six-ring molecules should be taken into account.

Concerning the *length of terminal chains* the phase behaviour of many homologous series of bent-core mesogens responds to the following general rules: Short-chain members can exhibit calamitic phases (N, SmA, SmC) sometimes also an intercalated SmC phase (B$_6$). Increasing the number of methylene groups in the hydrocarbon chains columnar phases occur. For the longer-chained homologues SmCP$_A$ phases are found in most cases. As usual, not all of the mentioned phases have to exist in such a phase sequence. The occurrence of the nematic phase for short-chained members can be attributed to the fact that in these cases the microsegregation of the aliphatic part is less pronounced.

There are first hints, however, that the lengthening of the terminal chains can also promote the conventional SmA phases although the shorter homologues exhibit a SmCP phase, only. One example is related to the compounds **7** (Table 1-7).

The interesting tetramorphism SmC$_S$P$_A$–Col–SmC$_S$P$_A$–SmA was proved for the longer homologues (n: 14, 16, 18). In contrast, the octyloxy and the dodecyloxy homologues (compounds **7a, 7b**) exhibit a SmCP$_A$ phase only, whereas the SmA phase is missing [43].

A similar relationship is reported for five-ring bent-core mesogens derived from 4-chlororesorcinol with lateral halogen substituents at the terminal rings [35]. It is seen from Figure 1-26 that for the octyloxy derivatives the attachment of halogene atoms at position A of the outer rings (A = F, Cl, Br) results only in SmCP phases.

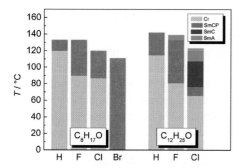

Figure 1-26. The formation of non-polar smectic phases with growing length of the terminal chains in presence of lateral halogen atoms (A = H, F, Cl, Br) (reprinted with permission from ref. 35, copyright, 2005, The Royal Society of Chemistry)

In dependence on the volume of the lateral substituents the clearing temperatures are lowered more or less. Starting from the dodecyloxy homologues, with increasing volume of the halogene atoms at the outer rings SmA and SmC phases exist additionally to the polar $SmCP_A$ phase. That means, the longer chained compounds are able to form conventional smectic phases although the shorter homologues exhibit only banana phases. It can be assumed that also in this case lateral groups enhance the intermolecular distance and therefore disturb the sensitive polar inter-action between the bent-core molecules, especially, if the bending angle is clearly higher than 120°, that means at the borderline between calamitic and bent-core mesogens.

The investigation of high-molecular compounds containing bent-core mesogens is also of interest in relation to this chapter, because SmA phases but also „banana-phases" have been reported for these materials. The formation of polar phases in such constraint systems depends on the chemical structure and length of the spacer as well as on the position at which the bent-core moiety is connected with the spacer group. The terminal connection of two bent-core mesogens, for example, results in dimeric compounds exhibiting „banana-phases" [109, 110]. The linking of two bent-core units using the 4-position of the central phenyl ring by means of short spacer yields dimers exhibiting conventional SmA phases only, because a polar packing is impossible for steric reasons [111].

For dendrimers the situation is more complicated. A polar smectic phase was reported by Tschierske et al. [112] for branched molecules containing four bent-core units. Increasing the numbers of generation of the dendrimer, however, results in compounds which show only a non-polar phase [109, 113].

For side-group polymers, the phase behaviour depends in addition on the number of bent-core units at the polymer backbone, which influences the flexibility or rigidity of the polymer and therefore the degree of freedom to form a polar packing. If in corresponding polysiloxanes each silicon atom is substituted with a bent-core moiety, a polar phase is missing [109]. In the case of strongly diluted polymers, however, the formation of a $SmCP_F$ phase could be observed [114]. It can be expected that by variation of this concept compounds should result which exhibit phase sequences with polar smectic and conventional smectic phases.

Although banana-shaped mesogens have been investigated for only about 10 years a lot of new findings have been obtained in this relatively short time which shows new fundamental aspects of the molecular self-organization in liquid crystals. As shown in this chapter that concerns for example new aspects of structure-property-relationships, polar structures and chiral structures formed by achiral molecules, field-induced switching of chirality or field-induced transition from non-polar to polar phases. It is of topical interest that also "conventional phases" (isotropic-liquid, nematic, smectic A, smectic C, and crystalline phases) formed by bent-core mesogens can exhibit exciting properties, which are clearly related to the non-linear shape of the molecules. Also in near future one can expect new surprising results, possibly concrete forms of practical application.

9.   ACKNOWLEDGEMENT

For financial support the DFG (Graduiertenkolleg 894 „Selbstorganisation durch koordinative und nichtkovalente Wechselwirkungen") is gratefully appreciated. The authors especially wish to thank Dr. Siegmar Diele (Martin-Luther-Universität Halle-Wittenberg) for helpful discussions and Dr. Martin W. Schröder for technical assistance.

10.   REFERENCES

1. Cotter M (1970) In: Luckhurst GR, Gray GW (eds) Molecular Physics of Liquid Crystals. Academic Press, London New York San Francisco, p 169
2. Gelbart WM (1982) J Phys Chem 86:4298
3. Tschierske C (1998) J Mater Chem 8:1485
4. Weissflog W (1998) In: Demus D, Goodby J, Gray GW, Spiess HW, Vill V (eds) Handbook of Liquid Crystals, Wiley VCH, Weinheim, p 855/chap XI/ vol 2B.
5. Nguyen HT, Destrade C, Malthete J (1998) In: Demus D, Goodby J, Gray GW, Spiess HW, Vill V (eds) Handbook of Liquid Crystals, Wiley VCH, Weinheim, p 865/chap XII/vol 2B.
6. Vorländer D, Apel A (1932) Ber Dtsch Chem Ges 65:1101

7. Akutagawa T, Matsunaga Y, Yasuhara K (1994) Liq Cryst 17:659; J Mater Chem 6:1231
8. Niori T, Sekine T, Watanabe J, Furukawa T, Takezoe H (1996) J Mater Chem 6:1231
9. Link DR, Natale G, Shao R, Maclennan JE, Clark NA, Körblova E, Walba DM (1997) Science 278:1924
10. Heppke G, Moro D (1998) Science 279:1872
11. Watanabe J, Niori T, Sekine T, Takezoe H (1998) Jpn J Appl Phys 37:L139
12. Pelzl G, Diele S, Weissflog W (1999) Adv Mat 11:707
13. Szydlowska J, Mieczkowski J, Matraszek J, Bruce DW, Gorecka E, Pociecha D, Guillon D (2003) Phys Rev E 67:31702
14. Takanishi Y, Izumi T, Watanabe J, Ishikawa K, Takezoe H, Iida A (1999) J Mater Chem 9:2771
15. Pelz K, Weissflog W, Baumeister U, Diele S (2003) Liq Cryst 30:115
16. Amaranatha Reddy R, Tschierske C (2006) J Mat Chem, 16:907
17. Takezoe H, Takanishi Y (2006) Jap J Appl Phys 45:597
18. Miyasato K, Abe S, Takezoe H, Fukuda A, Kuze (1983) Jpn J Appl Phys 22:L661
19. Kresse H, Schmalfuss H, Weissflog W (2002) Proceedings of SPIE 4759:222
20. Kresse H (2003) In: Haase W, Wrobel S (eds) Relaxation Phenomena in Dielectric, Magnetic and Superconducting Materials, Springer-Verlag, Heidelberg, 400–422.
21. Novotna V, Hamplova V, Kaspar M, Glogorova M, Pociecha D (2005) Liq. Cryst. 32:1115
22. Eremin A, Nadasi H, Pelzl G, Diele S, Kresse H, Weissflog W, Grande S (2004) Phys Chem Chem Phys 6:1290
23. Pelzl G, Diele S, Grande S, Jakli A, Lischka C, Kresse H, Schmalfuss H, Wirth I, Weissflog W (1999) Liq Cryst 26:401
24. Weissflog W, Lischka C, Diele S, Pelzl G, Wirth I, Grande S, Kresse H, Schmalfuss H, Hartung H, Stettler A (1999) Mol Cryst Liq Cryst 333:203
25. Diele S, Grande S, Kruth H, Lischka C, Pelzl G, Weissflog W, Wirth I (1998) Ferroelectrics 212:169
26. Nadasi H, Weissflog W, Eremin A, Pelzl G, Diele S, Das B, Grande S (2002) J Mater Chem 12:1316
27. Dong R Y, Fodor-Czorba K, Xu J, Domenici V, Prampolini G, Veracini CA (2004) J Phys Chem B 108:7694
28. Dong RY, Xu J, Benyei G, Fodor-Czorba K (2004) Phys Rev E 70:011708
29. Blanca Ros M, Serrano JL, Rosario de la Fuente M, Folcia CL (2005) J Mater Chem 15:5093
30. Mieczkowski J, Matraszek J (2005) Polish J Chem 79:179
31. Kovalenko L, Weissflog W, Grande S, Diele S, Pelzl G, Wirth I (2000) Liq Cryst 27:683
32. Wirth I, Diele S, Eremin A, Pelzl G, Grande S, Kovalenko L, Pancenko N, Weissflog W (2001) J Mater Chem 11:1642
33. Weissflog W, Kovalenko L, Wirth I, Diele S, Pelzl G, Schmalfuss H, Kresse H (2000) Liq Cryst 27:677
34. Weissflog W, Nadasi H, Dunemann U, Pelzl G, Diele S, Eremin A, Kresse H (2001) J Mater Chem 11:2748
35. Dunemann U, Schröder MW, Amaranatha Reddy R, Pelzl G, Diele S, Weissflog W (2005) J Mater Chem 15:4051
36. Eremin A (2003) PhD thesis, University of Halle

37. Weissflog W, Dunemann U, Schröder MW, Diele S, Pelzl G, Kresse H, Grande S (2005) J Mater Chem 15:939
38. Nakata M, Shao RF, Maclennan JE, Weissflog W, Clark NA (2005) Phys Rev Lett 96:067802
39. Schröder MW, Diele S, Pelzl G, Weissflog W (2004) Chem Phys Chem 5:99
40. Bedel JP, Rouillon JC, Marcerou JP, Nguyen HT, Achard MF (2004) Phys Rev E 69:61702
41. Amaranatha Reddy R, Schröder MW, Bodyagin M, Kresse H, Diele S, Pelzl G, Weissflog W (2005) Angew Chem 117 :784
42. Kovalenko L, Schröder MW, Amaranatha Reddy R, Diele S, Pelzl G, Weissflog W (2005) Liq Cryst 32:857
43. Shreenivasa Murthy HN, Bodyagin M, Diele S, Baumeister U, Pelzl G, Weissflog W (2006) J Mater Chem 16:1634
44. Weissflog W, Shreenivasa Murthy HN, Diele S, Pelzl G (2006) Phil Trans R Soc. A 364:2657
45. Schröder MW, Diele S, Pelzl G, Dunemann U, Kresse H, Weissflog W (2003) J Mat Chem 13:1877
46. Dunemann U, Schröder M W, Pelzl G, Diele S, Weissflog W (2005) Liq. Cryst. 32:151
47. Brand HR, Cladis PE, Pleiner H (1982) Macromolecules 25:7223
48. Eremin A, Diele S, Pelzl G, Nadasi H, Weissflog W, Salfetnikova J, Kresse H (2001) Phys Rev E 64:51707
49. Nadasi H (2004) PhD thesis, University of Halle
50. Shreenivasa Murthy HN, Sadashiva BK (2004) Liq Cryst 31:567
51. Sadashiva BK, Amaranatha Reddy R, Prathiba R, Madhusudana NV (2002) J Mater Chem 12:943
52. Amaranatha Reddy R, Sadashiva BK (2004) J Mater Chem 14:310
53. Amaranatha Reddy R, Sadashiva BK, Surajit Dhara (2001) Chem Commun 1972
54. Schröder MW, Pelzl G, Dunemann U, Weissflog W (2004) Liq Cryst 31:633
55. Eremin A, Diele S, Pelzl G, Weissflog W (2003) Phys Rev E 67:20702
56. Pelzl G, Eremin A, Diele S, Kresse H, Weissflog W (2002) J Mater Chem 12: 2591
57. Weissflog W, Sokolowski S, Dehne H, Das B, Grande S, Schröder MW, Eremin A, Diele S, Pelzl G, Kresse H (2004) Liq Cryst 31:923
58. Liao G, Stojadinovic S, Pelzl G, Weissflog W, Sprunt S, Jakli A (2005) Phys Rev E 72:21710
59. Gesekus G, Dierking I, Gerber S, Wulf M, Vill V (2004) Liq Cryst 31: 145
60. Shen D, Diele S, Pelzl G, Wirth I, Tschierske C (1999) J Mater Chem 9: 661
61. Amaranatha Reddy R, Sadashiva BK (2000) Liq Cryst 27:1613
62. Amaranatha Reddy R, Sadashiva BK (2004) J Mater Chem 14:1936
63. Amaranatha Reddy R, Sadashiva BK, Raghunathan VA (2004) Chem Mater 16:4050
64. Shreenivasa Murthy HN, Sadashiva BK (2004) Liq Cryst 31:1347
65. Niori T, Yamamoto J, Yokohama H (2004) Mol Cryst Liq Cryst 411:283
66. Kang S, Thisayukta J, Takezoe H, Watanabe J, Ogino K, Doi T, Takahashi T (2004) Liq Cryst 31:1333
67. Amaranatha Reddy R, Sadashiva BK, Baumeister U (2005) J Mater Chem 15:3303
68. Weissflog W, Lischka C, Diele S, Pelzl G, Wirth I (1999) Mol Cryst Liq Cryst 328: 101
69. Schröder MW, Diele S, Pelzl G, Pancenko N, Weissflog W (2002) Liq Cryst 29: 1039

70. Diele S, Pelzl G, Humke A, Wünsch S, Schäfer W, Zaschke H, Demus D (1989) Mol Cryst Liq Cryst 173:113
71. Dimitrowa K, Hauschild J, Zaschke H, Schubert H (1980) J Prakt Chem 322:933
72. Schröder MW (2006) PhD thesis, University of Halle
73. Schröder MW, Diele S, Pancenko N, Weissflog W, Pelzl G (2002) J Mater Chem 12:1331
74. Prathiba R, Madhusudana NV, Sadashiva BK (2001) Mol Cryst Liq Cryst 365:755;
75. Prathiba R, Madhusudana NV, Sadashiva BK (2000) Science 288:2184
76. Gorecka E, Nakata M, Mieczkowski J, Takanishi Y, Ishikawa K, Watanabe J, Takezoe H, Eichhorn SH, Swager TM (2000) Phys Rev Lett 85:2526
77. Rauch S, Selbmann C, Goc F, Heppke G, Dabrowski R (2005) Proceeding of the Arbeitstagung Flüssigkristalle, Paderborn, March 2005, P 45
78. Thisayukta J, Niwano H, Takezoe H, Watanabe J (2002) J Am Chem Soc 124:3354
79. Takanishi Y, Shin GJ, Jung JC, Choi SW, Ishikawa K, Watanabe J, Takezoe H, Toledano P (2005) J Mater Chem 15:4020
80. Jakli A, Cao W, Huang Y, Lee CK, Chien LC (2001) Liq Cryst 28:1279
81. Huang Y, Pedreira AM, Martins OG, Gigueiredo Neto AM, Jakli A (2002) PhysRevE 66:31708
82. Abeygunaratne S, Jakli A, Milkereit G, Sawade H, Vill V (2004) PhysRevE 69:21703
83. Yelamaggad CV, Prasad SK, Nair GG, Shashikala IS, Shankar Rao DS, Lobo CV, Chandrasekhar S (2004) Angew Chem Int Ed. 43:34229
84. Tamba MG, Kosata B, Pelz K, Diele S, Pelzl G, Vakhovskaya Z, Kresse H, Weissflog W, (2006) Soft Matter 2:60
85. Niori T, Yamamoto J, Yokohama H (2004) Mol Cryst Liq Cryst 400:475
86. Dozov J (2002) Europhys Lett 56:247
87. Memmer R (2002) Liq Cryst 29:483
88. Madsen LA, Dingemans TJ, Nakata M, Samulski ET (2004) Phys Rev Lett 92 :145505
89. Acharya BR, Primak A, Kumar S (2004) Phys Rev Lett 92:145506
90. Görtz V, Goodby J (2005) Chem Commun 3262
91. Sung HH, Lin HC (2004) Liq Cryst 31:831
92. Pleiner H, Brand HR, Cladis PE (2000) Ferroelectrics 243:291
93. Sparavigna A (2004) Rec Res Developments in Appl Phys 7:293
94. Torgova SI, Geivandova TA, Francescangeli O (2003) Pramana - J Phys 61:239
95. Stojadinovic S, Adorjan A, Sprunt S, Sawade H, Jakli A (2002) Phys Rev E 66: 60701
96. Semmler KJK, Dingemans TJ, Samulski ET (1998) Liq Cryst 24:799
97. Weissflog W, Schröder MW, Diele S, Pelzl G (2003) Adv Mat 15:630
98. Weissflog W, Naumann G, Kosata B, Schröder MS, Eremin A, Diele S, Vakhovskaya Z, Kresse H, Friedemann R, Ananda Rama Krishan S, Pelzl G (2005) J Mater Chem 15:4328
99. Helfrich W (1970) Phys Rev Lett 24:201
100. Thompson RT, Pintar MM (1976) J Chem Phys 65:1091
101. Reichert D, Hempel G, Zimmermann H, Schneider H, Luz Z (2000) Solid State Nuclear Magnetic Resonance 18:17
102. Stumber M, Zimmermann H, Schmitt H, Haeberlein W (2001) Mol Phys 99:1091
103. Kresse H, Salfetnikova J, Tschierske C, Dantlgraber D, Zhuchova T (2001) Liq Cryst 28:1575
104. Lansac Y, Maiti PK, Clark NA, Glaser MA (2003) Phys Rev E 67:011703

105. Mieczkowski J, Gomola K, Koseska J, Pociecha D, Szydlowska J, Gorecka E (2003) J Mater Chem 12:2132
106. Nguyen HT, Rouillon JC, Marcerou JP, Bedel JP, Barois P, Sarmento S (1999) Mol Cryst Liq Cryst 328:177
107. Rouillon JC, Marcerou JP, Laguerre M, Nguyen HT, Achard MF (2001) J Mater Chem 11:2946
108. Shreenivasa Murthy HN, Sadashiva BK (2005) J Mater Chem 15:2056
109. Kosata B, Tamba GM, Baumeister U, Pelz K, Diele S, Pelzl G, Galli G, Samaritani S, Agina EV, Boiko NI, Shibaev VP, Weissflog W (2006) Chem Mat 18:691
110. Dantlgraber G, Diele S, Tschierske C (2002) Chem Commun 2002:2768
111. Dehne H, Pötter M, Sokolowski S, Holzlehner U, Reinke H, Weissflog W, Diele S, Pelzl G, Wirth I, Kresse H, Schmalfuss H, Grande S (2000) Proceedings of the Freiburger Arbeitstagung Flüssigkristalle, Freiburg, March 2000, poster P 23
112. Dantlgraber G, Baumeister U, Diele S, Kresse H, Lühmann B, Lang H, Tschierske C (2002) J Am Chem Soc 124:14852
113. Kardas D, Prehm M, Baumeister U, Pociecha D, Amaranatha Reddy R, Mehl GH, Tschierske C (2005) J Mater Chem 15:1722
114. Keith C, Amaranatha Reddy R, Tschierske C (2005) Chem Commun 2005:2

CHAPTER 2

PHYSICAL PROPERTIES OF BANANA LIQUID CRYSTALS

ANTAL JÁKLI, CHRIS BAILEY AND JOHN HARDEN

Chemical Physics Interdisciplinary Program and Liquid Crystal Institute, Kent State University, Kent, OH, 44242, USA

1.   INTRODUCTION

Exactly a decade ago it was realized that not only rod-shape (calamitic) or discotic molecules can form liquid crystals, but bent-core (bow-like or banana-shape) molecules do, too [1]. Although the first synthesis of bent-shaped liquid crystals has been reported more than sixty years ago by Vorländer [2], they have not attracted much interest until the synthetic work of Matsunaga et al. [3] in the early 1990s. The discovery of the mesogenic properties of bent-core molecules has opened up a major new and exciting dimension in the science of thermotropic liquid crystals (LCS). Seminal findings – having broad implications for the general field of soft condensed matter – include the observation of ferroelectricity and spontaneous breaking of chiral symmetry in smectic phases composed of molecules that are not intrinsically chiral [4]. Typical bent-core molecules and their structure – property relations are discussed in detail in the previous chapter written by Weissflog.

In this chapter we will only review the first ten years of experimental physical results on bent core materials. We consider the molecules as bent rods and their packing and macroscopic phase behavior will be described based on the cartoons we present in Figure 2-2 to Figure 2-6.

1.1   Structural Cartoons

Considering their packing we assume that they tend to fill the space as effectively as possible. This steric requirement together with the bent-shape of the molecules immediately implies two important features.

A. Ramamoorthy (ed.), Thermotropic Liquid Crystals, 59–83.
© 2007 *Springer.*

Figure 2-1. Sketch of hypothetical nematic phase of bent-core molecules

(i) When translating a molecule in the "sea" of the neighbor molecules it is experiencing a periodic potential determined by the length *l* of the molecules. This means a translational symmetry breaking along the long axis of the molecules (the average direction of the hypothetical lines connecting to the ends of the molecules) and results in temporary periodic (smectic-type) clusters. In this case the nematic order is allowed only in a length scale that is much larger than that of the clusters as illustrated in Figure 2-1.

(ii) The tendency for layering combined with the close packing requirement results in a polar order along the kink direction. Provided that the molecules have a dipole moment along the kink direction (if the molecules are symmetric, the other components of the molecular dipoles average out anyway), this molecular dipole will not average out in a length scale of uniform domains. This means that the macroscopic electric polarization (volume density of the molecular dipoles) will not be zero in each uniform domain. This polarization $\hat{P}$ is a direct result of molecular packing, and as such is different by nature from that of the SmC* materials, where that molecular chirality and director tilt can result in polar order as shown first by R.B. Meyer [5].

Due to their shape bent-core molecules now can be characterized by 3 orthogonal unit vectors, $\hat{n}$, $\hat{m}$, and $\hat{p}$: $\hat{n}$ is the unit vector along the long axis, $\hat{m}$ is normal to the molecular plane, and $\hat{p}$ is along the kink direction, which therefore is parallel to the layer polarization $\vec{P}$.

Figure 2-2 shows the situation when $\hat{n}$ is parallel to the smectic layer normal $\hat{k}$. This is similar to the SmA phase of calamitic liquid crystals, except that now the layers are polar. This difference is designated by adding the letter P (for polar) to SmA. In this case one can have two distinct situations, the layer polarization $\vec{P}$ can be either parallel or antiparallel in the subsequent layers corresponding to ferroelectric (SmAP$_S$) or antiferroelectric (SmAP$_A$) subphases. Here and later

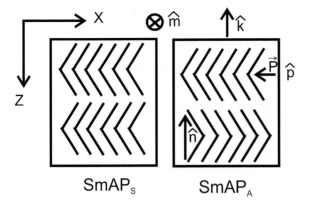

Figure 2-2. Schematic structures of the ferroelectric SmAP$_s$ and antiferroelectric SmAP$_A$ phases

throughout the paper we will denote the same directions by the subscript S for synchronous and the alternating directions by subscript A.

The situations when the molecular planes are tilted with respect to the layer normal, i.e., when $\hat{m}$ is not perpendicular to $\hat{k}$, are shown in the upper row of Figure 2-3. In the plane determined by the polarization $\vec{P}$ and the layer normal $\hat{k}$ (polar plane) this tilt is illustrated by a bar stuck to the end of the molecules which is closer to the observer. Depending on whether the tilt directions are parallel or antiparallel we can speak about synclinic and anticlinic situations. As mentioned above we will label them by the subscript S and A, respectively. Combining these different situations with the ferroelectric and antiferroelectric packing possibilities we have 4 different subphases: SmC$_S$P$_S$, SmC$_A$P$_A$, SmC$_S$P$_A$ and SmC$_A$P$_S$. Such a notation with some variations was introduced by Link et al. [6] and is widely used in the literature. Note that the SmCP layers have only a two-fold symmetry axis around $\hat{p}$ (C2 symmetry), i.e. they have the same symmetry as of chiral SmC* materials [5].

In principle, we can also envision that only the director $\hat{n}$ is tilted with respect to the layer normal $\hat{k}$. Successive smectic layers can be either *ferroelectric* (with the same direction of polar order) or *antiferroelectric* (with opposite directions). Likewise, successive layers can be either *synclinic* (with the same direction of molecular tilt) or *anticlinic* (with opposite tilt directions).Those situations are illustrated in the bottom row of Figure 2-3. To distinguish from the tilt of the molecular plane (which is denoted by the symbol C to express "clinic"), the tilt of the long axis will be called "leaning" and will be labeled with L. Just as in the SmCP cases we can have four distinct sub-phases *SmL$_S$P$_A$, SmL$_A$P$_S$, SmL$_A$P$_A$ and SmL$_S$P$_S$* depending on the subsequent tilt and polarization direction combinations. In such phases the polar axis is not parallel to the smectic layers and they have C$_s$ symmetry.

On the same grounds one also can imagine that both tilt directions are possible simultaneously [7]. Those "double – tilted" situations are labeled both with C and

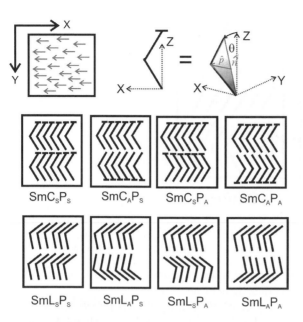

Figure 2-3. Possible single tilted bent-core smectic structures. Top row: Illustration of the fluid in plane order of the layer polarization, and 3 dimensional explanation of tilt. Middle row: 2 dimensional illustration of the 4 possible situations when only the molecular plane is tilted with respect to the layer normal; Bottom row: 2-dimensional illustration of the four possible situations when only the long axis is tilted (leaned) with respect to the layer normal

L, where the tilt independently can be the parallel or antiparallel to each other. Taking into account that the layer polarizations can either be parallel or antiparallel, we have altogether eight different subphases as illustrated in Figure 2-4.

Although in all situations the individual layer polarizations are not parallel to the smectic layers, the out of layer components average out in case of the anti-leaning structures, and their macroscopic symmetry would be the same as of the *SmAP* (for anticlinic situations, such as $SmC_A L_A P_A$ and $SmC_A L_A P_S$) or as of *SmCP* (for synclinic situations, such as $SmC_S L_A P_A$ and $SmC_S L_A P_S$). It is important to note that the *SmCLP* structures can be equivalently described by a tilt of the molecular plane and a rotation of the layer polarization $\vec{P}$ about the long axis $\hat{n}$ by an angle ϕ. If $\phi = 0$ or π then we have the SmCP case, $\phi = \pi/2$ or $3\pi/2$ correspond to the *SmLP* situations, otherwise we have the SmCLP phase. Double – tilted layers have triclinic symmetry, i.e., they are symmetric only with respect to a 360° rotation around the polar axis $\vec{P}$ (Schoenflis notation: C_1). Such a structure was theoretically predicted by de Gennes [8] and was labeled as SmC_G ("G" stands for generalized) without specifying bent-core molecules.

So far we have considered only those cases where the center of masses of molecules had periodicity only in one direction, i.e., when the material can be considered solid in 1 dimension and fluid in the other two directions. The

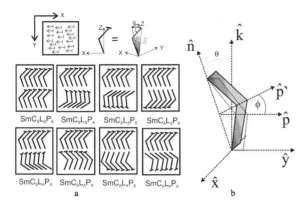

Figure 2-4. (a) Possible double – tilted bent-core smectic structures. Top row: Illustration of the fluid in plane orders of the layer polarization (either uniform or modulated), and 3 – dimensional explanation of tilt. Middle and bottom rows: 2 dimensional illustrations of the eight possible situations; (b) Graphical explanation of the SmCLP structures in term of two rotation angles θ and ϕ

theoretically possible 2-dimensional solid structures are illustrated in Figure 2-5 to Figure 2-6. The situation shown in the upper row of Figure 2-5 assumes an intercalated layer structure. In this case the hydrophobic hydrocarbon chains have to the in close proximity with the aromatic bent-cores, which usually are not favored entropically, and lead to a nano-segregation as shown in the bottom row of Figure 2-5.

Other two possibilities for two – dimensional solid structures are illustrated in Figure 2-6. In the first case we assume disc-shape aggregates of several molecules (four

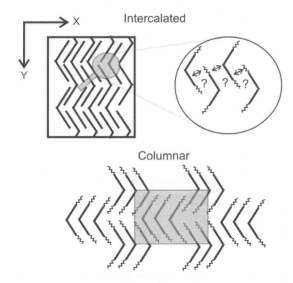

Figure 2-5. Schematic representation of the intercalated layer structures

Columnar hexagonal phase of disc shape aggregates

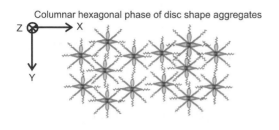

Antiferroelectric columnar structure

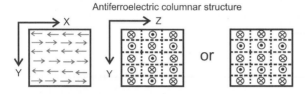

Figure 2-6. 2-dimensional periodic structures formed by local disc like aggregation of bent-core molecules (a) or due to antiferroelectric in-plane ordering (b)

in the drawing). To describe the orientations of the individual molecules in the aggregates we shall characterize it by the triplet of $\hat{p}, \hat{n}, \hat{m}$ unit vectors. We can express the aggregate with four triplets with coordinates in the X,Y,Z lab system with unit vectors $\vec{x}, \vec{y}$ and $\vec{z}$, as follows $\{[\vec{z}, \vec{y}, \vec{x}], [-\vec{z}, \vec{x}, -\vec{y}], [\vec{x}, \vec{y}, \vec{z}], [-\vec{x}, \vec{y}, -\vec{z}]\}$. These aggregates then form fluid stacks (columns) in the $\vec{z}$ direction and periodic structures in the XY plane. The last possibility to form 2-dimensional structures is illustrated in the bottom row of Figure 2-6. This corresponds to any local structures of *SmAP, SmCP, SmLP* or *SMCLP* illustrated in Figure 2-2 to 2-4, except that within each layers the polarization is assumed to alternate. This alternation imposes an entropic restriction to position of the center of molecules along the Y direction resulting in a rectangular periodicity in the YZ plane. In the neighbor rectangles the polarizations can either be parallel or antiparallel to each other corresponding to the P_S or P_A situations of Figure 2-2 to 2-4. Theoretically this last situation alone may have $2+4+4+8=18$ subphases

2.   EXPERIMENTAL RESULTS

2.1   Phase Structures

Experimental observations of the smectic and columnar phases of bent-core liquid crystal phases were pioneered by groups in Tokyo [9], Boulder [6], Berlin [10], Halle [11], Budapest [12], and several other places [13, 14]. In the last several years the research has quickly broadened with very intensive experimental studies carried out all around the world. Activities of this "banana mania" have been recently reviewed in several papers [15].

In the first few years of experimental studies seven different 'banana liquid crystal' textures have been observed and labeled as *B1, ..., B7* according to the chronological order of their observations [16]. The list of these phases, based on textural observations and X-ray measurements, are summarized in Table 2-1.

The **B1** phase is a columnar, non-polar non-tilted phase corresponding to the structure illustrated in Figure 2-5/b.

B2 is the most actively studied phase. It is identical to the SmCP structures shown in Figure 2-3. These materials show a variety of textures, electro-optical and electric polarization behavior, which we will detail later.

The **B3** phase represents a non-switchable (probably not polar) non-tilted smectic structure with in plane positional ordering. Due to the lack of the electro-optical response it is only seldom studied. Some studies indicate second harmonic generation (SHG activity) and therefore polar structure [18], however, these measurements could not be confirmed later [19].

The nature of the **B4** phase is a challenge of the current studies, although the first description of chirality in the banana phases was on the B4 phase. It is a solid phase, but it is currently not clear if this is some kind of crystal [20] or hexatic twist grain boundary (TGB) phase [21], or something else. X-ray diffraction measurements indicate broader peaks in the wide angle range than in the B3 phase that appears at higher temperatures. This would indicate less ordered packing in the short range, or layered packing with short range order. Freeze fracture electron microscopy studies suggest a structure similar to the lyotropic sponge phase [22]. Notably this structure is very similar to that proposed for an optically isotropic ferroelectric phase consisting of random SmC_AP_S nanodomains [23].

The **B5** name was originally assigned to a switchable phase, which appeared below the B2 phase and showed somewhat more complex X-ray profile [24] than of B2. However, investigations on freely suspended films showed that it has the same symmetry as of the B2 [16], so it seems that there is no need for a separate name.

The **B6** phase is an intercalated smectic structure such as shown in Figure 2-5/a without long range order in the YZ plane. In this phase the layer periodicity is smaller than half length of the molecules. As we mentioned before, this packing

Table 2-1. The different phases of bent-core molecules

Name	Structure	Polarity	Director tilt
B1	columnar	non-polar	non- tilted
B2 (*SmCP*)	smectic with no in-layer order	polar	tilted
B3(*SmIP*)	smectic with hexagonal in-layer order	polar?	tilted
B4	optically active solid	polar?	non-tilted
B5	highly viscous SmCP	polar	tilted
B6	interdigitated layers	non-polar	non-tilted
B7	columnar	polar	tilted
B7'	modulated layer structure ($SmCP^{17}$ or SmC_G	polar	Tilted

makes contacts between the aromatic rigid bent-core and the flexible tails, so it is not favored, explaining that experimentally it is found only in short chain materials [16].

The high temperature phase that appears directly below the isotropic phase showing characteristic helical filamentary growth in cooling [25] is denoted as **B7**. Later measurements showed that similar helical structures may appear in materials with different structures, and they correspond to at least two distinct phases. The exact structures of these phases are still under active debate, which we will detail later.

The polar smectic A (**SmAP**) [26] phase and nematic structures [27] were observed only after the Bi (i=1,...7) nomenclature was introduced and they have not been assigned with those symbols.

So far there is only one report[26] on non-tilted $SmAP_A$ phase corresponding to Figure 2-2/b. This phase does not show electro-optical switching, nor birefringence change, although polarization current measurements clearly show a large (almost $1000nC/cm^2$) antiferroelectric polarization value.

Nematic phases exhibiting purely orientational order are rather uncommon in bent-core compounds probably due to the tendency of the periodic positional ordering we mentioned above (see Figure 2-1). However, very recently, a number of new bent-core compounds with nematic phases have been synthesized [27–29]. In addition, their physical properties are currently being investigated [30–32]. Light scattering studies in the uniaxial nematic phase [31], and recent NMR measurements in the isotropic phase [33], reveal drastically slower fluctuations in bent-core compounds than observed in typical calamitics. This is because their viscosities are about 2 orders of magnitude larger than of normal calamitic materials [34]. This behavior is also consistent with the cluster concept illustrated in Figure 2-1 and corroborated by recent NMR measurements [35]. Simultaneously there has been a surge in theoretical studies [13, 36–38] predicting intriguing new thermotropic nematic and isotropic structures. These including uniaxial behavior [31], field induced biaxiality [32] and biaxial phases [39].

2.2   SmCP Materials

2.2.1   *Electro-optic properties*

Although the very first report on bent-core smectic materials showed ferroelectric switching without rotation of the optical axis during switching thus indicating SmAP structure corresponding to Figure 2-2/a [4] soon it became clear that they rather form tilted phases [40, 41] and show antiferroelectric switching [6, 12]. This is because of the typically much larger than $200\,nC/cm^2$ layer polarization, which favor antiferroelectric arrangement due to dipole – dipole interactions.

Interestingly it was observed that the switching between the field-induced ferro-electric states under typically above $3V/\mu m$ rectangular electric fields is either accompanied by rotation of the optic axis, or it takes place without any rotation of the optical axis. Usually both types of textures are present in one sample, typically under cooling from the isotropic phase mainly the non-rotating domains showed up, whereas when heating from the lower temperature B3 phase, mainly the rotating type domains formed. This strange behavior was first explained by Link et al. [6] by realizing that the domains that do not show rotation of the optic axis under rectangular fields have SmC_SP_A, whereas the other domains have SmC_AP_A structures (see cartoons in Figure 2-3).

2.2.2 Layer chirality

Importantly it was also realized by the Boulder group [6], that the combination of polar packing and the tilt of the molecular planes gives the smectic layers a chiral structure, which is usually referred as **layer chirality.** Since the molecules (at least those studied first) do not contain any chiral carbons, smectic layers can form two different structures that are non-superposable mirror images of each other. To distinguish between these two structures, we can define the chiral order parameter as [42]

$$(2\text{-}1) \qquad \chi = 2\left[\left(\hat{k} \times \hat{n}\right) \cdot \hat{p}\right]\left[\hat{k} \cdot \hat{n}\right],$$

where $\hat{k}$, $\hat{n}$ and $\hat{p}$ are unit vectors normal to the smectic layers, along the director and the layer polarization, respectively. This order parameter has the correct symmetry as a pseudoscalar, and it is invariant under the symmetries $\hat{n} \rightarrow -\hat{n}$ and $\hat{k} \rightarrow -\hat{k}$. Right- and left-handed mirror image layers have opposite signs of χ, which ranges from -1 to $+1$. These possibilities for polar and tilt order lead to two possibilities for the layer chirality. If the smectic phase is ferroelectric [43–45] and synclinic (SmC_SP_S), or if it is antiferroelectric [40, 41, 46] and anticlinic (SmC_AP_A), then the material has a homogeneously chiral structure, with the same sign of χ in all layers. By contrast, if the smectic phase is ferroelectric and anticlinic (SmC_AP_S), or if it is antiferroelectric and synclinic (SmC_SP_A), then the material has alternating right- and left-handed chiral layers, with alternating positive and negative values of χ. This latter possibility is generally called a "racemic" structure, although we might use the alternative term "antichiral" [47] to emphasize the rigid alternation from layer to layer. All these possibilities with their typical textures are shown in Figure 2-7.

The textures of racemic SmC_SP_A phase usually consist of fan-shaped domains decorated with a few micron wide stripes, which consist of oppositely tilted synclinic domains. In homochiral SmC_AP_A structure the optical axis is parallel to the layer normal regardless of the handedness of the domains. The antiferroelectric *(AFE)* arrangement can be easily switched to ferroelectric *(FE)* by applying an external electric field of typically larger than 3-5V/μm and f < 10kHz electric fields. The racemic *FE* state is anticlinic (SmC_AP_S) and the optical axes are parallel to the layer normal regardless of the sign of the electric field. The homochiral *FE* state is synclinic (SmC_SP_S) and the optical axes make angles $\pm\theta$ with the layer normal, depending on the sign of the electric field.

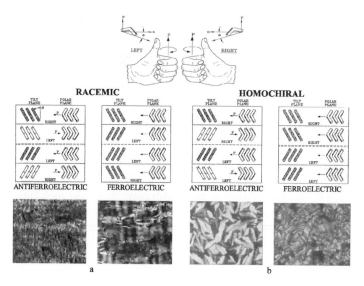

Figure 2-7. Definition of the layer chirality. Left (right) – handed layer: the average molecular axis is oriented (anti)clockwise from the layer normal. This can be envisioned relative to a left (right) hand: if the thumb of a left (right) hand is pointing in the direction of layer polarization **P**, the direction of the curling of the fingers represents the direction of the deviation of the molecular axis from the layer normal **n**. (a): Orthogonal views of the racemic SmCP phase of non-chiral banana-shaped molecules and polarizing microscope textures viewed in tilt plane in antiferroelectric (AFF) and ferroelectric(FE) states (left and right columns). The "Polar plane" contains the layer normal and the layer polarization (P), whereas the "Tilt plane" is perpendicular to P. The molecular plane is tilted with respect to the layer normal. The single dashed line (- - - - -) indicates synclinic interfaces in anticlinic state, whereas double dashed lines (=====) represents defect walls separating synclinc layers with opposite tilt directions. (b): Orthogonal views of the chiral SmCP phase of non-chiral banana-shaped molecules and polarizing microscope textures viewed in tilt plane in antiferroelectric (AFF) and ferroelectric (FE) states (left and right columns). The "Polar plane" contains the layer normal and the layer polarization (P), whereas the "Tilt plane" is perpendicular to P. The molecular plane is tilted with respect to the layer normal. In (a) and (b) (- - - - -) indicates synclinic interfaces in anticlinic state, whereas (====) represents defect walls separating synclinc layers with opposite tilt directions. The pictures represent 100-μm $\times$ 70-μm areas

2.2.3   Light scattering properties

In addition to these differences between anticlinic and synclinic structures in the behavior of the optical axis during switching by electric fields, there is a difference in their light scattering properties, too. The synclinic structures (*AFE* state of the racemic and *FE* state of the chiral structures) scatter light, whereas the anticlinic structures (*FE* state of the racemic and *AFE* state of the chiral structures) are transparent [46]. This is due to the fact that the differently tilted synclinic domains are separated by defect walls, which are observable even without polarizers. Antiferroelectric and ferro-electric textures of the chiral state viewed without polarizers are shown in Figure 2-8. In the anticlinic AFE state (a) only focal conic defects are present due to the imperfect layer alignment, whereas in the synclinic FE state (b) the texture is full of defect walls separating domains with different handedness. Defects cause light – scattering because

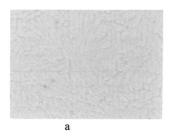
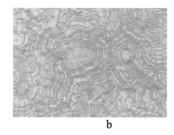

a   b

Figure 2-8. Textures of 4-μm cells of a banana-smectic in chiral state viewed without polarizers. The pictures represent 100-μm × 70-μm areas. Permission for Reprint, courtesy Society for Information Display

the refractive index of the defect is different from the uniform areas. Focal conic defects of Figure 2-8/a cause only a few percentage of scattering, whereas the defect walls of Figure 2-8/b scatter out 50% of incoming light. Such a behavior may have practical applications as fast light shutters.

2.2.4   Birefringence

Another important difference between the anticlinic and synclinic states is the birefringence. Anticlinic structures have low, whereas the synclinic phases have large birefringence. Accordingly when we switch between the antiferroelectric and ferroelectric states we can also switch the birefringence. Especially interesting is the case when the anticlinic state is optically isotropic that can be reversibly switched to birefringent state. Such a situation actually was observed recently [48, 51] on an ester-based banana material [49], which has orthoconic structure (i.e., nearly 45° tilt angle) in the antiferroelectric state.

Electro-optical observations on 5μm films of this material are summarized in Figure 2-9. The film is completely dark at zero fields between crossed polarizers even for oblique light incidence. The transmittance is linearly increasing with fields in the deformed antiferroelectric range (V < 40V), then it sharply increases at the transition to the ferroelectric state (V ∼ 50V), and it remains constant under higher fields. The corresponding birefringence, as judged from the birefringent color, increases from $\Delta n = 0$ at $V = 0$ to $\Delta n \sim 0.19$ at V>50V [48]. The director and layer structures in the antiferroelectric and ferroelectric states in positive and negative fields are sketched in the bottom of Figure 2-9.

By simple superposition of the dielectric tensors of the uniaxial arms, which are linked together with opening angle Φ it was shown [48] that the anticlinic antiferroelectric state is optically isotropic if the tilt angle is $\theta = 45°$ *and* $\Phi = 109.5°$ Actually for the experimentally studied material the angles are $\theta = 47°$ and $\Phi \sim 106°$, which results fairly good optical isotropy. (Note that anticlinic ferroelectric SmC$_A$P$_S$ states can also be optically isotropic if the $\theta = 45°$ *and* $\Phi = 109.5°$ condition is fulfilled.)

The optically isotropic structures are in fact unusually frequent in case of bent-core compound compared to the calamitic LCs, and were found also on materials

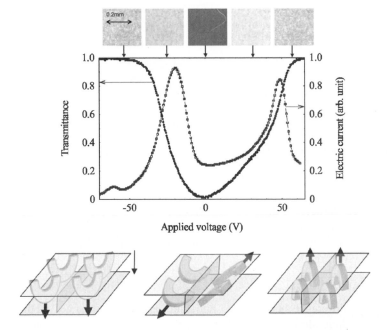

Figure 2-9. Voltage dependence (triangular waveform, f=31Hz) of the transmittance and the polarization current of a 4μm PBUBB cell in between crossed polarizers. At zero fields the material is in the optically isotropic antiferroelectric state. At increasing fields birefringence increases smoothly. Textures at different voltages At increasing fields the birefringence increases from zero to $\Delta n = 0.19$, and the color changes from black to blue The pictures represent 0.2mm × 0.2mm areas

with opening angles far from the tetrahedratic angle both in antiferroelectric [58, 57] and ferroelectric [23] phases. In these cases the absence of birefringence were explained assuming arbitrarily oriented sub-visible size birefringent domains [57]. Such materials are usually referred as "dark conglomerates [20, 63].

2.2.5   *Optical activity*

Chirality is usually seen optically in the form of rotation of the optical axis of a linearly polarized light crossing a chiral material (optical activity). In isotropic liquids optical activity requires chiral molecules, which results in typically of about 1 degree rotation of a light crossing 1 cm slab. In liquid crystals molecular chirality leads to helical structure, which enhances the optical activity so that the optical rotation (OR) can be as large as 100 deg/μm in some short pitch cholesteric or SmC* materials.

In achiral rod-shape molecules one does not expect any optical activity, but as we have seen achiral bent-core liquid crystals can have chiral layer structures. An interesting question therefore if we see optical activity in those materials. Observing planar textures of bent-core liquid crystals between slightly uncrossed polarizers, it is indeed often found [40, 50–59] that the texture splits into darker and brighter domains. For polarizers uncrossed in the opposite sense the darker and brighter domains exchange and for crossed polarizers they have the same brightness. This shows optical activity with OR $\sim 0.1 - 1 \deg/\mu m$. Except for the example demonstrated in Ref. [51], optical rotations were seen only on optically isotropic samples, although not all optically isotropic samples show observable optical rotation [60], or not in any alignment [52]. The reason for the optical activity is presently under active debate.

First it was argued to be due to a chiral molecular configuration characteristic of the particular type of bent-shape molecules, such as twisted or propeller shape (conformational chirality). The concept of conformational chirality was supported by simulations by Earl et al. [61], and was demonstrated by the observation that doping calamitic cholesteric liquid crystal by achiral bent-core molecules can lead to a decrease of the helical pitch, indicating an enhanced rotatory power of the mixture [62]. Unfortunately there is no proof that the decrease of the pitch is not due to a decrease of the twist elastic constant caused by the addition of bent-core units. Although the conformational chirality is usually not questioned in the solid B4 phase [20], its role has been questioned by Walba et al. [20] by arguing that these chiral conformations have very short lifetime, therefore they average out in fluid smectic, such as SmCP or SmC_G phases.

As an alternative to conformational chirality, Ortega et al.[57] modeled the SmC_AP_A with a locally achiral dielectric tensor where optical axis rotating with a pitch of two layers. This indicates that layer scale structural chirality can lead to observable optical rotation. Recently Hough and Clark have extended this model to other SmCP subphases, such as SmC_SP_S, SmC_AP_S and SmC_SP_A. They have shown [63] that the optical rotation of the unhelixed SmC_SP_S is comparable to that of SmC_AP_A. In addition they have also shown that even the racemic SmC_AP_S and SmC_SP_A may also present optical activity for lights not parallel or perpendicular to the smectic layers. It was found that the calculated optical rotation values have the same order of magnitude that is usually observed. They may also account for the observations [23, 64] where optical activity was observed only after repeated heating – cooling cycles to the crystalline, or B4 phase [65, 66] when surface memory effects may result in tilted layers with respect to the substrates. One therefore may be inclined to believe that there is no need of conformation chirality to explain the optical activity in the fluid tilted smectic phases, because the layer scale chirality can explain all observations. However, experimentally it is usually observed that the size of the domains with opposite optical rotation is typically in the range of $100-300\mu m$, whereas the size of domains with uniform layer chirality is only about $2-10\mu m$, i.e. an order of magnitude smaller. Since the optical activity is detected typically on optically isotropic samples [63, 65] where the homochiral domains are not visible the difference between the homochiral and optically active domains was not obvious until recently when the optically active domains were observed even in the SmC_SP_S state [51] (Figure 2-10). It is clearly seen that the domain boundaries do not correlate to those of different layer-chirality, which challenges the layer-scale chirality concept. In all cases we have observed an evolution of the optically

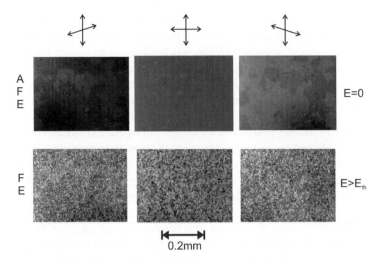

Figure 2-10. Textures between slightly uncrossed polarizers of the transformed state in antiferroelectric (upper row) and in the field-induced ferroelectric state (lower row). Arrows indicate the polarizer – analyzer configurations for the textures beneath

active domains after several cooling-heating cycles, and we suspect that they always related to surface memory effects. The optically active domains form only in the crystalline or close to crystalline structures such as B4 and B3 phases and those domains are imprinted in the surfaces.

2.2.6   Molecular chirality

Chirality can also be introduced when one or more chiral carbons are incorporated in the molecules, for example in the hydrocarbon terminal chains [67, 68], within the bent-core [69], or by addition of chiral dopants [6, 70]. It was noted during the early research that the handedness of the homochiral structures is very sensitive to chiral dopants [6], or even on chiral surfaces [71]. On the other hand, it was observed that banana-smectics made of enantiomeric chiral molecules form synclinic – antiferro- electric [44] and anticlinic ferroelectric [67] domains. This combination of tilt and polar order implies that the phase is racemic, with a rigid alternation of right- and left-handed chiral layers. This shows that the molecular chirality has no or minor effect on deciding about anticlinic or synclinic packing (which is mainly determined by entropic reasons), but it can bias the otherwise degenerate tilt directions.

Experiments by Binet et al. [72], carried out on chiral bent core materials containing biphenyl (BP) cores with S or R hydrocarbon chains, and on achiral biphenyl core molecules with chiral dopants, reveal the effect of the molecular chirality on the polarization structure. In addition the polarization P_b due to the closed packing of the bent-shape molecules, another polarization, P_c is introduced due to the chiral and tilted molecular structure. It was found [72] that in the antiferroelectric racemic domains at low fields, P_b of the synclinic – racemic domains averages out but, due to the synclinic order and of the chiral molecules, a P_c normal to the tilt plane similar to the SmC* phase is possible. In this case, a relatively low electric field is able to unwind the helical structure, but would not be able to switch the antiferroelectric P_b, which requires a higher threshold. Upon this antiferroelectric to ferroelectric transition, the synclinic structure becomes anticlinic and P_c vanishes, leaving only a P_b. Although the textural observations were consistent with this picture it was not possible to measure P_c separately, because in case of the enantiomeric molecules isolated metastable chiral (anticlinic antiferroelectric) domains are also present, where both P_c and P_b average out below the transition to the ferroelectric state where the effective polarization becomes $P_b + P_c$. Very recently however the chirality – induced polarization P_c could be separated from the bent-core packing related polarization P_b [73]. These studies were carried out on binary mixtures of an achiral fluorinated (B-2F) and a chiral material (B-Ch) where a chiral cholesterol unit was attached to one end of the bent core. At low B-Ch concentrations P_c could be switched and measured below P_b was switched. It is found that P_c is an order of magnitude smaller than P_b, but larger than of the pure B-Ch. This is because the molecular dipoles of *B-2F* are much larger than of *B-Ch* due to the presence of the highly polar fluorine atoms.

2.2.7   Chirality transformations

Although in the majority of the banana smectics the chirality is conserved during switching between AFE and FE states it was observed that in some materials strong fields cause a gradual change of the layer chirality [74]. In some examples both the opaque racemic *AFE* and the transparent chiral *AFE* states were found to be stable and could be interchanged. [74, 75], or the sign of the layer chirality even could be flipped [76]. Racemic domains can be rendered chiral by surface interactions, too [77]. An example for structural transformations is shown in Figure 2-11 [78].

Antiferroelectric racemic

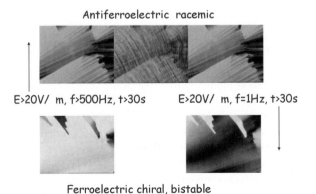

E>20V/ m, f>500Hz, t>30s   E>20V/ m, f=1Hz, t>30s

Ferroelectric chiral, bistable

Figure 2-11. Example of textural transformation observed under different electric fields on a 10μm film of a biphenyl based material

As pointed out by Lansac et al. [79] the synclinic SmC_SP_A enables the out-of-plane fluctuations, thus decreasing the entropy. This effect becomes less important at lower temperatures explaining that the synclinic phases often transform to the anticlinic SmC_AP_A state at lower temperatures, or appears in heating from the B_3 phase [10]. Based on this notion the switching between racemic and chiral states was explained as follows [75]. Assume that the ground state is antiferroelectric (racemic structure) and we apply strong rectangular electric fields that is switching the materials between the ferroelectric states. In case of racemic structure the structure would be anticlinic, which is disfavored entropically. To find the lower free-energy state the system, therefore has to drift to the synclinic ferroelectric state, which is chiral. After field removal this chiral state becomes antiferroelectric and anticlinic, so it can be only metastable. The more stable racemic state can reform either by nucleation process, or it can be driven back to the synclinic racemic state under triangular electric fields, because during switching it stays longer in the antiferroelectric than in the ferroelectric state [75].

All these chirality transformations mean that during the switching the molecules not strictly rotate about the layer normal around the tilt cone, but in some extent also around their long axis. As far as we know this model was first put forward in the 2nd Banana workshop held in Boulder, Colorado [70] and later discussed in several papers. [58, 76]. The rotation of the director around the tilt cone preserves the layer chirality, whereas during rotation around the long axis the chirality changes signs. This rotation is faster than the rotation around the cone, and is permissible only in racemic domains. Indeed it is usually observed that racemic switching is faster than chiral switching [74].

2.3   The B7 Materials

2.3.1   *Textures*

As mentioned before the B7 label denotes materials with peculiar helical super-structures [25],[80–82] including helical filaments [83] and banana leaf– shaped domains [84]. The helical filaments with a proposed telephone – wire type structure, where the smectic layers form concentric cylinders as shown in Figure 2-12.

Other type of interpretation of the textures was proposed by Nastishin et al. [85], who suggested that the B_7 phase is a smectic and columnar phase at the same time. The geometry of the helical filaments (ribbons) is that one of the central region of a screw disclination with a giant Burgers vector split into two disclination lines of strength $1/2$, which bound the ribbon.

The spontaneous chirality that result in the telephone wire type textures at the transition, persist in the fully formed B7 phase, resulting in various astonishing helical superstructures [80], as illustrated in Figure 2-13.

Since the first observations many other materials with similar textures have appeared in the literature [86, 87, 93] however, it is not clear that all the materials with *B*7 textures have really the same structure. For example Bedel et al. found that materials studied in Ref. [86] are not miscible with the reference *B*7 compound [25].

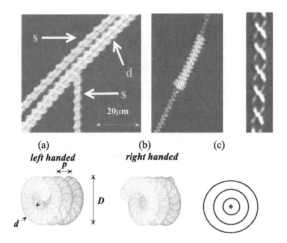

Figure 2-12. Textures of helical filaments forming on cooling from the isotropic phase [83] (upper row) and the proposed telephone cord structure consisting of smectic layers

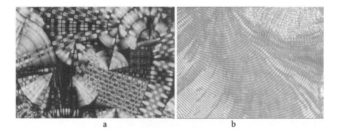

Figure 2-13. Helical superstructures in the B7 phases. (a) the original B7 material [80] (b) a binary mixture of a fluorinated B7' and the original B2 material [92]

Ortega et al. [88] Observed *B*7-like textures in a compound classified in principle as *B*1 [89] after a prolonged electric-field treatment. Other contradictory reports on the features of the *B*7 phase are also pointed out by several authors [90, 91, 43]. In view of all these facts we clearly have to distinguish between at least two different B7 materials.

2.3.2   *Phase structure*

X-ray measurements also clarified that B7 materials have at least two different phase structures. Some of them (for example, the first B7 material [25]) have distinct sharp peaks in the low angle range [16] which is characteristic to columnar structure. Other B7-type materials (we denote them by *B*7') have strong commensurate reflections – which indicate a layer structure – and small incommensurate satellites, which are very close to each other and hard to resolve with normal X-ray technique. These peaks were attributed to a one-dimensional undulation of the smectic layers with a relatively long wavelength modulation in the layer structures [17]. B7 and B7'can be also distinguished by electro-optical and polarization current measurements. The reference B7 – type compounds do not switch electro-optically and do not show polarization peaks at least up to about $40V\mu m^{-1}$ electric fields, whereas the B7' materials usually can be switched under high ($E > 10V/\mu m$) electric fields and posses either ferroelectric [45], or antiferroelectric [93] polarization peaks. Interestingly the threshold for switching decrease with decreasing temperatures [45, 93], which is opposite to the behavior of single tilted smectic materials, such as SmCP or SmC*. Based on these observations, and on field induced polar shifts of transition temperature [92] it has been suggested that the B7' materials have double tilted SmCLP (SmC_G) [92] structures, i.e., they have out–of–plane polarization components (see Figure 2-4).The other model, on the basis of observed layer modulation, assumes modulated in-layer polarization structures [17] in which the spontaneous polarization forms splay domains.

2.4   Free-Standing Filaments

Filaments usually can be pulled only in materials, which become solid during the fiber extruding process, like spider silk, [94] polymer fibers or silica glass fibers. It is also known that columnar liquid crystalline phases of disc-shaped molecules are one-dimensional fluids and they can form stable filaments [95]. The stability of the columnar strands can be explained by taking into account the bulk energy due to compression of the columnar structure, in addition to the surface contribution of the Newtonian fluids. Low molecular weight liquid crystals of rod shape molecules do not form fibers and it was observed [96] that nematic and smectic A bridges collapse at slenderness ratios of $R \approx \pi$ and at $R = 4.2$, respectively. Smectic liquid crystals of rod-shape molecules generally tend to form thin films, [97] similar to cell membranes. Repeated efforts in the past 20 years have failed to produce stable smectic filaments.

In the effort to make free-standing films of the first B7 material it was found that instead of films they form free-standing strands [25] just like columnar liquid crystals of disc-shape molecules [95]. Such observation was confirmed with other B7 materials [98], and later it was found that the B7' and B2 type banana-smectics also form strands of fibers [99]. Although the most stable fibers are the B7 materials with slenderness ratio as large as 5000, the B7' and B2 fibers also have aspect ratios over 1000 and 100, respectively. The values are orders of magnitudes larger than of the Rayleigh-Plateau limit [100] of Newtonian isotropic fluids and of nematic and smectic liquid crystals of rod-shape molecules [96].

B7 and B2 type fibers are completely stable, and are much thinner (1.5–10μm) than conventional extruded fibers (10–100μm). Interestingly metastable fibers could be pulled even in the nematic phase of bent-core materials and their slenderness ratios are much larger than those observed in nematic liquid crystals of rod-shape molecules [96]. This behavior is typical for viscoelastic fluid

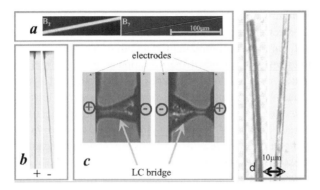

Figure 2-14. Freely suspended B2 and B7'and crystalline strands and bridges. a) Fibers viewed between crossed polarizers. Left side: a bundle of fibers in the B2 phase. The length of the entire fiber is 2mm. Right hand side: a single fiber in the B7' phase. The length of the whole fiber is 1cm. b) Photomicrograph of a B7 fiber with 2μm diameter at one edge when positive and negative voltages (200V) applied (total length is 1mm). c) 'Push-pull' effect in a B7' bridge under ±100V applied. The pictures represent 125μm × 125μm areas. d) The B7 fiber in the crystal phase at room temperature detached from the support. Reprinted figure with permission from A. Jákli, D. Krüerke, G.G. Nair, Phys. Rev. E Vol. 67, 051702, (2003), Copyright 2003 by the American Physical Society

fibers [101], and indicates that nematic bent-core liquid crystals may be considered as fluid agglomerates of smectic clusters [31].

The fibers possess helical properties: they form a helix when shrinking, or particles attached to it rotate during pulling and shrinking. Under dc electric fields applied along bundles of filaments some of the individual helical fibers unwind, then rupture, whereas others remain stable. This indicates polar nature of the individual fibers: those that rupture may have polarization antiparallel to the field; those that stabilize are polarized parallel to the field. Under alternating fields periodic flow may occur along the fibers. The direction of the flow depends on the sign of the electric fields, but in the different strands it can be opposite. In the *B*2 bundles the flow effects average out and the overall diameter does not change. In the *B*7 fibers, however the flow direction seem to be the same in all strands and a net flow is observed. Relatively thick and short B7' bridges posses' strange push-pull effects, as shown in Figure 2-14/c: one polarity can pull all the material from one side to the other and the other polarity can reverse it. Such effects were not observed in the *B*2 bridges indicating structural differences.

In the long single *B*7 fibers a transversal vibration could be observed under electric fields applied along the fibers. Interestingly the vibration takes place with the frequency of the electric field: the string is bent for one polarity of the voltage and becomes straight again for the other polarity. An example is shown in Figure 2-14/b, where the maximum deviation from the straight direction is about 5 degrees. The string is anchored at the end plates and moves sideways in the middle. The sensitivity of the length of the fiber to the sign of the field confirms that the fibers are polar. When the polarization is antiparallel to the applied field the electric field facilitates the fiber formation, which (at constant end-to-end distances) means that the fiber should be bent. When the polarization is antiparallel to the electric field the fiber formation is not favored just minimizing the length of the fiber. When the fiber snaps back after field reversal, or when the vibration is induced by a pulse we observe a small vibration with a natural frequency of about 10 Hz for strings of a $D = 2\mu m$ diameter and $L = 0.5mm$ length. From this the surface tension of the filaments were estimated as $\gamma \sim 30\,\text{mN/m}$, which is in good agreement with a more recent and more precise results of $\gamma = 26\,\text{mN/m}$ obtained by Eremin et al. [102] in a measurement wherein they have pulled a B7 filament with the cantilever of an Atomic Force Microscope and measured the force.

It is very important to note that the fibers remain stable even when cooled to the crystal phase, thus one can detach the fiber from the support and study them by methods that are not available when they sit between the supports. Parts of typical fibers in the *B*2, *B*7 and *B*7' fibers and bridges, in between the supports and when detached in their crystalline phase, are shown in Figure 2-14/d.

Recently Stannarius et al. [103] has found that the geometrical structures of two phases of the same material can be practically identical, whereas filaments are stable only in the higher temperature mesophase. They also conclude that probably the spontaneous curvature of the layers, which might be caused by the out of plane polarization component of the polarization favors the formation of the cylindrical fibers rather than the free-standing films.

Based on the experimental results it was proposed [99] that the fibers have a jelly-roll type structure, i.e. it consists of concentric smectic layers. Although this model was basically supported by recent freeze fracture measurements [17] quenched fibers also show grooves (periodic modulation) of the layers. These grooves have been explained in terms of the polarization modulation, however they cannot account for the stability of the fibers. Presently we are working on a unified model that combines the layer modulation model with an additional out of plane polarization, which seems to be able to explain the stability of the fibers [104].

As a summary, we have reviewed the main experimental results and concepts concerning the physical properties of the liquid crystals of bent-core molecules. We demonstrated that a number of seminal findings and basically new concepts emerged in the field in the last decade. However scientists regularly explore "unusual", "surprising" and not understood phenomena, and there are more unexplained observations than well understood ones. We are completely sure that the next decade of the physics of bent-core liquid crystals will bring new physics, and will be as rich in beautiful observations as was the first decade.

3. ACKNOWLEDGEMENTS

This work was supported by the Focused Research Grant: Ferroelectric phenomena in soft matter systems NSF-DMS-0456232.

4. REFERENCES

1. Y. Matsunaga, S. Miyamoto, *Mol. Cryst. Liq. Cryst.*, **237**, 311–317, (1993); H. Matsuzaki, Y. Matsunaga, *Liq. Cryst.*, **14**, 105–120, (1993)
2. D. Vorländer, A.Apel, "Die Richtung der Kohlenstoff-Valenzen in Benzolabkömmlingen (II.)." *Berichte der Deutschen Chemischen Gesellschaft*, 1101–1109, (1932)
3. Y. Matsunaga, S. Miyamoto, *Mol. Cryst. Liq. Cryst.*, **237**, 311-317, (1993); H. Matsuzaki, Y. Matsunaga, *Liq. Cryst.*, **14**, 105–120, (1993)
4. T. Niori, T. Sekine, J. Watanabe, T. Furukawa, H. Takezoe, *J. Mater. Chem.* **6(7)**, 1231–1233, (1996); T. Sekine, T. Niori, M. Sone, J. Watanabe, S.W. Choi, Y. Takanishi, H. Takezoe, *Jpn. J. Appl. Phys.*, **36**, 6455 (1997); D.R. Link, G. Natale, R. Shao, J.E. Maclennan, N.A. Clark, E. Körblova, D.M. Walba, *Science* **278**, 1924–1927 (1997)
5. R.B. Meyer, L. Liebert, L. Strzelecki, P. Keller, *J. Phys.* (France) **36**, L69-L74 (1975)
6. D.R. Link, G. Natale, R. Shao, J.E. Maclennan, N.A. Clark, E. Körblova, D.M. Walba, *Science* **278**, 1924–1927 (1997)
7. H.R. Brand, P.E. Cladis, H. Pleiner, *Eur. Phys. J. B*, **6**, 347 (1998)
8. P.G. de Gennes, "The Physics of Liquid Crystals", Clarendon Press, Oxford (1975)
9. Niori, T., Sekine, T., Watanabe, J., Furukawa, T., Takezoe, H., *J. Mater. Chem.* **6(7)**, 1231–1233, (1996); Sekine, T., Niori, T., Sone, M., Watanabe, J., Choi, S.W., Takanishi, Y., Takezoe, H., *Jpn. J. Appl. Phys.*, **36**, 6455 (1997)
10. G. Heppke, A. Jákli, D. Krüerke, C. Löhning, D. Lötzsch, S. Paus, R. Rauch, K. Sharma, ECLC'97 Abstract book, 34 (1997); Macdonald, R., Kentischer, F., Warnick, P., Heppke, G., Phys. Rev. Lett., **81**, 4408 (1998)
11. Weissflog, W., Lischka, C., Scharf, T, Pelzl, G, Diele, S., Kurth, H., Proc. Spie: Int. Soc. Opt. Eng., **3319**, 14 (1998), Diele, S., Grande, S., Kurth, H., Lischka, C., Pelzl,G., Weissflog, W., Wirth, I., *Ferroelectrics*, **212**, 169 (1998); Pelzl, G., Diele, S., Grande, S., Jákli, A., Lischka, C., Kresse, H., Schmalfuss, H., Wirth, I., Weissflog, W., Liq. Cryst., **26**, 401 (1999) ; G. Pelzl, S. Diele, A. Jákli, CH. Lischka, I. Wirth, W. Weissflog, *Liquid Crystals*, **26**, 135–139 (1999)
12. Jákli, S. Rauch, D. Lötzsch and G. Heppke, *Phys. Rev. E*, **57**, 6737–6740 (1998), A. Jákli, Ch. Lischka, W. Weissflog, S. Rauch, G. Heppke, *Mol. Cryst. Liq. Cryst*, **328**, 299–307 (1999)
13. H.R. Brand, P.E. Cladis, H. Pleiner, *Eur. Phys. J. B*, **6**, 347 (1998); P.E. Cladis, H.R. Brand, H. Pleiner, *Liquid Crystals Today*, **9**, 1 (1999) ;
14. Nguyen, H.T., Rouillon, J.C., Marcerou, J.P., Bedel, J.P., Barois, P., Sarmento, S., *Mol. Cryst. Liq. Cryst.*, **328**, 177 (1999)
15. R. Amaranatha Reddy, C. Tschierske, J. Mater. Chem. (2006); preprint DOI: 10.1039/b504400f; M.B. Ros, J.L. Serrano, M. R. de la Fuente, C. L. Folcia, J. Mater. Chem., 15, 5093 (2005)
16. G. Pelzl, S. Diele, W. Weissflog, *Adv. Mater.*, **11**, 707–724 (1999)
17. D.A. Coleman, J. Fernsler, N. Chattham, M. Nakata, Y. Takanishi, E. Körblova, D.R. Link, R.-F. Shao, W. G. Jang, J.E. Maclennan, O. Mondainn-Monval, C. Boyer, W. Weissflog, G. Pelzl, L-C. Chien, J. Zasadzinski, J. Watanabe, D.M. Walba, H. Takezoe, N.A. Clark, *Science*, **301**, 1204–1211 (2003)

18. T. Sekine, T. Niori, J. Watanabe, H. Takezoe, *Jpn. J. Appl. Phys.*, 36, L1201 (1997); S.W. Choi, Y. Kinoshita, B. Park, H. Takezoe, T. Niori, J. Watanabe, *Jpn. J. Appl. Phys.*, **37**, 3408 (1998)

19. F. Kentischer, R. Macdonald, P. Warnick, G. Heppke, *Liq. Cryst.*, **25**, 341 (1998)

20. D. Walba, L. Eshdat, E. Körblova, R. K. Shoemaker, *Crystal Growth and design*, 5 (6), 2091 (2005)

21. G. Heppke, D. Moro, *Science*, **279**, 1872–1873 (1998)

22. L.E. Hough, M. Spannuth, H.J. Jung, J. Zasadzinski, D. Krüerke, G. Heppke, D. Walba, N.A. Clark, *Bull. Am. Phys. Soc.*, **50**, 223 (2005)

23. G. Liao, S. Stojadinovic, G. Pelzl, W. Weissflog, S. Sprunt, A. Jákli, *Physical Review E*, **72**, 021710 (2005)

24. S. Diele, S. Grande, H. Kruth, C. Lischka, G. Pelzl, W. Weissflog, I. Wirth, *Ferroelectrics*, **212**, 169 (1998)

25. G. Pelzl, S. Diele, A. Jákli, CH. Lischka, I. Wirth, W. Weissflog, *Liq. Cryst.*, **26**, 135–139 (1999)

26. A. Eremin, S. Diele, G. Pelzl, H. Nadasi, W. Weissflog, J. Salfetnikova, H. Kresse, *Phys. Rev. E*, **64**, 051707-1-6 (2001)

27. J. Matraszek, J. Mieczkowski, J. Szydlowska, E. Gorecka, *Liq. Cryst.*, **27**, 429–436 (2000); I. Wirth, S. Diele, A. Eremin, G. Pelzl, S. Grande, L. Kovalenko, N. Pancenko, W. Weissflog, *J. Mater. Chem.*, **11**, 1642–1650 (2001); W. Weissflog, H. Nádasi, U. Dunemann, G. Pelzl, S. Diele, A. Eremin, H. Kresse, *J. Mater. Chem*, **11**, 2748–2758 (2001);

28. E. Mátyus, K. Keserü, *J. Mol. Struct.*, **543**, 89 (2001);

29. T.J. Dingemans, E.T. Samulski, *Liq. Cryst.*, **27**, 131–136 (2000)

30. K. Fodor-Csorba, A. Vajda, G. Galli, A. Jákli, D. Demus, *Macromolecular Chemistry and Physics*, **203**, 1556–1563 (2002); S. Demel, C. Slugovc, F Stelzer, K Fodor-Csorba, G. Galli, *Macromol. Rapid. Commun.* 2003, 24, 636

31. S. Stojadinovic, A. Adorjan, H. Sawade, A. Jakli, and S. Sprunt, "*Phys. Rev. E, Rapid Communications* **66**, 060707-1-4 (2002)

32. J. Olivares, S. Stojadinovic, T. Dingemans, S. Sprunt, and A. Jakli, *Phys. Rev. E* **68**, 041704-1-6 (2003)

33. V. Domenici, M. Geppi, C. A. Veracini, R. Blinc, A. Lebar, and B. Zalar, *J. Phys. Chem. B*, **109**, 769 (2005).

34. D. Wiant, J. T. Gleeson, N. Eber, K. Fodor-Csorba, A. Jakli, and T. Toth-Katona, *Phys. Rev. E* **72**, 041712 (2005); D. Wiant, S. Stojadinovic, K. Neupane, S. Sharma, K. Fodor-Csorba, A. Jákli, J. T. Gleeson, and S. Sprunt, *reprint: electronic-Liquid Crystal Communications*, **http://www.e-lc.org/docs/2005_11_23_10_05_03**

35. V. Domenici, C.A. Veracini, B. Zalar, *Soft Matter*, **1**, 408–411 (2005)

36. A. Roy, N.V. Madhusudana, P. Toledano, A.M. Figureiredo Neto, *Phys. Rev. Lett.*, **82**, 1466–1469 (1999);

37. T.C. Lubensky, L. Radzihovsky, *Phys. Rev. E*, **66**, 031704-1-27 (2002); L. Radzihovsky and T. C. Lubensky, *Europhys. Lett.* **54**, 206–212 (2001).

38. H. R. Brand, H. Pleiner, and P. E. Cladis, Eur. Phys. J. E 7, 163–166 (2002).

39. Aharya, B; Primak, A; Kumar, S; *Phys. Rev. Lett.*, **92**, 145506 (2004); L.A. Madsen, T.J. Dingemans, M. Nakata, E.T. Samulski, *Phys. Rev. Lett.*, **92** (14), 145505 (2004)

40. G. Heppke, A. Jákli, D. Krüerke, C. Löhning, D. Lötzsch, S. Paus, S. Rauch and K. Sharma, Abstracts, European Conference on Liquid Crystals, Zakopane, Poland, ISBN# 83-911181-8-1 (1997), p. 34.;

41. W. Weissflog, Ch. Lischka, I. Benné, T. Scharf, G. Pelzl, S. Diele and H. Kruth, Abstracts, European Conference on Liquid Crystals, Zakopane, Poland, ISBN# 83-911181-8-1 (1997) p.201; S. Diele, S. Grande, H. Kruth, Ch. Lischka, G. Pelzl, W. Weissflog, I. Wirth, *Ferroelectrics*, **212**, 169–177 (1998)

42. J. Xu, R. L. B. Selinger, J. V. Selinger, and R. Shashidhar, *J. Chem. Phys.* **115**, 4333–4338 (2001).

43. D.M. Walba, E. Körblova, R. Shao, J.E. Maclennan, D.R. Link, M.A. Glaser, N.A. Clark, *Science*, **288**, 2181–2184 (2000)

44. E. Gorecka, D. Pociecha, F. Araoka, D.R. Link, M. Nakata, J. Thisayukta, Y. Takanishi, K. Ishikawa, J. Watanabe, H. Takezoe, *Phys. Rev.E*, **62**, R4524-R4527 (2000)

45. S. Rauch, P.Bault, H. Sawade, G. Heppke, G.G. Nair, A. Jákli, *Phys. Rev. E*, **66**, 021706, (2002)

46. A. Jákli, D. Krüerke, H. Sawade, L-C. Chien, G. Heppke, *Liq. Cryst.*, **29**, 377–381 (2002)

47. J.V. Selinger, *Phys. Rev. Lett.*, **90**, 165501 (2003)

48. A. Jákli, K. Fodor-Csorba, IMID'03 Digest, 1108–1111 (2003)

49. K. Fodor-Csorba, A. Vajda, G. Galli, A. Jákli, D. Demus, *Macromolecular Chemistry and Physics*, **203**, 1556–1563 (2002)

50. J. Thysayukta, Y. Nakayama, S. Kawauchi, H. Takezoe, J.Watanabe, J. Am. Chem Soc., 122, 7441 (2000); J. Thysayukta, H. Takezoe, J.Watanabe, *Jpn. J. Appl. Phys.*, **40**, 3277–3281 (2001); T. Imase, S. Kawauchi, J. Watanabe, *J. Mol. Str.*, **560**, 275 (2001)

51. A. Jákli, Y.M. Huang, K. Fodor-Csorba, A. Vajda, G. Galli, S. Diele, G. Pelzl, *Advanced Materials* **15**, (19) 1606-1610 (2003);

52. G. Liao, S. Stojadinovic, G. Pelzl, W. Weissflog, S. Sprunt, A. Jákli, *Physical Review E*, **72**, 021710 (2005)

53. K. Kumazawa, M. Nakata, F. Araoka, Y. Takanishi, K. Ishikawa, J. Watanabe, H. Takezoe, *J. Mater. Chem.*, **14**, 157–164 (2004)

54. T. Sekine, T. Niori, M. Sone, J. Watanabe, S.W. Choi, Y. Takanishi, H. Takezoe, *Jpn. J. Appl. Phys.*, **36**, 6455 (1997)

55. G. Pelzl, A. Eremin, S. Diele, H. Kresse, W. Weissflog, *J. Mater. Chem.*, **12**, 2591 (2002); W. Weissflog, S. Sokolowski, H. Dehne, B. Das, S. Grande, M.W. Schröder, A. Eremin, S. Diele, G. Pelzl, H. Kresse, *Liq. Cryst.*, **31**, 923 (2004)

56. R. Amaranatha Reddy, B.K. Sadashiva, S. Dhara, Chem. Commun, 1972 (2001); H. N. Shreenivasa Murthy, B.K. Sadashiva, *Liq. Cryst.*, **29**, 1223 (2002)

57. C.L. Folcia, J. Etxebarria, N. Gimeno, M.B. Ros, *Phys. Rev. E* **68**, 011707 (2003)

58. J. Etxebarria, C.L. Folcia, J. Ortega, J. Ortega, M.B. Ros, *Phys. Rev. E*, **67**, 042702 (2003)

59. P. Pyc, J. Mieckowski, D. Pociecha, E. Gorecka, B. Donnio, D. Guillon, *J. Mater. Chem.*, **14**, 2374 (2004)

60. M. Y. M. Huang, A. M. Pedreira, O. G. Martins, A. M. Figueiredo Neto, and A. Jákli, *Phys. Rev. E*, **66**, 031708 (2002)

61. D.J. Earl, M.A. Osipov, H. Takezoe, Y. Takanashi, M. R. Wilson, *Phys. Rev. E* , **71**, 021706-1-11 (2005)

62. J. Thisayukta, H. Niwano, H. Takezoe, J. Watanabe, *J. Am. Chem. Soc.*, **124**, 3354 (2002); b) E. Gorecka, M. Cepic, J. Mieczkowski, M. Nakata, H. Takezoe, B. Zeks, *Phy. Rev. E.*, **67**, 061704 (2003); c) M. Nakata, Y. Takanishi, J. Watanabe, H. Takezoe, Phys. Rev. E, **68**, 04710 (2003)

63. L.E. Hough, N.A. Clark, *Physical Review Letters*, **95**, 107802-1-4 (2005)
64. H. Niwano, M. Nakata, J. Thisayukta, D.R. Link, H. Takezoe, J. Watanaba, *J. Phys. Chem.*, **108**, 1489–1496 (2004)
65. J. Thisayukta, H. Takezoe, J. Watanabe, *Jpn. J. Appl. Phys.*, **40**, 3277 (2001)
66. H. Kurosu, M. Kawasaki, M. Hirose, M. Yamada, S. Kang, J. Thisayukta, M. Sone, H. Takezoe, J. Watanabe, *J. Phys. Chem.*, **108**, 4674 (2004)
67. M. Nakata, D.R. Link, F. Araoka, J. Thisayukta, Y. Takanishi, K. Ishikawa, J. Watanabe, H. Takezoe, *Liq. Cryst.*, **28**, 1301–1308 (2001)
68. K. Kumazawa, M. Nakata, F. Araoka, Y. Takanishi, K. Ishikawa, J. Watanabe, H. Takezoe, *J. Mater. Chem.*, 2004, **14**, 157; b.) C.-K. Lee, S.-S. Kwon, T.-S. Kim, E.-J. Choi, S.-T. Shin, W.-C. Zin, D.-C. Kim, J.-H. Kim, L.-C. Chien, *Liq. Cryst.*, **30** (12), 1401 (2003).
69. G. Gesekus, I. Dierking, S. Gerber, M. Wulf and V. Vill, L, *Liq. Cryst.*, **31**, 145 (2003); b) J. P. F. Lagerwall, F. Giesselmann, M. D. Wand, D. M. Walba, *Chem. Mater.*, **16**, 3606 (2004); c) R. Amarnath Reddy, B. K. Sadashiva, U. Baumeister, *J. Mater. Chem.*, **15**, 3303 (2005)
70. Lecture at 2nd banana workshop, Boulder, August 23. 2002, http://anini.colorado.edu/bananas/jakli.pdf
71. A. Jákli, G.G. Nair, C.K. Lee, L.C. Chien, *Phys. Rev. E*, **63**, 061710-1-5 (2001)
72. C. Binet, S. Rauch, Ch. Selbmann, Ph. Bault, G. Heppke, H. Sawade, A. Jákli, *proceedings of German Liquid Crystal Workshop*, Mainz (2003)
73. A. Jákli, G. Liao, U.S. Hiremath, C. V. Yelamaggad, Physical Review E, *Phys. Rev. E*, **74**, 041706 (2006)
74. G. Heppke, A. Jákli, S. Rauch, H. Sawade, *Phys. Rev. E.*, **60**, 5575–5579 (1999)
75. A. Jákli, CH. Lischka, W. Weissflog, G. Pelzl, S. Rauch, G. Heppke, *Ferroelectrics*, **243**, 239–247 (2000)
76. M.W. Schröder, S. Diele, G. Pelzl, W. Weissflog, *ChemPhysChem*, **5**, 99–103 (2004)
77. A. Jákli, G.G. Nair, C.K. Lee, L.C. Chien, *Phys. Rev. E*, **63**, 061710-1-5 (2001)
78. P Bault, Ch Selbman, S Rauch, H Sawade and G Heppke, *Biphenyl–Based Banana Shaped Compounds*, P.611, ILCC 2003, Edinburgh
79. Y. Lansac, P. K. Maiti, N. A. Clark, and M. A. Glaser, *Phys. Rev. E* **67**, 011703 (2003)
80. W. Weissflog, C. Lischka, I. Benne, T. Scharf, G. Pelzl, S. Diele and H. Kruth, *Proc. SPIE*, 1998, **3319**, 14
81. (a) R. Amaranatha Reddy and B. K. Sadashiva, *Liq. Cryst.*, 2002, **29**, 1365; (*b*) R. Amaranatha Reddy and B. K. Sadashiva, *Liq. Cryst.*, 2003, **30**, 273; (*c*) H. N. Shreenivasa Murthy and B. K. Sadashiva, *Liq. Cryst.*, 2003, **30**, 1051; (*d*) H. N. Shreenivasa Murthy and B. K. Sadashiva, *J. Mater. Chem.*, 2003, **13**, 2863; (e) S. Umadevi and B. K. Sadashiva, *Liq. Cryst.*, 32, 1233 (2005)
82. C.-K. Lee, L.-C. Chien, *Liq. Cryst.* **26**, 609 (1999).
83. A. Jákli, CH. Lischka, W. Weissflog, G. Pelzl, A. Saupe, *Liquid Crystals*, **27**, 1405–1409 (2000)
84. J. P. Bedel, J. C. Rouillon, J. P. Marcerou, M. Laguerre, H.T. Nguyen and M. F. Achard, *Liq. Cryst.*, 2000, **27**, 1411
85. Yu. A. Nastishin, M.F. Achard, H.T. Nguyen, M. Kleman, *Eur. Phys. J. E*, **12**, 581 (2003)
86. J. P. Bedel, J. C. Rouillon, J. P. Marcerou, M. Laguerre, H.T. Nguyen and M. F. Achard, *Liq. Cryst.*, 2001, **28**, 1285; (*c*) J. P. Bedel, J. C. Rouillon, J. P. Marcerou, M. Laguerre, H.T. Nguyen and M. F. Achard, *J. Mater. Chem.*, 2002, **12**, 2214

87. D. S. Shankar Rao, G. G. Nair, S. Krishna Prasad, S. Anita Nagamani and C. V. Yelamaggad, *Liq. Cryst.*, 2001, **28**, 1041
88. J. Ortega, J. Etxebarria, C. L. Folcia, J. A. Gallastegui, N. Gimeno, and M. B. Ros, Poster STR-P075 presented at the *20th International Liquid Crystal Conference, Ljubljana, Slovenia*, 2004.
89. J. Ortega, M. R. de la Fuente, J. Etxebarria, C. L. Folcia, S. Díez, J. A. Gallastegui, N. Gimeno, M. B. Ros, and M. A. Pérez-Jubindo, *Phys. Rev. E*, **69**, 011703 _2004
90. J. Mieczkowski, J. Szydlowska, J. Matraszek, D. Pociecha, E. Gorecka, B. Donnio, and D. Guillon, *J. Mater. Chem.*, **12**, 3392 _2002
91. G. Pelzl, M. A. Schröder, U. Dunemann, S. Diele, W. Weissflog, C. Jones, D. Coleman, N. A. Clark, R. Stannarius, J. Li, B. Das, and S. Grande, *J. Mater. Chem.*, **14**, 2492 _2004
92. A. Jákli, D. Krüerke, H. Sawade, G. Heppke, "*Phys. Rev. Lett.*, **86**, (25), 5715–5718 (2001)
93. A. Jákli, G.G. Nair, H. Sawade, G. Heppke, *Liq. Cryst.*, **30** (3), 265–271 (2003)
94. P.J. Willcox, S.P. Gido, W. Muller, D. Kaplan, *Macromolecules*, **29**, 5106-5110 (1996); D.P. Knight, F. Vollrath, *Proc. R. Soc. Lond. B* **266**, 519–523 (1999); F. Vollrath, D.P. Knight, , *Nature*, **410**, 541–548 (2001)
95. D.H. Van Winkle, N.A. Clark, *Phys. Rev. Lett.*, **48**, 1407–1410 (1982)
96. M.P. Mahajan, M. Tsige, P.L. Taylor, C. Rosenblatt, *Liq. Cryst.*, **26**, 443–448 (1996)
97. C.Y. Young, R. Pindak, N.A. Clark, R.B. Meyer, *Phys. Rev. Lett.*, **40**, 773 (1978)
98. D.R. Link, N. Chattham, N.A. Clark, E. Körblova, D.M. Walba, p.322 Abstract Booklet FLC99, Darmstadt (1999)
99. A. Jákli, D. Krüerke, G.G. Nair, *Phys. Rev. E*, **67**, 051702 (2003)
100. J.W. Strutt (Lord Rayleigh), *Proc. Lond. Math. Soc.*, **10**, 4–13 (1879)
101. M. Yao, S.H. Spielberg, G.H. McKinley, *J. Non-Newtonian Fluid Mech.*, **89**, 1–43 (2000)
102. A. Eremin, A. Nemes, R. Stannarius, M. Schulz, H. Nádasi, W. Weissflog, *Phys. Rev. E.*, **71**, 031705 (2005);
103. R. Stannarius, A. Nemes, A. Eremin, *Phys. Rev. E*, **72**, 020702 (R) (2005)
104. Bailey, A. Jakli, to be published

CHAPTER 3

ATOMISTIC-RESOLUTION STRUCTURAL STUDIES
OF LIQUID CRYSTALLINE MATERIALS
USING SOLID-STATE NMR TECHNIQUES
Solid-state NMR of liquid crystals

AKIRA NAITO[1] AND AYYALUSAMY RAMAMOORTHY[2]

[1]*Graduate School of Engineering, Yokohama National University, 79-5 Tokiwadai, Hodogaya-ku,
Yokohama 240-8501, Japan. E-mail: naito@ynu.ac.jp*
[2]*Biophysics Research Division and Department of Chemistry, The University of Michigan, Ann Arbor,
MI 48109-1055, USA. E-mail: ramamoor@umich.edu*

1. INTRODUCTION

In the recent years, an amazing number of functional liquid crystalline molecules has
been successfully designed and developed for commercial applications. However,
further developments in this exciting area of science are unfortunately limited by
the difficulties in obtaining high-resolution structural details of these molecules.
For example, it is essential to understand the atomistic-level three-dimensional
structure of a liquid crystalline molecule in order to understand its properties,
which make them functional, such as the three-dimensional order/disorder, intermolecular packing/interactions and dynamics in different phases. Therefore,
there is considerable current interest in determining high-resolution structures of
functionally important liquid crystalline materials. While there are a number of
physical techniques that have been used to obtain low-resolution structures, NMR
spectroscopy has been the most powerful technique in providing atomistic-level
resolution structures [1–3]. It is possible to obtain such high-resolution structures
using X-ray crystallography if a high quality single crystal of the system can
be obtained. However, it is extremely difficult to obtain single crystals of most
functional molecules particularly in the most interesting nematic phase. Structures
obtained in a crystalline phase of a material need not be useful to understand
the molecular behavior in its nematic phase. In addition, crystal structures do
not provide molecular dynamics information. Similarly, structures solved from

85

A. Ramamoorthy (ed.), Thermotropic Liquid Crystals, 85–116.
© 2007 *Springer.*

solution samples of these materials using NMR experiments are also of limited use. Fortunately, solid-state NMR spectroscopy is capable of providing high-resolution molecular structure, dynamics, order/disorder and geometry in all non-isotropic phases (including crystalline, smectic, nematic and amorphous) of liquid crystalline materials even in a mixed state [3].

NMR spectral lines of non-isotropic solids are broader than that of isotropic solutions. The line broadening is mainly due to anisotropic chemical shift, dipole-dipole and quadrupole (only for nuclei with spins ≥ 1) interactions [2, 4]. These interactions can be suppressed by performing magic angle spinning (MAS) [5, 6] experiments to provide 'solution-like' narrow spectral lines but without denaturing the sample [2, 4]. Multiple radio frequency pulses can also be used to suppress the dipole-dipole interaction [2, 4, 7]. In addition, spectral resolution can be enhanced by the detection of dilute nuclei like 13C instead of protons. However, this approach reduces the sensitivity of the experiment as 13C is a less sensitive nucleus as compared to 1H. To enhance the sensitivity, a cross-polarization (CP) experiment is performed to transfer the proton magnetization to 13C via 1H-13C dipolar inter-action [8, 9]. A combination of CP and MAS are routinely used to record 13C chemical shift spectra of TLCs by decoupling protons. The isotropic chemical shift values obtained from high-resolution spectra are useful to identify the chemical groups in a compound and to further characterize its physicochemical properties [10–13].

We briefly mention about a recent solid-state NMR study on columnar mesophases of octa-alkyloxy tetrabenzo-orthocyclophanes (TBC) to show the usefulness of CPMAS applications [14]. Unlike most discotic mesognes, TBCs consists of a molecular core that is not flat and not even rigid and therefore it can acquire several flexible conformations as shown in Figure 3-1. The orthocyclophane core exists in a sofa conformation with the molecules stacked on top of each other to form columns. This study utilized 13C and 2H NMR experiments to charac-terize the reorientation of pseudorotation/rotation within the core of two homologs, namely TBC-9 and TBC-12. TBC-9 exhibits a single rectangular mesophase (Dr) while TBC-12 exhibits a low temperature Dr and a high temperature hexagonal columnar phase (Dh).

Carbon-13 CPMAS spectra of solid and columnar phases of TBC-9 and TBC-12 obtained at various temperatures are given in Figure 3-2. The chemical shift frequencies and line shapes of the spectral lines are different for different phases. While the spectra of both compounds in the solid phase are similar, the spectral lines of TBC-12 are relatively less revolved. Unlike the spectra of other columnar mesogens where the motion in the mesophase significantly narrows the spectral lines, the spectral lines of TBCs are narrower for the solid phase than that of columnar phases due to the inhomogenous stacking of molecules in the columnar phase. The spectra consist of peaks from aromatic, α-methylene and other aliphatic carbons. The spectra of solid phases in the aromatic region confirm the presence of a single sofa conformation. The mesophase spectra are consistent with a single sofa conformer that undergoes a slow pseudorotation in the NMR timescale.

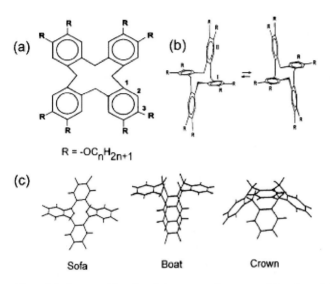

$R = -OC_nH_{2n+1}$

Sofa   Boat   Crown

Figure 3-1. Structure (a), sofa-sofa interconversion process (b), and possible conformations of the octa-alkyloxy orthocyclophane (c). Reprinted from reference 14, with permission from the *Journal of Physical Chemistry*

Use of higher magnetic fields (like 900 MHz) also enhances the sensitivity and resolution of an NMR experiment [15]. On the other hand, the chemical shift anisotropy (CSA) span increases with the magnetic field strength and therefore the sample has to be spun faster at the magic angle to suppress the line broadening CSA interaction. While spinning speeds up to 20 kHz are commonly used to obtain 13C CPMAS spectra of TLCs, it is now possible to spin the sample as fast as 75 kHz. Even though the suppression of line broadening interactions is essential for routine studies, CSA and dipolar or quadrupolar couplings are useful to understand chemical bonding, dynamics, order and geometry of molecules in solids. Studies have shown that multidimensional MAS techniques can be used to obtain high-resolution spectra that contain 'solution-like' spectra (constituting isotropic chemical shifts and scalar couplings) in one frequency dimension and 'solid-like' spectra (constituting CSA, dipolar or quadrupolar couplings) in other dimensions [16–19]. The measurement and analysis of these anisotropic interactions provides a wealth of information about TLCs [20].

Another approach to obtain high-resolution spectra of TLCs is to uniformly align the molecules in the magnetic field and use a CP experiment to record the chemical shift spectrum of S (less sensitive nucleus like 13C) spins by decoupling I (high sensitive nucleus like 1H) spins. Since TLC molecules in the nematic phase spontaneously align in the presence of an external magnetic field, high-resolution spectra can be obtained even though the phase of the material is not isotropic. This unique feature enables the measurement of NMR parameters like chemical shift, quadrupolar coupling and dipolar coupling, which can be used to determine

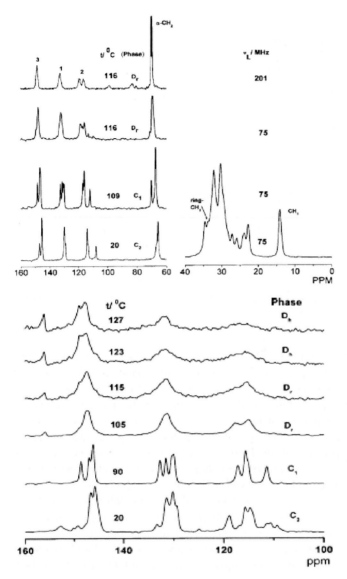

Figure 3-2. Magic angle spinning carbon-13 chemical shift spectra of TBC-9 (top) and TBC-12 (bottom) at different phases and temperature. Reprinted from reference 14, with permission from the *Journal of Physical Chemistry*

the structure, order, dynamics and geometry of molecules in the nematic phase. However, overlap of these anisotropic interactions could lower the resolution of a 1D chemical shift spectrum; on the other hand, multidimensional experiments on a magnetically aligned static sample can be used to resolve the overlapping spectral lines as discussed below. For example, obtaining a high-resolution 13C chemical shift spectrum of a fluorinated compound is difficult without the use of simultaneous 1H and 19F decoupling, which is a technically demanding approach. C-F interactions lead to broadening of 13C spectral lines and therefore limit the measurement of C-F couplings. A recent study investigated the effects of 13C-19F dipolar couplings on the resolution of the 13C chemical shift spectrum of a fluori- nated liquid crystalline compound in the nematic phase (Figure 3-3) [21]. This study demonstrated that using a suitable 1H decoupling pulse sequence it is possible to obtain high-resolution 13C chemical shift spectra and measure 13C-19F dipolar couplings in the nematic phase under static experimental condition. The efficacies of a CW (continuous wave), TPPM [22] and SPINAL [23] decoupling sequences were compared. As shown in Figure 3-3, SPINAL64 was found to provide superior resolution. Since C-F distances are known from the molecular structure, the measured C-F dipolar couplings and 13C chemical shifts can be used to determine the order parameters at each carbon site of the molecule in the nematic phase.

Microscopic orders of liquid crystalline samples have been studied successfully using one-dimensional NMR experiments under static condition that provide 13C chemical shifts and quadrupole couplings in the ^{2}H NMR spectra of ^{2}H-labeled samples [24, 25]. It is also possible to perform experiments at various temperatures to characterize the dynamics of the system. The frequency, lineshape and relaxation are highly dependent on the phase of the sample. Multidimensional experiments on an aligned nematic phase sample can be used to measure these anisotropic interactions for various sites of a molecule, which can provide information on the relative orientations of chemical groups and therefore a high-resolution image of the entire molecule. Typically, 2H quadrupole coupling, 13C CSA, and 1H-13C dipolar couplings are used in such studies. However, in principle, ^{1}H-1H dipolar couplings can also be used for a similar purpose. A direct analysis of a proton spectrum of a liquid crystalline sample, however, has been difficult, because the strongly coupled dipolar spin network causes splitting of individual proton resonances into complex multiplicities of resonance lines, resulting in a highly overlapped one-dimensional NMR spectrum. A successful analysis has usually been performed only after a reduction of the number of protons and a simplification of spin network by partial deuteration [26, 27].

For separating ^{13}C-^{1}H dipolar couplings of individual carbon nuclei in a liquid crystal, typically, two-dimensional separated-local-field (SLF) experiments are performed to obtain 13C chemical shift and 1H-13C dipolar couplings [28–30]. A number of sophisticated laboratory frame and rotating frame SLF sequences have been reported in the literature to determine the microscopic order of liquid crystals. This SLF technique was further modified by combining an off magic angle spinning and a separated local dipolar field (SLF/VAS) [31, 32] and by switching off magic angle spinning to magic angle spinning [33] to obtain scaled ^{13}C-^{1}H dipolar inter- actions in various liquid crystalline samples with an excellent resolution. Among all these SLF experiments, PISEMA [30] [34–38] has provided the narrowest lines in rigid systems [39–44]. Laboratory frame experiments like PDLF [45–47] have

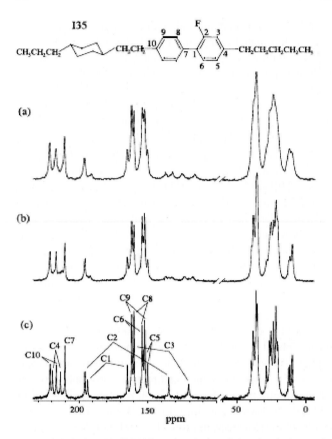

Figure 3-3. Carbon-13 CPMAS spectra (a-c) and the chemical structure of nematic 135 (top). The spectra were obtained using CW (a), TPPM (b) and SPINAL64 (c) decoupling of protons at 40 °C (10 °C above the crystal to nematic phase transition) using a cross-polarization sequence. The C-F interactions degrade the resolution in spectra (a) and (b), while the SPINAL64 enhances the spectral resolution and enables the measurement of C-F dipolar couplings (c). C-F dipolar splittings are indicated in (c). Reprinted from reference 21, with permission from the *Chemical Physics Letters*

recently been shown to provide high-resolution dipolar coupling spectra in less rigid systems. In this Chapter, we present various such SLF methods that have been used in the studies of TLCs. Another Chapter in this book illustrates the applications of some of the SLF methods like PDLF [45–47], PISEMA [30] and HIMSELF [48, 49] to study columnar molecules.

State-correlated 2D NMR (SC-2D NMR) spectroscopy [50] has been developed as an alternative approach to elucidate microscopic order of liquid crystalline samples [51–53], give correlation between native and denatured states of proteins [54], and to reveal correlation between the solid and liquid states of camphor using a CO_2 laser as a heat source [55]. This technique turned out to be useful to observe ^{1}H dipolar coupling patterns of ^{1}H NMR spectra with high-resolution.

In this technique, the local dipolar interaction of individual protons in the liquid crystalline state can be obtained via resonances in the isotropic phase [51–53]. A phase transition from a nematic to an isotropic phase is completed rapidly within the spin-lattice relaxation times of ^{1}H nuclei by applying a pulsed microwave. By this method, homonuclear dipolar interactions associated with individual protons can be separately observed without applying a multiple pulse sequence, and hence a detailed information on the microscopic order parameters, geometry of different chemical groups and 1H spin networks can be obtained. In particular, recent technical improvements in the microwave temperature jump probe have realized a transition in even less than 10 ms [52, 53], and enable us to obtain simpler dipolar patterns. In this Chapter, SC-2D NMR experiments between the nematic and the isotropic phase of liquid crystalline samples are described, in which well separated dipolar pattern for individual protons are observed as cross sectional spectra. Besides, this technique can also provide spin diffusion pathways among proton spin networks. These pieces of information provide insights into the microscopic order of liquid crystalline materials.

2. SLF

Separated-local-field experiments are commonly used to measure the heteronuclear dipole-dipole interactions (called 'local-field' in the molecule) in solids [2, 4, 30], These are two-dimensional pulse sequences as depicted in Figure 3-4 that consists of preparation, evolution and detection periods. A 90° pulse on the I nuclear spin (abundant nuclei like protons) channel followed by the cross-polarization sequence is the typical preparation period of the pulse sequence. This preparation step generates transverse magnetization of S nuclei (less sensitive nuclei like 13C) that evolves during the evolution period (t1) of the sequence under a suitably dressed Hamiltonian. There are a number of pulse sequences that differ in the

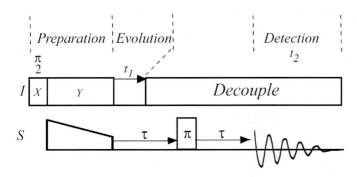

Figure 3-4. A laboratory-frame two-dimensional separated-local-field pulse sequence that correlates the S spin chemical shift in the t_2 period with the *I-S* dipolar couplings in the t_1 period. A multiple pulse sequence is usually applied in the t_1 period to suppress *I-I* interactions that degrade the resolution in the *I-S* dipolar coupling dimension

type of Hamiltonians that are operative during the t1 period. In most solids, the dominant 1H-1H dipolar couplings need to be suppressed in the t1 period in order to obtain high-resolution I-S dipolar coupling spectral lines. Therefore, various multiple pulse sequences have been used in the t1 period. In the detection period (t2), S spin magnetization is acquired by decoupling I spins. Therefore, the resultant 2D spectrum consists of chemical shift of S nuclei in w2 and I-S dipolar couplings in w1 dimension. The I-S dipolar couplings are scaled by the pulse sequence used to suppress I-I interactions in the t1 period.

In the laboratory frame SLF experiments (Figure 3-4), S spins are allowed to evolve under the S spin chemical shift and I-S dipolar couplings in the t1 period. The evolution under the S spin chemical shift in t1 is refocused after t1 typically using a spin echo sequence (tau-180°-tau) pulse while protons are decoupled. On the other hand, in a rotating frame experiment, I and S spins are locked by rf pulses during t1 and the amplitudes of spin-locks are matched to a set resonance condition so that I and S spins can exchange the magnetization via the dipole-dipole interaction (Figure 3-5). Multiple radio frequency pulse sequences are used to spin-lock I spins as they suppress I-I dipolar couplings [30]. In these SLF experiments, the S spin magnetization evolves under several I-S dipolar couplings and therefore the spectral line shapes in the w1 dimension can be complicated to extract the individual I-S dipolar coupling values. For example, if a carbon is dipolar coupled to N protons, 2^N spectral lines are expected in the dipolar coupling spectrum obtained using a laboratory-frame SLF experiment. To overcome this difficulty, there are SLF pulse sequences that utilize the evolution of the I spin magnetization in t1 followed by a transfer of I magnetization to S nuclei. These sequences are called 'proton-detected

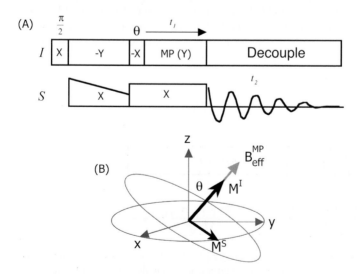

Figure 3-5. (A) A rotating-frame SLF pulse sequence and (B) the effective field direction of the multiple pulse (MP) sequence during t1

local-field' (but protons are not directly detected) or 'proton-encoded local-field' (PDLF) (Figure 3-6) [45–47]. Several studies have demonstrated the advantages of this sequence in the measurement of H-C dipolar couplings from samples that are not labeled with 13C isotope. In a PDLF spectrum, the w1 dimension will have a single dipolar coupling split doublet due to 1H-13C interaction as the proton magnetization evolved only under a single H-C dipolar coupling. We briefly highlight some of the structural studies on TLCs using SLF sequences below.

Figure 3-7 shows a 1D 13C chemical shift and a 2D PDLF spectrum of 5CB liquid crystalline compound in the nematic phase under static conditions [56]. Since the spectral lines are well resolved, C-H dipolar couplings can be directly measured from the 2D spectrum for each carbon site of the molecule.

Figure 3-8 shows a 1D 13C chemical shift and 2D PITANSEMA [39] spectra of a thiophene-based liquid crystal [41]. Unlike the laboratory frame experiments, the rotating frame experiments suppress small I-S couplings as the Hamiltonians do not commute. While this property simplifies the dipolar spectra but at the loss of small dipolar couplings. Among the rotating frame experiments, PISEMA [30] and its variant broadband-PISEMA [37] provides the highest spectral resolution. However, HIMSELF sequences based on BLEW [48] or WIM [49] provide the best

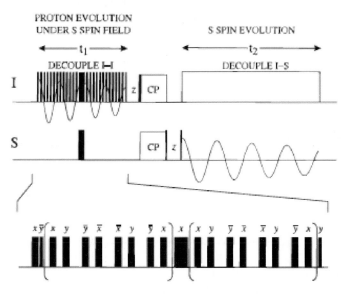

Figure 3-6. A laboratory-frame proton encoded local field (PDLF) two-dimensional pulse sequence. After the evolution under the *I-S* dipolar coupling, the *I* spin magnetization is transferred to *S* spins via cross-polarization, and *S* magnetization acquired by decoupling *I* spins. An MREV-8 pulse sequence is used to suppress *I-I* interactions and a pair of 180° pulses are used to refocus the chemical shift. Since a regular CP could transfer magnetization from various *I* spins to *S* spin via *I-I* interactions, a short contact should be used or a magic angle irradiation (or a Lee-Goldburg sequence) can be used as an *I* spin-lock (called Lee-Goldburg-CP) [30]. Reprinted from reference 47, with permission from the *Journal of Physical Chemistry*

a

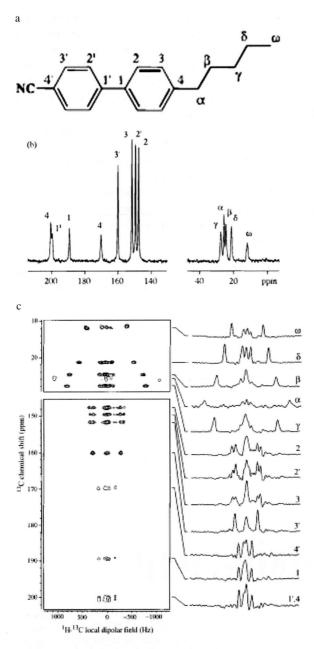

Figure 3-7. (a) Chemical structure of a liquid crystalline compound, 4-*n*-pentyl-4'-cyanobiphenyl (5CB). (b) Carbon-13 chemical shift spectrum of 5CB in the nematic phase obtained via cross-polarization under static condition. Thirteen carbon sites are well resolved. (c) A 2D proton-encoded local field spectrum of 5CB in the nematic phase under static conditions. H-C dipolar coupling spectra for each carbon site are shown. Reprinted from reference 47, with permission from the *Journal of Physical Chemistry*

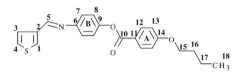

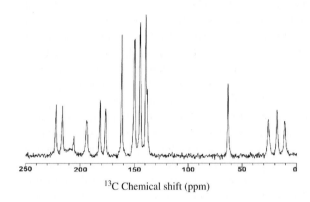

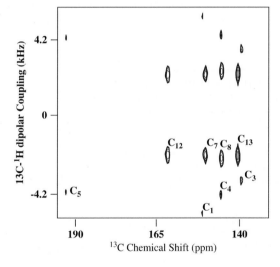

Figure 3-8. Chemical structure of a thiophene-based thermotrophic liquid crystalline compound (top). Carbon-13 chemical shift spectrum of this compound in the nematic phase obtained via cross-polarization under static condition (middle). A 2D PITANSEMA spectrum of this compound in the nematic phase under static conditions (bottom). H-C dipolar coupling spectra for each carbon site are shown. More details about the PITANSEMA pulse sequence can be found elsewhere [39]. Reprinted from reference 41, with permission from the *Journal of the American Chemical Society*

resolution spectra for semi-solids with significant dynamics like in TLCs and other liquid crystalline systems. The details of this aspect of the pulse sequences can be found in a recent publication [57, 58].

3.   STATE-CORRELATED 2D NMR SPECTROSCOPY

The pulse sequence used in the state-correlated two-dimensional (SC-2D) NMR spectroscopy is essentially the same for the radio frequency part as that of a 2D exchange or an NOE experiment (Figure 3-9). The first 90° pulse prepares the transverse magnetization from thermal equilibrium. During the evolution period, the temperature of the sample is kept constant to maintain the nematic phase of the sample so that the ^{1}H spins evolve under strong dipolar interactions between proton nuclei in the nematic phase. After the t_1 period, a second 90° pulse is applied to store the magnetization along the z-axis. During the transition period, a pulsed microwave is applied for a short time to increase the sample temperature so that the nematic phase can be transformed into an isotropic phase. Any remaining transverse magnetization is assumed to diphase within a couple of milliseconds during the transition period under strong dipolar interactions. To study the spin diffusion processes, a mixing time, τ_m, is inserted in the beginning of the transition period. After the third 90° pulse, the free induction decay is acquired in the detection period, t_2, during which the system experiences magnetic interactions in the isotropic phase. Free induction decay signals recorded as functions of t_1 and t_2 are double

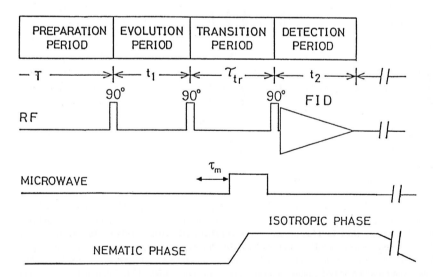

Figure 3-9. Pulse sequence for the state-correlated 2D NMR (SC-2D NMR) experiments. A mixing time τ_m is inserted into the beginning of a transition period to examine the spin diffusion properties. Reprinted from reference 53, with permission from the *Journal of Chemical Physics*

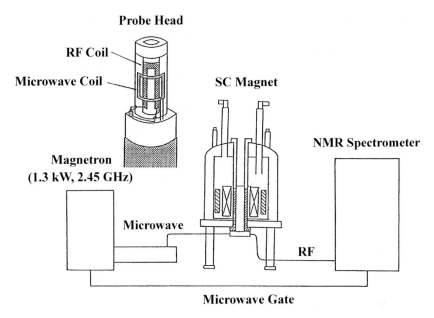

Figure 3-10. Block diagram of the temperature jump NMR spectrometer equipped with a microwave transmitter. Microwave and radiowave coils in the probe head are also shown

Fourier-transformed to generate the 2D NMR spectrum that correlates the nematic and isotropic phases.

The microwave pulse is applied through a microwave circuit built into a JEOL high-resolution ^{1}H NMR probe for liquids (Figure 3-10). A flat copper ribbon, 3 mm in width, is used for the microwave coil, which is wound inside the radio wave coil coaxially to reduce arcing during the microwave irradiation (Figure 3-10). The microwave circuit is tuned properly to 2.45 GHz by using a sweep generator. NMR spectra were recorded at 399.8 MHz on a JEOL GX 400 pulse FT NMR spectrometer, equipped with a pulsed microwave transmitter (IDX, Tokyo Electric Co. Ltd.) capable of transmitting 1.3 kW pulsed microwave at a frequency of 2.45 GHz. After the temperature jump, the sample was cooled down to the nematic phase with a help of a JEOL gas flow temperature controller in order to repeat the experiment for signal averaging.

4. SC-2D NMR SPECTRA OF LIQUID CRYSTALLINE SAMPLES

4.1 Theoretical Background of SC-2D NMR and Analysis of Order Parameters in 4'-methoxybenzylidene-4-acetoxyaniline

4.1.1 *SC-2D NMR experiments of APAPA*

Static one-dimensional ^{1}H NMR spectra of 4'-methoxybenzylidene-4-acetoxyaniline (APAPA) in the isotropic phase at 110 °C and in the nematic phase at

108 °C are shown in Figures 3-3 top and bottom, respectively. The proton signals in the isotropic phase are very narrow, and multiplet patterns due to spin-spin couplings are clearly observed. Therefore, all the peaks were unambiguously assigned to the individual protons as shown in Figure 3-11 (top). In contrast, the signals in the nematic phase at 108 °C were very broad and were overlapped with each other as shown in Figure 3-11 (bottom). It was, therefore, not to assign the NMR patterns to individual protons by a simple inspection of the one-dimensional NMR spectrum.

A nematic/isotropic phase correlated 2D NMR (SC-2D NMR) spectrum of APAPA (Figure 3-12) was obtained by using the pulse sequence shown in Figure 3-1 with $\tau_m = 0$ and a microwave pulse of 10 ms duration. The temperature before the transition was kept at 108°K, and a repetition time of 120 s was used for the system to be completely back to the initial nematic phase. The NMR spectrum in the nematic phase appeared in the $\omega 1$ dimension and that in the isotropic phase in the $\omega 2$ dimension. The cross sections along the $\omega 1$ axis clearly showed different dipolar coupling pattern for individual resonances separated in the $\omega 2$ axis.

Particularly, a triplet pattern with the intensity ratio of 1:2:1 and with a splitting frequency of 10.2 kHz appeared for the α–methyl protons. This indicates the presence of three protons with equal values of dipolar coupling. On the other hand, a broad singlet pattern was observed for the α'-methyl protons, with minor satellite signals in both sides of the central peak. This observation indicates that the dipolar coupling constant of the α'-methyl protons is much smaller than that of the α methyl protons. The satellite lines are attributable to those of the aromatic protons, and are caused by exchange of longitudinal magnetization due to cross relaxation and/or spin diffusion during the transition period under the influence of strong dipolar interactions in the nematic phase.

Doublet patterns were observed for 2,6 and 3,5 protons with a splitting frequency of 9.7 kHz. Doublet patterns were also seen in 2',6' and 3',5' protons, although they were mixed with the dipolar patterns of the α' methyl and 7' protons by spin exchange. It is noted that the doublet patterns for 2',6' and 3',5' protons have splitting frequencies slightly different from those of 2,6 and 3,5 protons.

To characterize the origin of the spectral mixing, SC-2D NMR spectra were recorded with varying τ_m values as shown in Figure 3-13. It is clear that the mixing patterns are dependent on τ_m values. When τ_m was chosen to be 100 ms, the central lines due to the α' protons grew considerably in the resonance lines of 2,6 and 3,5 protons. On the other hand, mixing of signals from 2', 6' and 3', 5' protons did not significantly change even for $\tau_m = 600$ ms because an equilibrium state has already been established at $\tau_m = 100$ ms. It is also noted, that the sum of cross sections, shown in the top of the 2-D spectrum, is invariant during the course of the τ_m variation. These results indicate that spectral mixing occurs more efficiently within the aromatic rings, which are considered to constitute a core group of the liquid crystalline molecule, than between the aromatic and the α' methyl protons. Further, spectral mixing between aromatic and the α methyl protons is very slow, and hence spectral pattern is different from the other even at $\tau_m = 600$ ms.

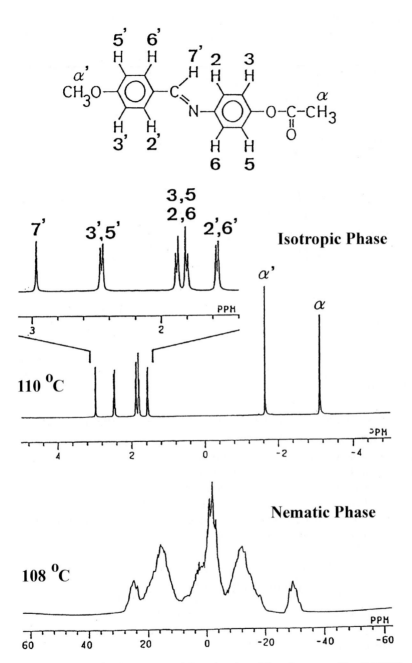

Figure 3-11. Static ¹H NMR spectra of 4'-methoxybenzylidene-4-acetoxyaniline (APAPA) measured at 110 anf 108 °C. Reprinted from reference 53, with permission from the *Journal of Chemical Physics*

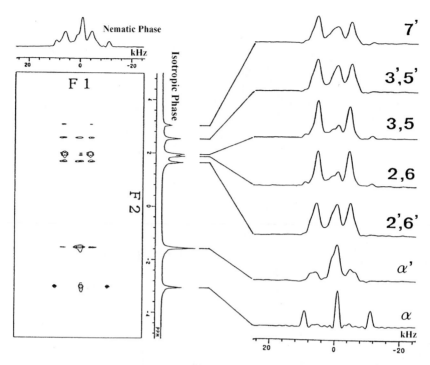

Figure 3-12. A nematic /isotropic phase correlated 2D NMR spectrum (SC-2D NMR) of APAPA with cross sectional patterns by applying the microwave pulses of 10 ms durations. Reprinted from reference 53, with permission from the *Journal of Chemical Physics*

4.1.2   Analysis of the cross sectional spectra in a SC-2D NMR experiment

The cross sections of the SC-2D NMR spectra of APAPA in terms of a limited number of spin interactions (subspin system) in the nematic phase are considered. First, it is assumed that the signals in the F2 dimension are well separated so that transverse magnetization of individual protons I_y^k ($k = 1, 2 \ldots\ldots n$) are detected separately. To obtain the kth cross section, we follow the discussion by Schuff et al [59]. for the case of a dipolar-coupled spin system in the crystal, and calculate the t_1 variation of the component of kth spin by

$$<I_y^k(t_1)> = \mathrm{Tr}\{\rho(t_1)I_y^k\}$$

$$(3\text{-}1) \qquad = \mathrm{Tr}\{\exp(-iH_1t_1)\rho(0)\exp(iH_1t_1)I_y^k\},$$

With

$$\rho(0) = \Sigma_1 I_y^1$$

$$H_1 = -\Sigma_i \nu_L iIzi - 2\Sigma_{i,j}A_{i,j}[I_{zi}I_{zj} - (I_{+i}I_{-j} + I_{-i}I_{+j})/4]$$

$$A_{ij} = (\gamma^2 h/4\pi r_{ij}^3) < 3\cos^2\theta_{ij} - 1 > .$$

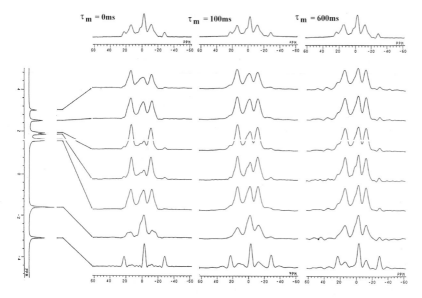

Figure 3-13. Dependence of the cross sectional patterns on the mixing time tm in the SC-2D NMR spectra of APAPA. Reprinted from reference 53, with permission from the *Journal of Chemical Physics*

Here $\rho(0)$ is the density matrix at the start of the evolution period, which evolves under the Hamiltonian in the nematic phase, H_1, to $\rho(t_1)$ at the end of the evolution period t_1. We consider the Zeeman interaction, ν_L, and the secular part of the homonuclear dipolar interaction in H_1, in which θ_{ij} represents the angle between the internuclear vector connecting i and j nuclei and the static external magnetic field. The cross section of the SC-2D NMR spectrum for kth proton, $S_k(\omega_1)$, is then obtained by Fourier transforming, $I_y^k(t_1)$., i.e.

(3-2) $S_k(\omega_1) = \int dt_1, I_y^k(t_1). \exp(-i\omega_1 t_1).$

Figure 3-14 shows simulated SC-2D NMR spectra for the α methyl protons and the 2,3,5,6 ring protons, both of which were calculated as a 7 spin system consisting of the α methyl protons and the 2,3,5,6 ring protons, with effective dipolar coupling parameters of $A_{\alpha\alpha} = -3410$, $A_{\alpha3} = 400$, $A_{\alpha2} = 50$, $A_{32} = 3250$, $A_{35} = -400$, and $A_{36} = -100\,Hz$. Calculations with a Lorentzian half width of $500\,Hz$ (panel b) showed good agreement with the experimentally obtained cross sectional spectra (panel c) except for the central portion of the cross section for the 2,3,5,6 protons (panel c, right), which would originate dominantly from the α' methyl protons by spin diffusion in a finite transition time of the experiment. While this result points to the necessity of further shortening of the transition time to completely isolate each other, it is also recognized in the calculated spectra that, even at the zero transition time for which the calculations are performed, some mixing should occur in the cross sectional patterns. Furthermore, when a Lorentzian half width of $100\,Hz$

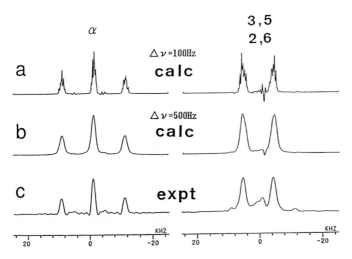

Figure 3-14. Experimental and simulated cross sectional patterns for the α-protons and the 3,5/2,6 protons in the SC-2D NMR spectrum of APAPA. (a) The experimental pattern. (b) the simulated patterns calculated based on Eq. (3-1) and (3-2), with the parameters of $A_{\alpha\alpha} = -3410$, $A_{\alpha3} = 400$, $A_{\alpha3} = 50$, $A_{32} = 3250$, $A_{35} = -400$, and $A_{36} = -100\,\text{Hz}$, and a Lorentzian half width of 500 Hz. (c) the simulated pattern with the same parameters as in (b) but with a half width of 100 Hz. Reprinted from reference 53, with permission from the *Journal of Chemical Physics*

was applied, complex fine structures appeared in the calculated spectra (panel a). These calculated fine structures for the α methyl protons matches well with the fine structures appearing in the central portion of the one-dimensional spectrum measured separately under high resolution (Figure 3-11). This result indicates that these fine structures are real. An interesting point is that some emissive patterns also appear in the calculated spectra (Figure 3-14(a) right). This feature is a common characteristic of a two-dimensional correlation spectrum of a strongly-coupled spin system, such as a J-coupled spin system in solution [60] or a dipolar-coupled spin system in solid [59]. Unfortunately, the insufficient resolution in the present SC-2D NMR spectrum did not allow a direct experimental observation of the emission.

4.1.3 Interpretation of dipolar interactions of the aromatic and methyl protons

In the cross section of the aromatic protons, doublet patterns with a splitting of 9.7 kHz were commonly observed. The frequency of splitting of two strongly-coupled aromatic protons is given by

(3-3) $\Delta v = 3\gamma^2 h/(4\pi r^3)(3\cos^2\beta - 1)S$

With

$$S = <(3\cos 2\zeta - 1)/2>.$$

S is the order parameter and gives a long-range order in terms of the molecular axis. β is the angle of the director inclined to the proton internuclear vectors. S is a statistical value and ζ represents the angle between the static magnetic field and the director vector. In this expression, the molecular axis and the director of the liquid crystal are considered to be parallel. When we adopt r = 2.45 A and β = 10°, as calculated from X-ray diffraction data, with $\Delta v = 9.7$ kHz, we obtain S = 0.41 for the S-value of the aromatic protons of APAPA at 108 °C.

Cross-section of the SC-2D NMR spectrum of APAPA indicates a large difference in dipolar couplings between two types of methyl protons. In general, protons of a methyl group with mutually equivalent dipolar coupling constants will show a triplet pattern with an intensity ratio of 1:2:1, whose splitting frequency is given by

$$(3\text{-}4) \qquad \Delta v = (-1/2)3\gamma^2 h/(4\pi r^3)(3\cos^2\beta' - 1)S$$

Here the value of -1/2 comes from the fact that proton internuclear vector of methyl protons rotates rapidly about the C3 axis of the methyl group making an angle of 90°. β' is the angle of the C3 axis and the molecular axis. For a methyl proton, Δv was determined from Figure 3-12 to be -10.2 kHz. The internuclear distance was estimated to be 1.66 Å using neutron diffraction data. Putting those values together with the order parameter of the molecular axis, S = 0.41, as evaluated from the splitting of the aromatic protons of APAPA, into equation [27], the angle β' was determined to be 29°. On the other hand, the Δv value of the α' methyl protons was estimated to be close to zero from Figure 3-12, and hence, the β' value must be close to the magic angle 54°. Therefore, it is concluded that the large difference of the dipolar coupling constants between the α and α' methyl protons are attributed to the differences in the angle between the molecular axis and the C3 axis.

4.1.4 *Interpretation of mixing of the cross sectional patterns*

The fact that mutually different cross sectional patterns were observed in the SC-2D NMR spectrum where τ_m was chosen to be 0 ms indicates that both inter- and intramolecular cross relaxation rates and spin flip-flop rates between interacting pairs of protons are relatively slow. This can be understood if one considers that dipolar interactions are partially averaged out by fast translational and rotational molecular motions in the liquid crystalline phase in contrast to the solid phase.

However, spectral mixing does occur among different cross-sections, and more so with larger τ_m values. To understand the degree of mixing among the cross-sectional patterns of an SC-2D NMR spectrum more quantitatively, we must find an equation to govern the exchange of longitudinal magnetizations within the homonuclear spin system during the mixing time. The result of Figure 3-4 indicates that the exchange occurs in the liquid crystalline phase and suggests that it must arise from the static and time-dependent parts of the dipolar coupling Hamiltonian. It can be analyzed by starting with a Solomon type equation for a dipolar-coupled homonuclear spin system, which can be expressed in a compact form as

$$(3\text{-}5) \qquad d<\mathbf{I}_z>/dt = \mathbf{R} <\mathbf{I}_z>.$$

Here I_z denotes the longitudinal magnetization vector of the homonuclear spin system with components $I_{z1}, I_{z2}, \cdots$. For individual protons and $\mathbf{R}$ denotes the relaxation matrix. For an exact estimation of the relaxation matrix elements, we require details not only on the conformation of the molecule but also on the molecular motion, which are presently unknown. Thus, we take a phenomenological approach under a reasonable simplification. We neglect the spin-lattice relaxation rates entirely, which are on the order of a second at 400 MHz proton resonance frequency and for the first approximation unimportant for the consideration of spectral patterns, and take only the exchange of magnetizations into account. Then, the elements of the relaxation matrix should follow the relation

$$(3\text{-}6) \qquad R_{ij} = -\sum_j R_{ij}(i \neq j)$$

with $R_{ij} = R_{ji}$ $(i \neq j)$.

Now the element R_{ij} simply denotes the rate of exchange of magnetizations between two protons i and j due to mutual dipolar coupling. In the SC-2D NMR experiments using the pulse sequence of Figure 3-1, the transverse magnetization $<I_y(t_1)>$ that has evolved for t_1 under the nematic phase Hamiltonian is transformed into $<I_z(t_1)> = \sum_k <I^k_z(t_1)>$ by the second 90° pulse. Then during the mixing time τ_m, the longitudinal magnetization further evolves by exchange obeying the following equation

$$(3\text{-}7) \qquad <I^z(t_1, \tau_m)> = \exp(\tau_m \mathbf{R}) < I^z(t_1) > .$$

Of which the component of the kth proton is given by

$$(3\text{-}8) \qquad < I^k_z(t_1, \tau_m) > = \sum_l [\exp(\tau_m \mathbf{R})]_{kl} < I_{lz}(t_1) > .$$

The matrix $\exp(\tau_m \mathbf{R})$ can be obtained numerically after diagonalizing R. Then, the longitudinal magnetization $<I_{kz}(t_1, \tau_m)>$ can be transformed into $<I_k(t_1, \tau_m)>$ by the third 90° pulse. Finally, the resonance lines of the kth proton $S_k(\omega_1, \tau_m)$ mixed with other proton resonances can be obtained by the Fourier transformation of $\sum_l [\exp(\tau_m \mathbf{R})]_{kl} <I'_z(t_1)>$, i.e.,

$$(3\text{-}9) \qquad S_k(\omega_1, \tau_m) = \sum_l [\exp(\tau_m \mathbf{R})]_{kl} \int dt_1 < I'_y(t_1) > \exp(i\omega t_1).$$

In the actual calculation of the mixing pattern of the α methyl protons, seven protons were included and the exchange rates for the individual proton pairs were assumed to be determined from Figure 3-5. Because the 3',5'/2',6',7 and a' methyl protons are far from the α methyl protons and only the mixing with the 3,5/2,6 protons was observed in Figure 3-12, a subspin system consisting of α methyl and 3,5/2,6 protons was considered in the simulation. Figure 3-15 shows the experimentally obtained and the best-fit simulated cross sections for protons with the mixing time of 0, 100, and 600 ms. The results of the calculation indicate that the cross relaxation

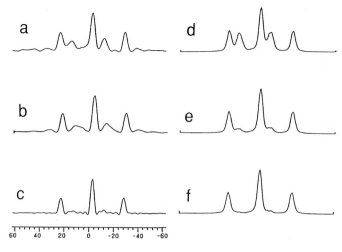

Figure 3-15. Experimental and simulated patterns of SC-2D NMR spectra for the α-protons of APAPA. Experimental patterns with (a)$\tau_m = 600$ ms, (b) $\tau_m = 100$ ms, and (c) $\tau_m = 0$ ms. Simulated pattern with (d) $\tau_m = 600$ ms, (e)$\tau_m = 100$ mS, and (f)tm = 0 ms. Simulation was performed based on Eqs. (3-5)–(3-7) with elements of the exchange matrix of $k_{\alpha\alpha} = 3.41$, $k_{\alpha3} = 0.40$, $k_{\alpha2} = 0.05$, $k_{32} = 3.25$, $k_{35} = 0.40$, and $k_{36} = 0.10$ s^{-1}. Reprinted from reference 53, with permission from the *Journal of Chemical Physics*

rate from the aromatic to a proton is only 0.40 s^{-1}, far slower than that expected in solids. This slow rate of exchange clearly explains the reason why proton NMR spectra in nematic liquid crystals are inhomogeneously broadened unlike those in solids.

4.2   SC-2D NMR Spectra of EBBA

Figure 3-16 shows the contour plot of the SC-2D NMR spectrum of 4'-ethoxybenzylidene-4-n-butylanilin (EBBA) between the nematic and isotropic phases, carried out using the pulse sequence of Figure 3-9. This figure clearly demonstrates that the SC-2D NMR experiment was successful in the liquid crystal sample of EBBA. It is recognized that the dipolar coupling spectra of individual protons are separated in the ω1 dimension due to the high-resolution in the ω2 dimension rendered by the isotropic phase.

Figure 3-16 also shows cross-sections of individual types of protons along the ω1 axis as a stacked plot. The local dipolar coupling fields of nine types of protons are resolved, except that of 2,6 and 3,5 protons, whose chemical shift differences are too small even in the isotropic phase, overlapped completely. It is recognized that the local dipolar coupling fields for the methyl protons of the ethoxy group (β' protons) contain a triplet pattern with 1:2:1 intensity ratio and a splitting of 9.6 kHz characteristic of a triangle network of protons as discussed for APAPA. Besides, each triplet lines were further split into small couplings due to the couplings from α' protons. On the other hand, the δ methyl protons show

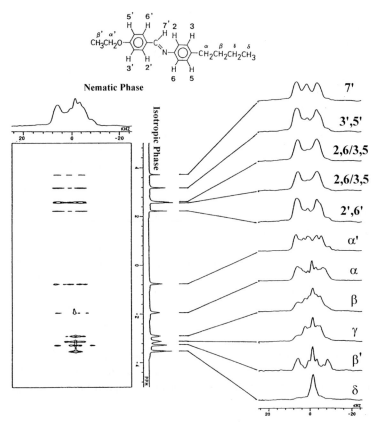

Figure 3-16. A nematic/isotropic phase correlated 2D NMR spectrum (SC-2D NMR) of EBBA with cross sectional patterns by applying a microwave pulse of 10 ms duration

a singlet pattern indicating that the splitting constant is very small. The aromatic protons in the phenyl groups (3,5/3',5' and 2,6/2',6' protons) are considered to give mainly doublet patterns with a splitting of 12.8 kHz, as were reported previously using deuteration of the alkyl group [26]. The main features of the spectra of the methylene and methyl protons in the alkyl group (α, β, γ, and δ protons) reflect the mobility of the groups. Namely the most mobile δ methyl group shows a singlet pattern while the linewidth increases as the protons are close to the core group with the splitting ranging from 8 to 15 kHz corresponding to the increasing order from δ to α protons. Thus this small splitting constant of the δ methyl protons turned out to be due to the higher mobility of the alkyl group in contrast to the case of the α' methyl protons, besides whose splitting constant is small because the C3 axis is making an angle of 54° with the molecular axis. The splitting of the doublets for methylene protons was observed for the α proton, which is a little larger than those of aromatic protons. Doublet patterns for the methylene protons have not been reported so far. It is of interest to note that the methane proton (7' proton) shows a mixture of doublet and singlet patterns and the singlet pattern should arise from the 7' proton.

Although a triplet and singlet patterns were clearly observed for the β' and δ methyl protons, respectively, in the cross-sections of the SC-2D NMR spectrum as shown in Figure 3-17 ($\tau_m = 0$ ms), doublet patterns of other protons mixed to the lineshapes of both the β' and δ protons at $\tau_m = 200$ ms (Figure 3-17). This fact indicates that the spin states of methyl protons and those of other protons exchange with each other by mutual dipolar interactions. Phenomenologically, this is similar to the appearance of a cross peak in a 2D exchange or 2D NOE spectrum. This phenomenon is expected to be efficient in the nematic phase since the intramolecular dipolar interactions are relatively strong. Because the intermolecular dipolar inter-actions are averaged out more effectively than the intramolecular ones due to the rapid intermolecular rearrangement, the mixing rate between intermolecular protons should be slower than that between intramolecular protons particularly in nematic phase. Therefore, this mixing of the spectra should represent pathways of intramolecular rather than those of intermolecular spin diffusion. Actually, Figure 3-17 clearly show that mixing among aromatic protons is the most efficient and that between methyl protons and methylene protons is less efficient. Particu-larly, the triplet pattern of the methyl protons of the ethoxy group is clear, because they are fairly isolated from the aromatic and butyl protons within the molecule. On the other hand, δ protons mixed rapidly with other protons to grow into a doublet. On the other hand, β' protons still remain as triplet, leading that ethoxy group is more isolated from aromatic protons. When τ_m is set to 600 ms, most of the protons

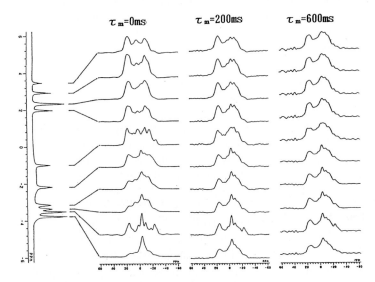

$\tau_m = 0$ ms $\tau_m = 200$ ms $\tau_m = 600$ ms

Figure 3-17. Dependence of the cross sectional patterns on the mixing time tm in the SC-2D NMR spectrum of EBBA

show now same lineshapes with each other to indicate that it is long enough to establish an equilibrium among all protons within the molecule of EBBA in the nematic phase.

4.3   SC-2D NMR Spectra of Mixed System of 5OCB and PCH3

Figure 3-18 shows SC-2D NMR spectra of 4-(n-pentyloxy)-4'-cyanobiphenyl (5OCB) between the nematic and the isotropic phases for three different tempera-tures of the nematic phase, i.e., 67.1, 65.1, and 63.1°C. Compared to a relatively short microwave pulse (5 ms) that attained the phase transition from 67.1°C, longer microwave pulses of duration 20 and 25 ms were required to attain the transition from the lower temperatures. The complex asymmetric one-dimensional ^{1}H NMR spectrum below Tc for the whole proton system is now correlated with the individual proton transitions above Tc, which were already assigned as shown in Figure 3-18. The patterns of the correlated signals are best displayed as cross sections along the ω_1 axis for those values of ω_2 corresponding to frequencies of transition of individual protons in the isotropic phase. It is noted that each cross-section shows a nearly symmetric pattern with respect to its center of resonance and has its own characteristic shape that indicates that the cross-section is practically free from mixing with others, while its center of resonance is characteristically displaced from each other owing to its own chemical shift value in the nematic phase. It is also noted that the splittings at the low temperature are larger than those at a higher temperature.

Figure 3-19 shows an SC-2D NMR spectrum of 1-(4'-cyanophenyl)-4-propylcyclohexane (PCH3) (Tc = 45.0 °C, the temperature before the transition was 43.1°C) obtained with a microwave pulse duration of 5 ms, the signal assignments in the isotropic phase being made by an independent solution experiment (DQF-COSY). It is noticed that the cross sections of the ring protons and those of the aliphatic protons (both cyclohexyl and propyl) differ considerably, but cross sections are quite similar among the ring protons themselves and among the aliphatic protons themselves, indicating that nearly complete mixing of the subspectra took place among each group of protons.

Figure 3-20 shows an SC-2D spectrum of a 1:1 mixture (by weight) of 5OCB and PCH3 at 50.1°C (Tc = 51.0 °C) measured with a microwave pulse duration of 7 ms. Although the separation of signals is not completed even in the isotropic phase, a number of lines were assignable to individual proton transitions of 5OCB and PCH3 by performing a DQF-COSY experiment on the mixture, as indicated in Figure 3-20.

It is fairly straightforward to evaluate the largest splitting for individual protons directly from the cross-sectional patterns (subspectra) of 5OCB as shown by arrows in Figure 3-18 top. The major doublet splittings of about 10 kHz for the ring protons could be attributed to the dipolar couplings between the nearest-neighbor protons only 2.45 Å apart, parallel to the rings, i.e., pairs of 2-3, 5-6, 2'-3', and 5'-6' protons. Likewise, the major doublet splittings for the methylene protons and the

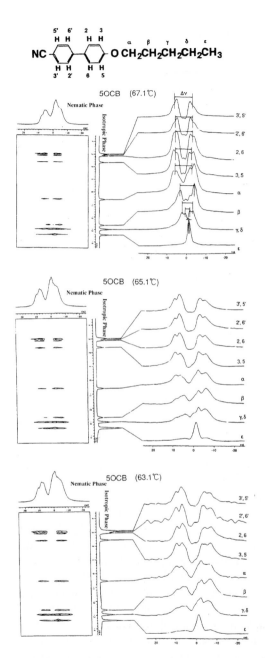

Figure 3-18. Nematic/isotropic phase correlated 2D ¹H NMR spectra of 4-(n-pentyloxy)-4'-cyanobiphenyl (5OCB), together with cross sections. Temperature were jumped from (top) 67.1, (middle) 65.1, and (bottom) 63.1 °C with microwave pulses of duration 5, 20, and 25 ms, respectively. Reprinted from reference 52, with permission from the *Journal of Physical Chemistry*

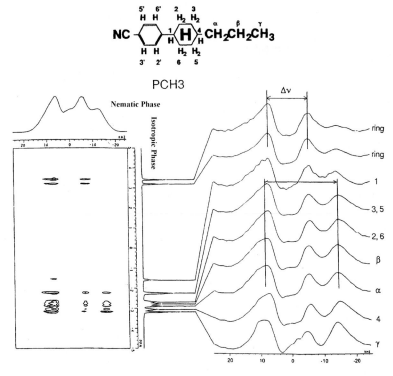

Figure 3-19. Nematic/isotropic phase correlated 2D ^{1}H NMR spectrum of 1-(4'-cyanophenyl)-4-propylcyclohexane (PCH3), together with cross sections. Temperature was jumped from 43.1 °C with a microwave pulse of duration 5 ms. Reprinted from reference 52, with permission from the *Journal of Physical Chemistry*

unresolved triplet splitting for the methyl protons can be attributed to the mutual dipolar coupling (geminal couplings) within the same methylene protons and within the same methyl protons, respectively.

The major doublets in the experimental cross sections of the ring protons such as those in Figure 3-18 middle and bottom show splittings into further doublets with much smaller coupling constants of 2-3 kHz. Indeed, each proton of 5OCB should be coupled to several other protons, and therefore in principle each of the cross sectional spectra should show a multiple pattern whose complete analysis requires computer simulation with all these coupling constants of subspin systems taken into account as discussed above. Unfortunately, the resolution of the experimental cross sectional spectra of Figure 3-18 is not high enough to render analysis of these couplings by spectral simulation. Therefore, we limit our discussion only to the secondary splittings of 2-3 kHz, which can be estimated directly on the ring proton spectra of Figure 3-18.

Under the assumption that the biphenyl ring flips rapidly about its para-axis, calculation using the C-C inter-ring distance of 1.493 and 1.52 Å for biphenyl

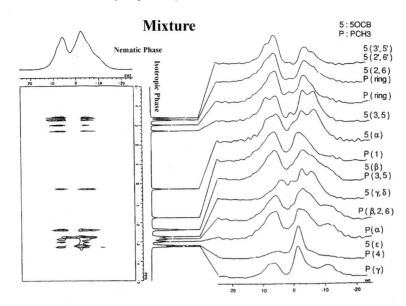

Figure 3-20. Nematic/isotropic phase correlated 2D ^{1}H NMR spectrum of a 1:1 mixture (by weight) of 4-(n-pentyloxy)-4'-cyanobiphenyl (5OCB) and 1-(4'-cyanophenyl)-4-propylcyclohexane (PCH3). Temperature was jumped from 50.1 °C with a microwave pulse of duration 7 ms. Reprinted from reference 52, with permission from the *Journal of Physical Chemistry*

and 4,4'-difluoro-biphenyl, respectively, standard C-C distance of 1.400 Å and C-H distance of 1.082 Å and standard angle of 120° for the biphenyl part [61] predict splitting due to coupling between the proton pairs across the rings (2-6, 3-5, 2'-6', 3'-5') to be only 1. kHz and much less for other pairs (such as 2-5 and 3-6). On the other hand, similar calculation for the nearest inter-ring proton pairs, i.e., 2-6' and 2'-6, predicts the experimentally observed values of 2-3 kHz, for a single conformation with the rotation angle of 47-52° between the two rings of the biphenyl group. Thus the secondary splittings in the subspectra of the aromatic protons can be assigned to the nearest inter-ring proton interactions. Indeed, our experience of the SC-2D NMR study of liquid crystals so far suggests that the double-doublet nature of the cross section of the ring protons is observed only in liquid crystals with a biphenyl moiety but not with phenyl rings separated by heteroatoms (e.g., APAPA and EBBA as shown in previous section). The fact that the secondary splittings were observed not only in (2,6) and (2',6') protons but also in (3,5) and (3',5') protons having no direct inter-ring interactions is probably due to strong cross relaxation within the ring protons. The determinated tortional angle 47-52° is larger than those (31.8–35.4°) reported for pure biphenyl or meta or para-halogenated biphenyls at 30 °C but is smaller than those reported for ortho-halogenated biphenyls (68.1–77.4°) [62].

The local order parameters S_{kl} obtained from the experimental major splittings are plotted in Figure 3-21 for the nematic phase of 5OCB at three different temperatures.

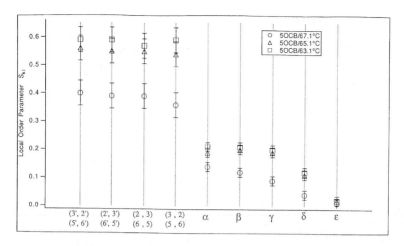

Figure 3-21. Local orders S_{kl} of 4-(n-pentyloxy)-4'-cyanobiphenyl (5OBC). Evaluation is made for each observed proton (k) and is ploted against each proton pair (k,l) as deduced from the dipolar coupling obtained from Figure 18 and Eq. (3-3). The error bars were estimated from line widths of the cross-sectional spectra in Figure 3-18. Reprinted from reference 52, with permission from the *Journal of Physical Chemistry*

At all the temperatures studied, evaluated local orders are, as expected, the same within the accuracy of measurements for any protons in the two rings. However, it becomes smaller and smaller as the position of the methylene protons becomes more distant from the ring along the aliphatic chain as same as the case of the alkyl group of EBBA. The latter observation is in qualitative agreement with what had been disclosed previously from ^{2}H NMR spectra of ^{2}H-labeled n-octylcyanobiphenyls [25]. Furthermore, the local orders of the two rings clearly depended on the temperature of observation. At 67.1°C, only 0.9 °C below Tc, the local order was the lowest, but it increased considerably at lower temperatures, as also shown previously from ^{2}H NMR [25]. Furthermore, if we compare the patterns of the cross sections among parts in Figure 3-18 (top to bottom), we note that more spectral mixing occurs among the cross sections as the microwave pulse duration, i.e., the transition period, becomes longer (compare $\alpha, \beta, \gamma, \delta$ and ε among Figure 3-18 top to bottom).

This spectral mixing occurs because of the exchange of longitudinal magnetizations during the transition period in which the sample partly spends in the liquid-crystalline phase. In principle, the exchange should occur by two mechanisms in the liquid crystalline phase; namely, cross-polarization or dipolar coupling oscillation between the coupled protons due to the static part of the dipolar interaction and cross relaxation between the coupled protons due to the zero-frequency component of the fluctuating part of the dipolar interaction. In the present case, the dipolar coupling oscillation may not be apparent, since it will rapidly damp out in less than milliseconds because of dipolar couplings with other protons and relaxation processes, leaving mainly the effect of cross relaxation. Since the molecules under study constitute dipolar-coupled multiproton systems, consecutive cross relaxations may occur leading to spin diffusion. If we examine the patterns of Figure 3-18 in more detail, it is noticed that the mixing is most complete within the ring protons and extends much less along the chain. This indicates that cross relaxation and spin diffusion is a sensitive parameter reflecting the dynamics of dipolar couplings among protons in the liquid-crystalline phase.

For PCH3, the subspectra of the ring protons are essentially a doublet with a splitting of 12.7 kHz, giving S = 0.56 as the order of the ring, whereas those of the aliphatic protons have another doublet component with a larger splitting in addition to the 12.7 kHz doublet which apparently comes from the ring protons by spin diffusion.

Major dipolar protons of 5OCB are compared between pure 5OCB and its 1:1 mixture with PCH3 (Figure 3-12). It is noticed that the local orders of the ring and the δ methylene protons of 5OCB in the mixture at 50.1°C are much smaller than that of pure 5OCB expected at the same temperature. In fact, the local orders of 5OCB in the mixture at 50.1°C are comparable to those of pure 5OCB at about 66°C. It can be compared that the dipolar splitting and the order parameters of the ring protons in the mixture with the corresponding values in pure 5OCB and PCH3 at a reduced temperature of Tex(K)/Tc(K) = 0.99. It is obvious that by the mixing the order of the ring of 5OCB increased while the order of the ring of PCH3 decreased.

The strong mixing of subspectra in pure PCH3 indicates a distinctly high efficiency of spin diffusion in PCH3, as compared to that in 5OCB and in other nematic phase liquid crystals so far examined by SC-2D NMR. This result seems to suggest that at 43.1°C PCH3 does not exist in the nematic phase but probably in the smectic phase in which intermolecular interaction is stronger and mobility is more severely restricted. It is also noticed that the cross-sections of PCH3 in the mixture are mutually far more distinct from each other than those of pure PCH3, indicating that the efficiency of spin diffusion in PCH3 is considerably reduced by the presence of 5OCB. These observations indicate that spin diffusion is quite sensitive to intermolecular interactions between liquid crystalline molecules and suggests that it can be used to detect direct intermolecular interactions in liquid crystals.

5. SUMMARY AND SCOPE

Solid-state NMR spectroscopy is a powerful tool in studying the structure, dynamics, order and geometry of liquid crystalline compounds at atomistic-level resolution in different phases at various temperatures. There are a plethora of two-dimensional techniques that has been utilized in such studies includes SLF, PDLF, PISEMA, PITANSEMA, HIMSELF, off-magic angle spinning, HETCOR, and correlation experiments under static and MAS conditions. Among the nuclei, 13C and 2H are commonly used in such experiments. However, labeling with 19F isotope will provide further insights into the structural features of the molecule and also on the intermolecular packing in three-dimension. Recent studies have shown that it is possible to use proton-detection experiments in solid-state [63]. While it is certainly beyond the scope of this chapter to cover all published solid-state NMR techniques, we have presented details on techniques that are highly valuable in the measurement of dipolar couplings and the determination of order parameters of liquid crystalline compounds. These techniques can also be applied to study other types of LCs including metallo-mesogens and polymeric liquid crystals. SLF experiments can also be performed under MAS [2, 4, 30, 35, 38] to determine dipolar couplings if magnetic alignment and/or resolution pose difficulties in studying static nematic materials.

2D SLF and SC-2D experiments allow the elucidation of local-order parameters without the need for deuteration and therefore are readily applicable to a wide range of liquid crystalline samples. SLF techniques are routinely utilized in the study of liquid crystals as they require simple experimental set up as compared to the SC-2D experiment. Further developments on improving the sensitivity and resolution of SLF methods are in progress in several laboratories. One direction is on the use of micro-size radio frequency coils that significantly reduces the sample quantity, enables samples to be spun at very fast spinning speeds (up to 75 kHz), and dramatically reduces the required rf power and enable studies on heat-sensitive liquid crystalline samples. Another important direction is the use of higher magnetic fields and proton detection approaches to enhance the sensitivity of NMR experiments. However, there are several unique advantages in using SC-2D method to study liquid crystals because this method can observe 1H NMR signals. Firstly, specific ^{2}H labeling is not necessary for assignments of the cross sectional spectra, since SC-2D NMR allows their automatic assignments through cross peaks to the signals in the isotropic phase which are readily assignable by the conventional 2D method. Secondly, information on local conformation, as exemplified above by the torsional angle of the two phenyl rings, could be obtained through the analysis of fine structures of the cross-sectional spectra which provide 1H-1H dipolar splittings. Finally, cross relaxation and spin diffusion can be a unique means to characterize the dynamics as well as intermolecular interactions in liquid crystalline molecules. Although the temperature range of measurement for SC-2D NMR spectroscopy of liquid crystals is limited to those close to Tc, the method has some unique advantages over the conventional ^{2}H NMR spectroscopy.

6.   REFERENCES

1. C.A. Fyfe, Solid State NMR for Chemists, CFC Press, Guelph, Ontario, Canada (1983).
2. K. Schmidt-Rohr and H.W. Spiess, Multidimensional Solid-State NMR and Polymers, Academic Press, New York (1994).
3. A. Ramamoorthy (Ed.), NMR Spectroscopy of Biological Solids, Taylor & Francis, CRC Press, New York (2006).
4. M. Mehring, High Resolution NMR in Solids, 2nd Edition, Springer-Verlag, New York (1983).
5. E.R. Andrew, A. Bradbury and R.G. Eades, Nature 182, 1659 (1958).

6. I.J. Lowe, Phys. Rev. Lett., 2, 285 (1959).
7. U. Haeberlen, High Resolution NMR in Solids, Selective Averaging, Academic Press, New York (1976).
8. A. Pines, M.G. Gibby and J.S. Waugh, J. Chem. Phys., 59, 569 (1973).
9. J. Schaefer and E.O. Stejskal, J. Am. Chem. Soc., 98, 1031 (1976).
10. J.W. Emsley (Eds.), Nuclear Magnetic Resonance of Liquid Crystals, Reidel Publishing Company, Dordrecht (1985).
11. R.Y. Dong, Nuclear Magnetic Resonance of Liquid crystals, 2nd Edition, Springer, New York (1997).
12. B.M. Fung, 13C NMR studies of liquid crystals, Prog. NMR Spectrosc., 41, 171 (2002).
13. T. Narasimhaswamy, N. Somanathan, D. K. Lee and A. Ramamoorthy, *Chem. Mater.*, **17**, 2013-2018 (2005).
14. R. Kannan, T. Sen, R. Poupko, Z. Luz and H. Zimmerman, J. Phys. Chem. B 107, 13033 (2003).
15. T. Fujiwara and A. Ramamoorthy, Annu. Rep. NMR Spectrosc., (2006).
16. Z.H. Gan, J. Am. Chem. Soc., 114, 8307 (1992).
17. A. Ramamoorthy, L.M. Gierasch and S.J. Opella, J. Magn. Reson., B110, 102 (1996).
18. J.Z. Hu, C.H. Ye, R.J. Pugmire and D.M. Grant, J. Magn. Reson., 145, 230 (2000).
19. M. Strohmeier and D.M. Grant, J. Magn. Reson., 168, 296 (2004).
20. J. Xu, F. Csorba, R.Y. Dong, *J. Phys. Chem. A.*, 109, 1998 (2005).
21. G. Antonioli, D.E. McMillan, P. Hodgkinson, Chem. Phys. Lett., 344, 68 (2001).
22. A.E. Bennett, C.M. Reinstra, M. Auger, K.V. Lakshmi, and R.G. Griffin, J. Chem. Phys., 103, 6951 (1995).
23. B.M. Fung, A.K. Khitrin, K. Ermolaev, J. Magn. Reson., 142, 97 (2000).
24. J.W. Emsley, G.R. Luckhurst, E.J. Parsona, and B.A. Timimi, Mol. Phys. 56, 767 (1985).
25. C.J.R. Counsell, J.W. Emsley, G.R. Luckhurst, and H.S. Sachdev, Mol. Phys., 63, 33 (1988).
26. Y.C. Lee, Y. Hsu, and Dolphin, Liquid Crystals and Ordered Fluids (Plenum, New York, p. 357 (1974).
27. J.S. Prasad, J. Chem. Phys. 65, 941 (1976).
28. R.K. Hester, J.L. Ackerman, B.L. Neff, and J.S. Waugh, Phys. Rev. Lett., 36, 1081 (1976).
29. A. Hohener, L. Muller, and R.R. Ernst, Mol. Phys. 38, 909 (1979).
30. A. Ramamoorthy, Y. Wei and D. K. Lee, Annu. Rep. NMR Spectrosc., 52, 1 (2004).
31. B.M. Fung, J. Afzal, T. L. Foss, and M.H. Chan, J. Chem. Phys., 85, 4808 (1986).
32. C.B. Frech, B.M. Fung, and M. Schadt, Liq. Cryst., 3, 713 (1988).
33. M. Hong, K. Schmidt-Rohr, and A. Pines, J. Am. Chem. Soc., 117, 3310 (1995).
34. C. H. Wu, A. Ramamoorthy and S. J. Opella, J. Magn. Reson., A109, 270 (1994).
35. A. Ramamoorthy and S. J. Opella, *Solid state NMR Spectrosc.*, 4, 387 (1995).
36. A. Ramamoorthy, C. H. Wu and S. J. Opella, *J. Magn. Reson.* **140**, 131 (1999).
37. K. Yamamoto, D. K. Lee, and A. Ramamoorthy, *Chem. Phys. Lett.*, **407**, 289–293 (2005).
38. K. Yamamoto, V.L. Ermakov, D. K. Lee, and A. Ramamoorthy, *Chem. Phys. Lett.*, **408**, 118–122 (2005).
39. D. K. Lee, T. Narasimhaswamy, and A. Ramamoorthy, Chem. Phys. Lett., 399, 359 (2004).
40. K. Nishimura and A. Naito, Chem. Phys. Lett., 402, 245 (2005).

41. T. Narasimhaswamy, D. K. Lee, K. Yamamoto, N. Somanathan, and A. Ramamoorthy, J. Am. Chem. Soc., 127, 6958 (2005).
42. T. Narasimhaswamy, D.K. Lee, N. Somanathan, and A. Ramamoorthy, *Chem. Mater.*, **17**, 4567–4569 (2005).
43. T. Narasimhaswamy, M. Monette, D.K. Lee, and A. Ramamoorthy, *J. Phys. Chem. B*, **109**, 19696–19703 (2005).
44. K. Nishimura and A. Naito, Chem. Phys. Lett., 419, 120 (2006).
45. P. Caravatti, G. Bodenhausen, and R.R. Ernst, *Chem. Phys. Lett., 89*, 363 (1982).
46. T. Nakai and T. Terao, *Magn. Reson. Chem.,* 30, 42 (1992).
47. K. Schmidt-Rohr, D. Nanz, L. Emsley, and A. Pines, J. Phys. Chem., 98, 6668 (1994).
48. K. Yamamoto, S. Dvinskikh, and A. Ramamoorthy, *Chem. Phys. Lett.*, **419**, 533–536 (2006).
49. S. Dvinskikh, K. Yamamoto, and A. Ramamoorthy, *Chem. Phys. Lett.*, **419**, 168–173 (2006).
50. A. Naito, H. Nakatani, M. Imanari, and K. Akasaka, J. Magn. Reson., 87, 429 (1990).
51. A. Naito, M. Imanari, and K. Akasaka, J. Magn. Reson., 92, 85 (1991).
52. K. Akasaka, M. Kimura, A. Naito, H. Kawahara, and M. Imanari, J. Phys. Chem., 99, 9523 (1995).
53. A. Naito, M. Imanari, and K. Akasaka, J. Chem. Phys., 105 4504 (1996).
54. K. Akasaka, A. Naito, and M. Imanari, J. Am. Chem. Soc., 113, 4688 (1991).
55. D.B. Ferguson, T.R. Kramiete, and J.F. How, Chem. Phys. Lett., 229, 71 (1994).
56. S. Caldarelli, M. Hong, L. Emsley, and A. Pines, J. Phys. Chem., 100, 18695 (1996).
57. S. Dvinskikh, K. Yamamoto, U. Duerr and A. Ramamoorthy, J. Am. Chem. Soc., 128, 6326 (2006).
58. S. Dvinskikh, K. Yamamoto, and A. Ramamoorthy, J. Chem. Phys., in press.
59. N. Schuff and U. Haeberlen, J. Magn. Reson., 52, 267 (1981).
60. G. Bodenhausen, R. Freeman, G.A. Morris, and D.L. Tyrner, J. Magn. Reson., 31, 75 (1978).
61. W. Niedelberger, P. Diel, and L. Lunazzi, Mol. Phys. 26, 571 (1973).
62. L.D. Field, and S.J. Sternhell, J. Am. Chem. Soc. 99, 5249 (1981).
63. K. Saalwachter and A. Ramamoorthy, (Ed. A. Ramamoorthy), NMR Spectroscopy of Biological Solids, Chapter 6, Taylor & Francis, CRC Press, New York (2006).

CHAPTER 4

SEPARATED LOCAL FIELD NMR SPECTROSCOPY IN COLUMNAR LIQUID CRYSTALS

SLF NMR spectroscopy in columnar liquid crystals

SERGEY V. DVINSKIKH

Institute of Physics, St. Petersburg State University, 198504 St. Petersburg, Russia

1.   INTRODUCTION

Many compounds composed of disc-shaped molecules exhibit stable thermotropic liquid crystalline phases referred to as discotic liquid crystals. The history of discotic liquid crystals goes back to the pioneering work of Chandrasekhar [1] on hexaesters of benzene. Over the past twenty-five years the properties of compounds composed of disc-shaped molecules have been investigated extensively [2–6]. The archetypal discotic molecule has a flat core with flexible side chains covalently linked to it. The molecular symmetry of these compounds is usually trigonal although some molecules with lower symmetry also form discotic liquid crystals. Examples of aromatic cores are displayed in Figure 4-1a.The phase sequence and the temperature range of the mesomorphic region depend on the size and symmetry of the core, the length and chemical structure of the side chains, and the type of the linkage between the core and the chains. Most of the discotic mesophases exhibits an architecture where the molecules are stacked into columns, which in turn form two-dimensional (2D) arrays with various symmetries, see Figure 4-1b. The columnar phases are characterized by both long-range orientational and positional order. In addition, discotic molecules form nematic liquid crystals [2–7] which, in analogy with the conventional nematics formed by rod-like mesogens, exhibit orientational order but where the positional order is absent.

  Columnar liquid crystals have some unique properties, which are starting to get exploited for commercial use. For technical applications, the most important aspect is the geometry of the mesophase that enables one-dimensional transport of charge within the columns [8–10]. There is a huge anisotropy in the conductivity between the direction parallel to the column axis and that perpendicular to it, caused by the

A. Ramamoorthy (ed.), Thermotropic Liquid Crystals, 117–140.
© 2007 *Springer.*

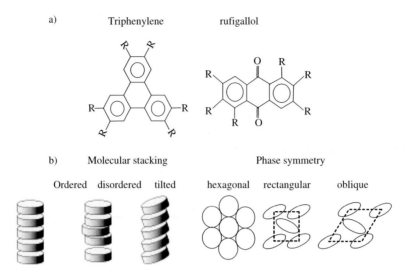

Figure 4-1. Discotic liquid crystals: a) examples of aromatic cores; side chains R are, for example, $-O-C_nH_{2n+1}, -OC(O)-C_nH_{2n+1}, -S-C_nH_{2n+1}$; b) structures of columnar phases. Columnar mesophases are labeled D_{xy}, where the index x indicates the symmetry of the two-dimensional unit cell of the mesophase: *h*, *r* and *ob* refer to the hexagonal, rectangular and oblique arrangements, respectively. The molecular organization within the columns is indicated by the second index, y, where ordered and disordered stackings are represented by the letters *o* and *d*. The samples of the discotic compounds used in the experiments described in this chapter, were 1,2,3,5,6,7-hexaoctyloxy-rufigallol (RufH8O), 2,3,6,7,10,11-hexa-n-oxy-triphenylene (THEn, n = 5, 6), 2,3,6,7,10,11-hexahexyl-thiotriphenylene (HHTT) and their ^{13}C and ^2H labeled analogs. These compounds have following phase diagrams

RufH8O: Solid $\xrightarrow{19°C}$ $D_{ho}(III)$ $\xrightarrow{31°C}$ $D_{ho}(II)$ $\xrightarrow{96°c}$ isotropic;

The5: Solid $\xrightarrow{63°C}$ D_{ho} $\xrightarrow{120°C}$ isotropic;

THE6: Solid $\xrightarrow{64°C}$ D_{ho} $\xrightarrow{97°C}$ isotropic;

HHTT: Solid $\xrightarrow{62°C}$ D_{ho} $\xrightarrow{70°C}$ D_{hd} $\xrightarrow{93°C}$ isotropic;

insulating effect of the side chains. Some of the compounds exhibit extremely high charge mobility, in fact higher than that in any other organic material [8, 11].

The important physical property of liquid crystals is the orientational ordering of the molecules. The orientational order is characterized by the average orientation of the molecular symmetry axes, which defines the director, **N** [12]. The orientational order is usually quantified by means of the ordering tensor with the elements defined by the averages of the Wigner functions, $\langle D_{mn}^L (\Omega_{MN}) \rangle$ [13], where L is the rank and the set of angles Ω_{MN} defines the relative orientation of the molecular coordinate system and the director. In general, five order parameters are required to completely describe the orientational order of a rigid molecule in a uniaxial liquid crystalline phase. For molecules with trigonal (or higher) symmetry only one order parameter is necessary

$$(4-1) \quad S = \langle D_{00}^2 (\Omega_{MN}) \rangle = \frac{1}{2}(3\langle \cos^2 \vartheta_{MN} \rangle - 1)$$

where ϑ_{MN} is the angle between **N** and the main molecular axis. Columnar phases typically exhibit high ordering with $S \approx 0.9$. For flexible molecules, local order parameters can be defined for individual segments.

Nuclear magnetic resonance (NMR) spectroscopy is a powerful experimental tool for studies of liquid crystals in general and discotic phases in particular. Deuterium (^{2}H) NMR has extensively been used for investigations of the molecular order, structure and dynamics in columnar mesophases [14–24]. Recently, molecular self-diffusion measurements employing ^{2}H NMR in columnar phases have been reported [25]. An inherent drawback of the method is that it requires isotopic labeling that can be both difficult and expensive. Moreover, the resonance assignment of ^{2}H spectra of molecules containing many different ^{2}H sites is not straightforward [26]. Furthermore, the ^{2}H NMR line shape reflects the motion of C-D vectors and is, therefore, rather insensitive to other details of molecular conformations.

Carbon-13 NMR spectroscopy has several advantages for studying columnar mesophases. Spectra with good signal-to-noise ratios may be obtained in samples with the natural isotopic abundance. Resonances from chemically non-equivalent sites are typically well resolved and can often be readily assigned. Anisotropic spin interaction, such as dipolar couplings, provide rich information on ordering, conformational structure, and phase transitions. In this chapter, the application to columnar mesophases of recently developed and advanced ^{13}C NMR methods for measurements of heteronuclear spin interactions are described.

2. ^{13}C-^{1}H SEPARATED LOCAL FIELD NMR SPECTROSCOPY

2.1 Heteronuclear Spin-Spin Interactions

Heteronuclear spin interactions include two terms, H_D and H_J, corresponding to the direct dipolar coupling due to the local magnetic field generated by the magnetic moments of the spins and the indirect J-coupling mediated by the bonding electrons, respectively [27]. The direct dipolar interaction is totally anisotropic. The indirect coupling contains a significant isotropic part or the scalar coupling, while the anisotropic contribution is usually much smaller and often neglected. In the presence of a strong external magnetic field along the z axis of the laboratory frame, the secular part of the Hamiltonian for heteronuclear interactions for a spin pair I and S is given by

$$(4\text{-}2) \qquad H_{IS} = H_D + H_J = 2\pi \left[2d_{IS} \times d_{00}^2 (\cos \vartheta_{PL}) + J_{IS} \right] \times I_z S_z$$

where ϑ_{PL} is the angle between the inter-nuclear vector (which defines an interaction principal frame axis **P**) and the laboratory frame z-axis along the magnetic field B_0, $d_{00}^2 (\cos \vartheta_{PL}) = (3\cos^2 \vartheta_{PL} - 1)/2$ is the reduced Wigner matrix element [13], the dipolar coupling constant is given by $d_{IS} = -(\mu_0/8\pi^2) \times (\gamma_I \gamma_S \hbar / r^3)$, and J_{IS} is the scalar coupling constant. For a typical one-bond C-H group, the coupling constants amount to $d_{CH} \approx -21\,\text{kHz}$ and $J_{CH} \approx 140\,\text{Hz}$. For a spin-1/2 pair in an oriented and static sample, the spectrum consists of a doublet with a splitting of

$$(4\text{-}3) \qquad \Delta\nu = |2d_{IS} \times d_{00}^2(\cos\vartheta_{PL}) + J_{IS}|$$

For a random distribution of ϑ_{PL} angles in a powder sample, a superposition of the doublets with different splittings gives rise to a Pake-type spectrum [28].

In the mesophase, anisotropic reorientations of the inter-nuclear vector (for directly bonded spins the distance r is constant) leads to partial averaging of the angular factor in Eq. (4-2). The averaging is over conformational transitions of flexible molecular parts, molecular rotations around the molecular symmetry axis, and fluctuations of the molecular axis around the director. These motions are, typically, much faster than the interaction time scale defined by $(2\pi \times \Delta\nu)^{-1}$. In an uniaxial liquid crystal, the orientational averaging of the bond direction with respect to $\mathbf{N}$ is conveniently described by a local order parameter $S_{local} = \langle d_{00}^2(\cos\vartheta_{PN})\rangle$, where ϑ_{PN} is the angle between the inter-nuclear vector $\mathbf{P}$ and the director $\mathbf{N}$. Thus, the dipolar splitting is given by

$$(4\text{-}4) \qquad \Delta\nu_{LC} = |2 \times \langle d_{IS}\rangle \times d_{00}^2(\cos\vartheta_{NL}) + J_{IS}|,$$

where $\langle d_{IS}\rangle = d_{IS} \times S_{local}$ is the residual dipolar coupling and ϑ_{NL} defines the $\mathbf{N}$ direction in the laboratory frame. For rigid molecular fragments with well-defined geometry, the local order parameter can often be expressed via the molecular order parameter S (defined in Eq. (4-1)): $S_{local} = S \times d_{00}^2(\cos\vartheta_{PM})$, where ϑ_{PM} is the angle between the bond vector and the molecular symmetry axis. Thus, the spectrum of an oriented liquid crystalline domain exhibits a doublet with a scaled splitting

$$(4\text{-}5) \qquad \Delta\nu_{LC} = |2d_{IS} \times S \times d_{00}^2(\cos\vartheta_{PM}) \times d_{00}^2(\cos\vartheta_{NL}) + J_{IS}|,$$

Although the J-coupling constant is generally small compared to d_{IS}, its contribution to the spectral splitting can often not be ignored in liquid crystals owing to the scaling of the dipolar term in Eq. (4-5) by the angular coefficient. The value of J_{CH} can readily be estimated from the isotropic solution spectrum and is practically not affected by the bulk medium.

Homonuclear proton decoupling is often a prerequisite for obtaining resolved $^{13}C - ^1H$ heteronuclear couplings. A side effect of homonuclear decoupling is that the heteronuclear interactions are apparently reduced by a scaling factor k specific of the decoupling sequence. The value of the scaling factor, $0 < |k| < 1$, is predicted by calculation of the average heteronuclear spin Hamiltonian [29].

2.2   The Effect of Magic-Angle Sample Spinning

The magic-angle spinning (MAS) technique was developed to suppress the effect of anisotropic interactions in NMR spectra. The result is an increase of the spectral resolution and sensitivity [30, 31]. The magic angle ϑ_m is defined by $3 \times \cos^2 \vartheta_m - 1 = 0$ and is $\approx 54.7°$. The term *fast* MAS is used when the spinning frequency v_r exceeds the width of the powder spectrum.

When the sample is spun rapidly at the magic angle with respect to the magnetic field, anisotropic interactions such as the chemical shift anisotropy and dipolar couplings average to zero over one rotational period. Thus, the powder pattern collapses into a single line at the isotropic chemical shift position [32]. This effect can be understood by considering that, for any orientation in the powder sample, the anisotropic spin interaction components perpendicular to the rotational axis are averaged to zero. Only the parallel component survives, which is scaled by the conventional angular factor $d_{00}^2(\cos \vartheta)$, where ϑ is the orientation of the rotational axis with respect to magnetic field. At the magic angle, $\vartheta = \vartheta_m$, this parallel component scales to zero. When homonuclear *I–I* dipolar interactions are present, suppression of the *I–S* dipolar interactions requires a spinning frequency exceeding not only the *I–S* dipolar couplings, but also the *I–I* dipolar couplings. If the spinning frequency is not high enough, these interactions can be removed by homonuclear and heteronuclear multiple-pulse decoupling sequences.

When spinning frequency v_r does not exceed the width of the powder pattern, the modulation of the NMR frequencies under MAS results in spinning sidebands in the spectrum at frequencies separated from the isotropic chemical shift by multiples of v_r. They extend over a range of about the powder pattern linewidth. The intensities of the sidebands can be used to determine the spin interaction parameters [33].

Fast MAS has the advantage of increased spectral resolution and signal intensity, but it also removes important information from the spectra by suppressing anisotropic interactions. It is, however, possible to selectively reintroduce the anisotropic spin interactions by suitably designed radio-frequency (rf) pulse sequences [34–36]. This technique, called recoupling, allows for combining the advantages of fast MAS and preserving the informative anisotropic interactions. In experiments, the anisotropic spin interactions are turned on and off at will during the separate time periods.

2.3   Measurements of ^{1}H-^{13}C Dipolar Couplings

General approach for measuring and assigning of the ^{13}C-^{1}H dipolar couplings is 2D separation of dipolar splittings according to ^{13}C chemical shifts. Since the dipolar couplings correspond to local magnetic fields, this class of experiments is referred to as separated local field (SLF) spectroscopy [37]. Technically, SLF experiments are designed to include two time intervals: during the first time period, the dipolar couplings are encoded in the NMR signal, while in the detection period the signals of different ^{13}C sites are separated on the basis of their differences in the ^{13}C chemical shifts.

Typical 1D cross-polarization (CP) ^{13}C spectra in columnar mesophase of the discotic sample THE6 are demonstrated in Figure 4-2. In both, static oriented and spinning samples, the spectra provide high chemical resolution.

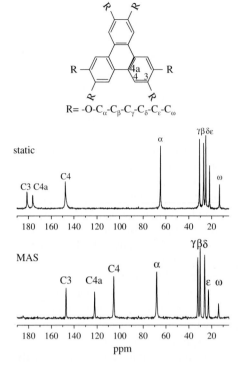

Figure 4-2. One-dimensional proton decoupled ^{13}C CP spectra of THE6 in the columnar D_{ho} phase at 85 °C. Line positions in static (top) and MAS spectrum (bottom) are determined by one of the principal values of the residual CSA tensor and by isotropic chemical shift, respectively

Three popular experimental protocols for SLF spectroscopy are outlined in Figure 4-3. They differ in the details of the preparation and evolution periods, while in all cases the ^{13}C signal is observed during the detection period t_2 as it evolves under the ^{13}C chemical shift interaction and in the presence of ^{1}H heteronuclear decoupling. These general protocols are applicable to both stationary and MAS samples.

In the conventional SLF sequence [37] shown in Figure 4-3a, the ^{13}C magnetization evolves under multiple ^{1}H$-^{13}$C heteronuclear dipolar couplings during the variable evolution period t_1. For many-proton spin systems this results in a crowded multiplet type dipolar spectrum, where each additional proton contributes with a successive first-order splitting.

In the proton-detected local field (PDLF) [38–40] experiment (Figure 4-3b), ^{1}H magnetization evolves under the local field of rare ^{13}C spins during t_1 and is subsequently transferred to ^{13}C spin for detection. Hence, PDLF dipolar spectrum is governed by two-spin dipolar interactions, which leads to a superposition of dipolar doublets.

In CP-based local field spectroscopy (CP-SLF) [41, 42], the dipolar couplings are monitored through the oscillations resulting from coherent polarization transfer

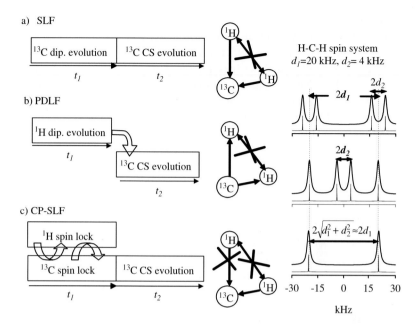

Figure 4-3. Experimental protocols of 2D experiments used for the measurements of heteronuclear dipolar couplings and the corresponding simulated dipolar spectra for three spin system C-H$_2$. (a) SLF; (b) PDLF; (c) CP-SLF

between ^{1}H and ^{13}C spins during the spin lock by Hartmann-Hahn matched rf fields (Figure 4-3c). The resolution of the large dipolar interaction in this approach is significantly enhanced due to effect of the truncation of the small couplings [43]. When the protons in the spin cluster are few, the dipolar resolution in all the methods is comparable. The simulated spectra for a three spin H-C-H system are demonstrated in Figure 4-3, right panel. For increasing number of remote protons, the resolution of the large coupling deteriorates in conventional SLF, while it is essentially unaffected in the PDLF and CP-SLF techniques. This is demonstrated below by experimental examples.

Efficient homonuclear ^{1}H decoupling during the evolution period is essential for obtaining high resolution of the heteronuclear dipolar couplings. The windowless sequences BLEW-n [44], LG [45], and FSLG [46, 47] have been used to achieve adequate resolution. The CP-SLF technique combined with FSLG decoupling is denoted PISEMA (*polarization inversion and spin exchange at the magic angle*) [48, 49]. Under MAS, the rf sequences applied during the dipolar evolution period typically combine heteronuclear recoupling with homonuclear decoupling.

2.3.1   Stationary samples

In contrast to conventional nematic liquid crystals, the columnar phases are usually highly ordered and viscous which often prevents spontaneous alignment in external fields, particularly, in the magnetic field of an NMR spectrometer. The samples can, however, be oriented in strong magnetic fields by slow cooling from the isotropic liquid. Columnar phases formed by discotic molecules typically align with the directors distributed in a plane perpendicular to the magnetic field. It is so because the net magnetic susceptibility of these mesophases is negative, which in turn is an effect of the strongly negative contribution from the aromatic core. For discotic molecules with long side chains, the positive contribution from the aliphatic part can balance the susceptibility from the core, resulting in a liquid crystalline sample with an isotropic director distribution.

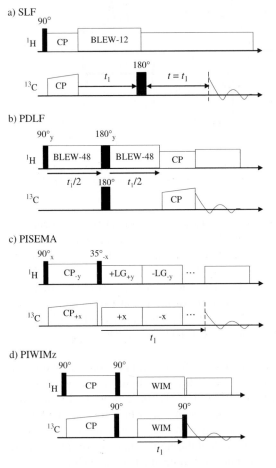

Figure 4-4. Examples of two-dimensional experiments for measurements of the ^{13}C-^{1}H spin coupling in static samples. In SLF technique (a) ^{13}C magnetization is encoded during t_1 period. In PDLF technique (b) the evolution of ^{1}H magnetization is encoded. In PISEMA (c) and PIWIMz (d) experiments the ^{13}C-^{1}H spin polarization exchange during t_1 is monitored. Homonuclear proton decoupling is achieved by BLEW-12, BLEW-48, FSLG, and WIM24 sequences, respectively. In detection period the heteronuclear proton decoupling is applied

The examples of the popular 2D SLF, PDLF and CP-SLF pulse sequences for measurements of ^{1}H-^{13}C dipolar couplings in columnar phases under stationary sample conditions are shown in Figure 4-4. The details can be found in Refs [48–53].

For relatively isolated C-H$_n$ ($n = 1, 2$) spin groups in columnar mesophases, well-resolved dipolar splitting are typically obtained by any of these techniques. That is demonstrated by the spectra for the directly bonded C-H pair in the aromatic core of RufH8O shown in Figure 4-5 [50]. Multiple dipolar couplings to the remote protons in the side chains, although weak, result in extra line broadening of the SLF spectrum and, to a smaller extend, of the PISEMA spectrum. The PDLF spectrum is essentially free of this effect and provides superior resolution.

Chain-per-deuterated THE5 is a good model sample for studying spin cluster with a limited number of protons (see molecular structure in Figure 4-6). For example, the core C$_4$ carbon dipolar coupled to two protons, H$_4$ and H$_5$, represents a relatively isolated H-C-H three spin system with one large (~ 5 kHz) and one small (~ 0.4 kHz) ^{1}H-^{13}C couplings. As shown in Figure 4-6, the resolution produced by all methods for this spin system is sufficient to observe all predicted splittings (cf. simulated spectra in Figure 4-3). Note, that in the PISEMA spectrum the smaller coupling is suppressed in the presence of the strong one.

$R = -C_8H_{17}$

PISEMA

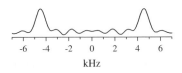

PDLF

SLF

-6 -4 -2 0 2 4 6

kHz

Figure 4-5. Experimental dipolar spectra obtained by different techniques, SLF, PDLF, and PISEMA, of the virtually isolated C-H spin pair provided by the signal of the C4 core carbon in the static and oriented RufH8O sample in the columnar phase at 85 °C. The frequency axes have been corrected for the scaling factors

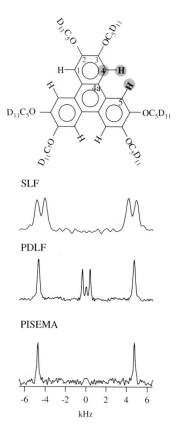

Figure 4-6. Experimental dipolar spectra of the virtually isolated H-C-H three spin system with one strong ($\approx 5\,\text{kHz}$) and one weak ($\approx 0.4\,\text{kHz}$) dipolar coupling provided by the signal of the C4 core carbon in static and oriented sample of chain-per-deuterated THE5 in columnar phase at 80 °C. The frequency axes have been corrected for the scaling factors

Example of well resolved dipolar spectra for a CH_2 group with two equal ^1H-^{13}C couplings is given by the α_3-methylene segment in RufH8O in the high temperature mesophase (Figure 4-7). The small splittings in the SLF spectrum and the inner doublet in the PDLF spectrum originate from a weak coupling of $\approx 300\,\text{Hz}$ to the core H_4 proton.

In general, for proton-rich samples, the PDLF method results in superior dipolar resolution. This is demonstrated in Figure 4-8, which compares dipolar spectra of all non-equivalent carbon sites in THE6. A resolution gain up to factor of 10 is obtained by PDLF as compared to SLF, and up to factor of 3 as compared to PISEMA and WIM. Similar observations have previously been made for calamitic liquid crystals in the nematic phase [50, 53]. A serious drawback of the original PISEMA technique, the excessive sensitivity of the dipolar splittings to proton chemical

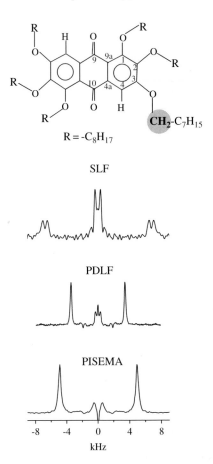

Figure 4-7. Experimental dipolar spectra of a methylene group with equal C-H dipolar couplings provided by the signal of the α_3-carbon in static and oriented RufH8O sample in columnar phase at 85 °C. The smaller splittings in the SLF spectrum and the inner doublet in the PDLF spectrum are due to a weak coupling to the core proton H4. The frequency axes have been corrected for the scaling factors

shift and resonance offsets, is alleviated in the offset-refocused modification of the method introduced in Refs [54, 55]. Table 4-1 compares the ^{1}H-^{13}C dipolar couplings estimated from the three different experiments. The results are consistent within experimental errors. Contributions from J_{CH} coupling were neglected in the analysis.

The measured dipolar couplings are often converted into bond order parameter S_{CH} useful for the structural and conformational studies [56–58]. The molecular order parameter S in THE6 mesophase is conveniently estimated from the heteronuclear dipolar splitting $\Delta\nu$ for the aromatic protonated C_4 carbon. Since the C-H bond vector for this site is perpendicular to the molecular rotation axis and the liquid-crystalline director is aligned perpendicular to the external magnetic field,

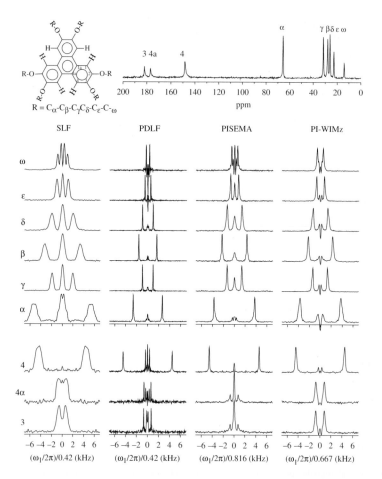

Figure 4-8. Experimental dipolar spectra for all non-equivalent carbons in the static and oriented THE6 sample in the coulmnar phase at 85° C. The 1D chemical shift spectrum is shown on the top. The frequency axes have been corrected for the scaling factors

the dipolar splitting is given by (cf. Eq. (4-5) with the angle values $\vartheta_{PM} = 90°$ and $\vartheta_{NL} = 90°$)

$$(4\text{-}6) \quad |\Delta v| = \frac{1}{2} k |S \, d_{CH}|$$

where k is the scaling factor of multiple pulse sequence. Using the C_4 splitting from the Table 4-1 (data from PDLF technique), S is estimated to 0.83. The S_{CH} values for the aliphatic chains are given in Table 4-2. Good agreement with previous results obtained in deuterated analogs [59] is found. Due to the high dipolar resolution the accuracy of the local order parameters is competitive to that available from the traditional ^{2}H NMR technique.

Table 4-1. Partially averaged dipolar couplings $\langle d_{CH}\rangle^{a)}$ (in kHz) in THE6 in the columnar phase at 85° C obtained by the SLF, PDLF, PISEMA, and WIM techniques

Carbon site	SLF	PDLF	PISEMA	WIM
ω	0.62	0.62	–	–
ε	1.04	1.02	1.04	0.97
δ	1.98	1.92	1.90	1.87
β	3.34	3.22	3.22	3.00
γ	1.96	1.92	1.88	1.82
α	5.30	5.30	5.30	5.07
C4	8.74	8.96	9.14	8.96
C4a	1.48	1.22	1.46	1.57
C3	1.28	1.24	1.46	1.50

$^{a)}$ J_{CH} couplings are not accounted for in the calculation of the dipolar couplings.

Table 4-2. Comparison of local order parameters extracted from dipolar and quadrupolar couplings in THE6 at 85° C

Carbon	$|S_{CH}|^{a)}$	$|S_{CD}|^{b)}$
α	0.24	0.24
β	0.15	0.16
γ	0.09	0.09
δ	0.09	0.09
ε	0.05	0.04
ω	0.03	0.02

$^{a)}$ $S_{CH} = \langle d_{CH}\rangle/d_{CH}$, where $d_{CH} = 21.5\,kHz$.
$^{b)}$ Calculated from data in Ref. [59].

2.3.2   MAS samples

The same general experimental protocols as described above for 2D SLF, PDLF and CP-SLF experiments can be applied to the samples under MAS condition. The design of the rf sequences employed during the evolution period is, however, more involved in this case. It is so, because active heteronuclear dipolar recoupling with simultaneous homonuclear proton decoupling is required. A wide range of dipolar recoupling NMR techniques have recently been developed [34–36]. Although most of these methods aim at applications to rigid solids, they can also be applied to mobile anisotropic systems either directly, or, after suitable modifications. Indirect J_{CH}-couplings can be safely ignored in these MAS experiments since they are suppressed during action of the recoupling sequences.

R-SLF and R-PDLF techniques. Levitt and co-workers have presented symmetry theorems that are useful in the design of rf schemes for selective recoupling and decoupling in MAS solid-state NMR [36]. For example, it was found that certain R-type sequences [60] lead to reactivation of heteronuclear dipolar couplings whereas homonuclear interactions are suppressed. It has also been demonstrated that the combination of R-type rf irradiation and 2D SLF spectroscopy (referred to as R-SLF spectroscopy [61] is an efficient approach for measuring and assigning heteronuclear dipolar interactions in solids [60, 62, 63]. Application of R-SLF was also successful for measurements of motionally-averaged ^{1}H-^{13}C interactions in columnar liquid crystals, especially for ^{13}CH groups [61, 62]. However, the method was proven to be less efficient for extracting dipolar couplings in ^{13}CH$_2$ methylene groups in systems exhibiting fast axial motions, as is common for mobile side chains in liquid crystals. The reason is that the motionally-averaged directions of the two CH bond vectors in the methylene moiety are collinear and, furthermore, that the two residual couplings are equal (Figure 4-9a). Therefore, the contribution from

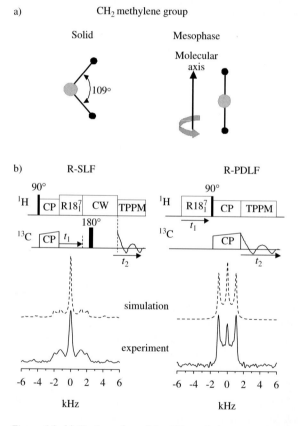

Figure 4-9. (a) Conformation of the CH$_2$ methylene group in the solid state and effective motionally averaged conformation in the mesophase. (b) 2D R-SLF and R-PDLF MAS experiments and corresponding spectra of a mobile CH$_2$ methylene group in the mesophase. The $R18_1^7$ pulse sequence [60] is applied in the evolution period to recouple heteronuclear dipolar interactions with simultaneous decoupling of homonuclear ^{1}H dipolar interactions. The simulated and experimental dipolar recoupled spectra of the α_3-methylene in RufH8O at 80 °C are shown by solid and dotted lines, respectively

such $^{13}CH_2$ spin system to the R-SLF powder spectrum is a 1:2:1 triplet with an orientation-dependent splitting. This results in a line shape, which is dominated by an uninformative zero-frequency peak superimposed on a broad and low-intensity base (see Figure 4-9b, left panel).

In contrast, the R-PDLF technique, which incorporates R-type recoupling in the PDLF protocol [61], produces high-resolution dipolar spectra irrespective of the ^1H multiplicity (see Figure 4-9b, right panel). The spectral simplification occurs due to the fact that each ^1H spin in a $^{13}CH_n$ group experiences the local dipolar field from only one ^{13}C spin and, hence, the PDLF spectrum is governed by simple two-spin interactions resulting in a doublet spectral structure. An additional advantage of the R-PDLF technique is the possibility to select certain ^1H-^{13}C spin pairs during the proton-to-carbon polarization transfer step. If J_{CH} couplings are exploited, polarization transfer occurs almost exclusively between directly bonded heteronuclei.

One limitation of heteronuclear dipolar recoupling based on R-type sequences is the relatively small dipolar scaling factor (e.g., 0.315 for the $R18_1^7$ sequence [60]), which may hamper measurements of small dipolar couplings in highly mobile systems. In this case, the CP-SLF approaches, which provide much higher scaling factors, can be advantageous.

CP-SLF technique. Efficient recoupling of heteronuclear dipolar interactions with simultaneous homonuclear proton decoupling can be achieved by combining the CP sequence with FSLG irradiation [45, 47], FSLG-CP [64, 65]. In contrast to the PISEMA technique [48] designed for stationary samples, FSLG-CP employs amplitude modulation of the ^{13}C rf fields synchronous with the phase switching (Figure 4-10a). It has been shown [64–66] that efficient dipolar recoupling is obtained if the amplitude modulation depth is equal to $2v_r$.

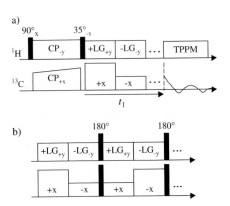

Figure 4-10. (a) 2D FSLG-CP MAS experiment. [64] Amplitude modulated FSLG-CP sequence is applied in the evolution period to recouple the heteronuclear dipolar interactions with simultaneous decoupling of homonuclear ^1H dipolar interactions. (b) Frequency offset refocused FSLG-CP sequence [54]

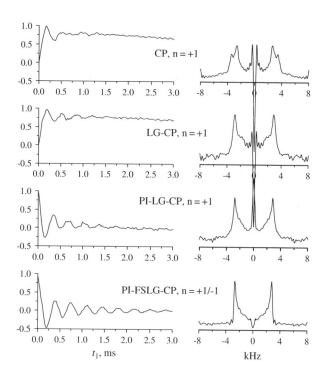

Figure 4-11. Recoupled dipolar spectra (right column) and corresponding time domain signals (left column) obtained with different CP-SLF approaches for the α_3-carbon of the RufH8O sample in the columnar phase at 80 °C (spinning frequency = 8 kHz). Polarization inversion (PI) prior the dipolar evolution period increases the dipolar oscillation amplitude

In Figure 4-11, recoupled dipolar spectra of the α_3-methylene group in the columnar phase of RufH8O obtained by different CP-SLF approaches are compared. The corresponding time-domain signals are also shown. Both continuous-wave on-resonance CP and off-resonance LG-CP result in distorted line shapes due to influence of ^1H homonuclear couplings and rf inhomogeneity. An extra splittings in the on-resonance CP spectrum is caused by a strong non-suppressed H-H dipolar coupling within the CH_2 group [67]. In contrast, the FSLG-CP spectral shape is very similar to the theoretical spectrum simulated with the assumption of the ideal rf fields and fully removed homonuclear proton interaction. The polarization inversion (PI) increases the dipolar oscillation amplitude.

It has been pointed in many investigations [43, 64, 65, 68] that off-resonance sequences like LG- and FSLG-CP are very sensitive to the choice of the ^1H carrier frequency. Therefore, it may be difficult to obtain accurate dipolar couplings for multiple sites in the molecule due to the ^1H chemical shift dispersion. To remedy this problem, combining CP with an on-resonance magic-echo sequence in ^1H

channel (thus replacing the off-resonance ^{1}H rf irradiation) has been suggested [69]. Another solution is to apply 180° refocusing pulses orthogonal to the spin-lock fields, as shown in Figure 4-10b. Simultaneous pulses in the two channels remove the effect the frequency offset, while preserve the heteronuclear couplings. This approach has successfully been demonstrated in stationary liquid crystals and rigid solids under MAS [54].

Yet another approach to deal with the ^{1}H offset effects, useful for mobile systems exhibiting small heteronuclear dipolar couplings (below $\sim 5\,\mathrm{kHz}$), is to perform *on-resonance* phase-alternated CP [67, 70]. For such systems, the ^{1}H-^{1}H interactions are either significantly reduced or exhibit inhomogeneous line broadening due to effect of the projection of the spin-spin vectors on common molecular rotational axis [71]. Hence, MAS is capable to efficiently suppress the homonuclear couplings. Thus, *on-resonance* ^{1}H rf irradiation during CP recoupling eliminates the sensitivity to the frequency offsets, while the phase alternation preserves the robustness with the respect to rf mismatch and inhomogeneity. The flip angle of the individual pulses must be set to multiples of 180°, and the amplitude modulation of one (or both) channels synchronized with the phase inversion is necessary to maintain the heteronuclear recoupling, similar to FSLG-CP under MAS. The scaling factor of this sequence, 0.707, is higher as compared to FSLG-CP, 0.577, which results in better resolution of small dipolar couplings.

Such an on-resonance technique, abbreviated APM-CP (*a*mplitude- and *p*hase-*m*odulated CP) is demonstrated for the HHTT compound in the columnar phase and is compared to FSLG-CP (Figure 4-12). The dipolar couplings in the APM-CP spectra are well resolved for all methylene sites. For the least mobile groups, α and β, outer shoulders in the spectral shapes are observed which are indicative of H-H dipolar coupling, as confirmed by numerical analysis. It is also clear that the apparent dipolar splittings are larger in APM-CP spectra owing to a higher dipolar scaling factor, as compared to FSLG-CP.

Comparison of the recoupling techniques. The dipolar recoupling approaches discussed above have the advantage of a reduced dependence on the crystallite orientations, a property referred to as γ-encoding [72]. Generally, γ-encoded recoupling schemes produce longer dipolar oscillation curves, due to minimizing the destructive interference from signals of differently oriented crystallites or domains in the sample. This results in a more accurate determination of dipolar couplings [36, 72]. Another useful feature of the R and FSLG sequences is a short rf cycle time, which enables sampling of large splittings without spectral aliasing. The CP-based sequences, in contrast to R sequences, do not require rotor synchronization which makes them flexible in terms of choosing the rf power and spinning speed.

Table 4-3 lists various recoupling techniques, which have been applied to liquid crystals. The methods are compared in terms of heteronuclear dipolar scaling factor, γ-encoding property, sensitivity to frequency offsets and chemical shift anisotropy of S (rare) and I (abundant) spins, tolerance to rf inhomogeneity, accomplishment of active homonuclear proton decoupling, length of rf cycle time (which limits the maximum spectral width in dipolar dimension), and requirement of rotor

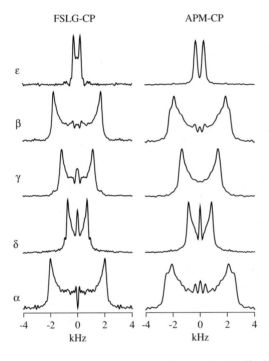

$R = C_\alpha\text{-}C_\beta\text{-}C_\gamma\text{-}C_\delta\text{-}C_\varepsilon\text{-}C_\omega$

Figure 4-12. Comparison of FSLG-CP and APM-CP ^{13}C-^{1}H dipolar recoupled spectra of methylene carbons in the HHTT sample in the columnar phase at 80 °C (spinning frequency = 8 kHz)

synchronization. The sequences are listed in order of decreasing scaling factor. Table 4-3 reveals that the FSLG-CP technique is the best scored one.

Figure 4-13 shows FSLG-CP and R-PDLF dipolar spectra of the α_3 methylene carbon in RufH8O-^{13}C$_\alpha$. The experiments were carried out under similar conditions. Lineshape simulations are also included. Both techniques result in well-resolved dipolar patterns, although the resolution in the R-PDLF spectrum is somewhat lower. This can partly be explained by the smaller dipolar scaling factor of the R sequence. A better suppression of homonuclear couplings in the FSLG sequence [47], as

Table 4-3. Comparison of heteronuclear recoupling techniques applied to liquid crystals

Method	Scaling factor	γ-encoding	S offset	S CSA	I offset	I CSA	RF inhom.	I homo--dec.	Short cycle time[a]	Asynch-ronous	Ref.
CW-CP	0.707	+	+	+	+	+	−	−	+	+	[42]
APM-CP	0.707	+	+	+	+	+	+	−	+	+	[70]
LG-CP	0.577	+	+	+	−	+	−	+	+	+	[73]
FSLG-CP refocused	0.577	+	+	+	−	+	+	+	+	+	[64]
FSLG-CP[b]	0.577	+	+	+	+	+	+	+	+	+	[54]
DROSS	0.39	−	+	−	+	+	−	−	−	−	[74]
R-PDLF	0.315 [c]	+	+	+	+	+	−	+	+	−	[61]
R-SLF	0.315 [c]	+	+	+	+	+	−	+	+	−	[60]

[a] Marked '+' if cycle time can be set to a fraction of the rotor period.
[b] FSLG-CP with ^{1}H offset refocusing [54].
[c] Using the $R18_1^7$ sequence [60].

a) MAS sample

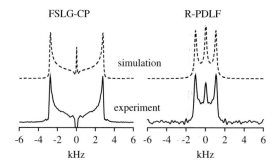

b) stationary sample

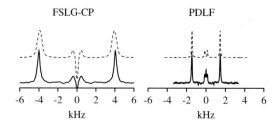

Figure 4-13. Comparison of FSLG-CP and PDLF ^{13}C-^{1}H dipolar recoupled spectra the of α_3-carbons in the RufH8O sample in the columnar phase at 85 °C. (a) MAS sample. (b) Oriented sample under stationary conditions. The following values were used for the simulations (dashed lines): $d_{CH(1)} = d_{CH(2)} = 7\,$kHz and $d_{HH} = 12\,$kHz

compared to the R sequence [60], may result in additional resolution enhancement of FSLG-CP spectra.

The dipolar splittings observed in the MAS spectra are consistent with those obtained in the oriented stationary sample of the same compound. The corresponding spectra for the stationary sample are included in Figure 4-13. Taking into account the scaling factors of the sequences, the splittings in MAS FSLG-CP, MAS R-PDLF, static FSLG-CP and static PDLF spectra result in dipolar couplings of 6.9, 7.0, 7.0, and 6.8 kHz, respectively, which agree within the experimental errors.

3.   SUMMARY

NMR spectroscopy is a particularly versatile experimental method for study of columnar systems. Investigations of columnar phases require the same type of NMR equipment as used in studies of the solid state. Deuterium NMR has over the years been the predominant method, employed in many investigations of the molecular structure, order, and dynamics in columnar mesophases. An attractive alternative to the ^{2}H technique is ^{13}C NMR spectroscopy. In this chapter, we have described applications to columnar phases of modern and advanced ^{13}C NMR methods for measuring the heteronuclear ^{13}C-^{1}H dipolar couplings. The major advantage of these techniques can be summarized as follows: i) high quality spectra may be obtained from natural isotopic abundance samples, ii) in-equivalent carbon sites are well resolved and signals can usually be easily assigned, and iii) the anisotropic interactions determining the line shape provide a wealth of information on molecular and liquid crystalline phase properties.

4.   ACKNOWLEDGEMENTS

This work was supported by the Swedish Research Council and the Russian foundation for basic research (Grant 04-03-32639).

5.   REFERENCES

1. S. Chandrasekhar, B. K. Sadashiva, and K. A. Suresh, Liquid crystals of disc-like molecules, *Pramana* **9** 471–480 (1977).
2. S. Chandrasekhar, Liquid Crystals of Disklike Molecules, *Adv. Liq. Cryst.* **5** 47–78 (1982).
3. S. Chandrasekhar and G. S. Ranganath, Discotic Liquid Crystals, *Rep. Prog. Phys.* **53** 57–84 (1990).
4. S. Chandrasekhar, Discotic Liquid Crystals, *Liq. Cryst.* **14** 3–14 (1993).
5. R. J. Bushby and O. R. Lozman, Discotic liquid crystals 25 years on, *Curr. Opin. Colloid Interface Sci.* **7** 343–354 (2002).
6. S. Kumar, Recent developments in the chemistry of triphenylene-based discotic liquid crystals, *Liq. Cryst.* **31** 1037–1059 (2004).
7. C. Destrade, P. Foucher, H. Gasparoux, N. H. Tinh, A. M. Levelut, and J. Malthete, Disc–like Mesogen Polymorphism, *Mol. Cryst. Liq. Cryst.* **106** 121–146 (1984).

8. D. Adam, P. Schuhmacher, J. Simmerer, L. Häussling, K. Siemensmeyer, K. H. Etzbachi, H. Ringsdorf, and D. Haarer, Fast photoconduction in the highly ordered columnar phase of a discotic liquid crystal, *Nature* **371** 141–143 (1994).

9. L. Schmidt-Mende, A. Fechtenkotter, K. Mullen, E. Moons, R. H. Friend, and J. D. MacKenzie, Self-organized discotic liquid crystals for high-efficiency organic photovoltaics, *Science* **293** 1119–1122 (2001).

10. N. Boden, R. J. Bushby, and J. Clements, Mechanism of quasi-one-dimensional electronic conductivity in discotic liquid-crystals, *J. Chem. Phys.* **98** 5920–5931 (1993).

11. G. B. M. Vaughan, P. A. Heiney, J. P. McCauley, and A. B. Smith, Conductivity and structure of a liquid-crystalline organic conductor, *Phys. Rev. B* **46** 2787–2791 (1992).

12. P. G. de Gennes and J. Prost, *The Physics of Liquid Crystals* (Clarendon, Oxford, 1993).

13. C. Zannoni, in: *The Molecular Dynamics of Liquid Crystals*, edited by G. R. Luckhurst and C. A. Veracini (Kluwer, Dordrecht, 1994), p. 11–40.

14. D. Goldfarb, I. Belsky, Z. Luz, and H. Zimmermann, Axial-biaxial phase transition in discotic liquid crystals studied by beuterium NMR, *J. Chem. Phys.* **79** 6203–6210 (1983).

15. D. Goldfarb, R. Poupko, Z. Luz, and H. Zimmermann, Deuterium NMR of biaxial discotic liquid crystals, *J. Chem. Phys.* **79** 4035–4047 (1983).

16. Z. Luz, D. Goldfarb, and H. Zimmermann, in: *Nuclear Magnetic Resonance of Liquid Crystals*, edited by J. W. Emsley (Reidel, Dordrecht, 1985), p. 343–377.

17. D. Goldfarb, R. Y. Dong, Z. Luz, and H. Zimmermann, Deuterium N.M.R. relaxation and spectral densities in the discotic mesophase of hexahexyloxytriphenylene, *Mol. Phys.* **54** 1185–1202 (1985).

18. J. Hirschinger, W. Kranig, and H. W. Spiess, A deuteron NMR study of axial motion and side chain conformation in the mesophase of discotic liquid crystal main-chain polymers, *Colloid Polym. Sci.* **269** 993–1002 (1991).

19. J. Leisen, M. Werth, C. Boeffel, and H. W. Spiess, Molecular Dynamics at the Glass Transition: One Dimensional and Two Dimensional Nuclear Magnetic Resonance Studies of a Glass-Forming Discotic Liquid Crystal, *J. Chem. Phys.* **97** 3749–3759 (1992).

20. M. Werth, J. Leisen, C. Boeffel, R. Y. Dong, and H. W. Spiess, Mobility changes of side chains ascribed to density modulations along columnar structures detected by 2D NMR, *J. Phys. II France* **3** 53–67 (1993).

21. A. Maliniak, S. Greenbaum, R. Poupko, H. Zimmermann, and Z. Luz, Deuterium and Carbon-13 NMR of the Solid and Discotic Phases of Three Benzenehexa-n-alkanoates, *J. Phys. Chem.* **97** 4832–4840 (1993).

22. S. Zamir, R. Poupko, Z. Luz, B. Huser, C. Boeffel, and H. Zimmermann, Molecular ordering and dynamics in the columnar mesophase of a new dimeric discotic liquid-crystal as studied by X-ray-diffraction and deuterium NMR, *J. Am. Chem. Soc.* **116** 1973–1980 (1994).

23. D. Sandström, M. Nygren, H. Zimmermann, and A. Maliniak, Deuterium NMR Investigation of a Discotic Mesogen Based on Hexasubstituted Truxene, *J. Phys. Chem.* **99** 6661–6669 (1995).

24. R. Y. Dong and C. R. Morcombe, Rotational dynamics of a discotic liquid crystal studied by deuteron spin relaxation, *Liq. Cryst.* **27** 897–900 (2000).

25. S. V. Dvinskikh, I. Furó, H. Zimmermann, and A. Maliniak, Molecular self-diffusion in a columnar liquid crystalline phase determined by deuterium NMR, *Phys. Rev. E* **65** 050702(R) (2002).

26. D. Sandström and H. Zimmermann, Correlation of deuterium quadrupolar couplings and carbon-13 chemical shifts in ordered media by multiple-quantum NMR, *J. Phys. Chem. B* **104** 1490–1493 (2000).

27. M. H. Levitt, *Spin Dynamics* (Wiley, Chichester, 2001).

28. G. E. Pake, Nuclear resonance absorption in hydrated crystals: Fine structure of the proton line, *J. Chem. Phys.* **16** 327–336 (1948).

29. U. Haeberlen, High Resolution NMR in Solids – Selective Avereging, *Adv. Magn. Reson.* **1** (1976).

30. E. R. Andrew, A. Bradbury, and R. G. Eades, NMR spectra from a crystal rotated at high speed, *Nature* **182** 1659 (1958).

31. I. J. Lowe, Free induction decays in rotating solids, *Phys. Rev. Lett.* **2** 285–287 (1959).

32. K. Schmidt-Rohr and H. W. Spiess, *Multidimensional solid-state NMR and polymers* (Academic Press, London, 1994).

33. J. Herzfeld and A. E. Berger, Sideband intensities in NMR spectra of samples spinning at the magic angle, *J. Chem. Phys.* **73** 6021–6030 (1980).

34. A. E. Bennett, R. G. Griffin, and S. Vega, Recoupling of Homo- and Heteronulear Dipolar Interactions in Rotating Solids, *NMR Basic Princ.Progr.* **33** 1–77 (1994).

35. S. Dusold and A. Sebald, Dipolar recoupling under magi-angle spinning conditions, *Annu. Rep. NMR Spectrosc.* **41** 185–264 (2000).

36. M. H. Levitt, in: *Encyclopedia of Nuclear Magnetic Resonance*, Vol. 9, edited by D. M. Grant and R. K. Harris (Wiley, Chichester, 2002), p. 165–196.

37. R. K. Hester, J. L. Ackerman, B. L. Neff, and J. S. Waugh, Separated local field spectra in NMR: Determination of structure of solids, *Phys. Rev. Lett.* **36** 1081–1083 (1976).

38. P. Caravatti, G. Bodenhausen, and R. R. Ernst, Heteronuclear solid-state correlation spectroscopy, *Chem. Phys. Lett.* **89** 363–367 (1982).

39. T. Nakai and T. Terao, Measurements of heteronuclear dipolar powder patterns due only to directly bonded couplings, *Magn. Reson. Chem.* **30** 42–44 (1992).

40. K. Schmidt-Rohr, D. Nanz, L. Emsley, and A. Pines, NMR measurement of resolved heteronuclear dipole couplings in liquid crystals and lipids, *J. Phys. Chem.* **98** 6668–6670 (1994).

41. L. Müller, A. Kumar, T. Baumann, and R. R. Ernst, Transient Oscillations in NMR Cross-Polarization Experiments in Solids, *Phys. Rev. Lett.* **32** 1402–1406 (1974).

42. P. Bertani, J. Raya, P. Reinheimer, R. Gougeon, L. Delmotte, and J. Hirschinger, $^{19}F/^{29}Si$ distance determination in fluoride-containing octadecasil by Hartmann-Hahn cross-polarization under fast magic-angle spinning, *Solid State Nucl. Magn. Reson.* **13** 219–229 (1999).

43. Z. Gan, Spin dynamics of polarization inversion spin exchange at the magic angle in multiple spin systems, *J. Magn. Reson.* **143** 136–143 (2000).

44. D. P. Burum, M. Linder, and R. R. Ernst, Low-power multipulse line narrowing in solid–state NMR, *J. Magn. Reson.* **44** 173–188 (1981).

45. M. Lee and W. I. Goldburg, Nuclear-magnetic-resonance line narrowing by a rotating rf field, *Phys. Rev.* **140** A1261–A1271 (1965).

46. M. Mehring and J. S. Waugh, Magic-angle NMR experiments in solids, *Phys. Rev. B* **5** 3459–3471 (1972).

47. A. Bielecki, A. C. Kolbert, H. J. M. de Groot, R. G. Griffin, and M. H. Levitt, Frequency-switched Lee-Goldburg sequences in solids, *Adv. Magn. Reson.* **14** 111–124 (1990).

48. C. H. Wu, A. Ramamoorthy, and S. J. Opella, High-resolution heteronuclear dipolar solid–state NMR spectroscopy, *J. Magn. Reson. Ser. A* **109** 270–272 (1994).

49. A. Ramamoorthy, Y. Wei, and D.-K. Lee, PISEMA solid-state NMR spectroscopy, *Annu. Rep. NMR Spectrosc.* **52** 1–52 (2004).

50. S. V. Dvinskikh, H. Zimmermann, A. Maliniak, and D. Sandström, Separated local field spectroscopy of columnar and nematic liquid crystals, *J. Magn. Reson.* **163** 46–55 (2003).

51. S. V. Dvinskikh, K. Yamamoto, and A. Ramamoorthy, Separated local field NMR spectroscopy by windowless isotropic mixing, *Chem. Phys. Lett.* **419** 168–173 (2006).

52. K. Yamamoto, S. V. Dvinskikh, and A. Ramamoorthy, Measurement of heteronuclear dipolar couplings using a rotating frame solid-state NMR experiment, *Chem. Phys. Lett.* **419** 533–536 (2006).

53. B. M. Fung, K. Ermolaev, and Y. Yu, ^{13}C NMR of liquid crystals with different proton homonuclear dipolar decoupling methods, *J. Magn. Reson.* **138** 28–35 (1999).

54. S. V. Dvinskikh and D. Sandström, Frequency offset refocused PISEMA-type sequences, *J. Magn. Reson.* **175** 163–169 (2005).

55. K. Yamamoto, D. K. Lee, and A. Ramamoorthy, Broadband-PISEMA solid-state NMR spectroscopy, *Chem. Phys. Lett.* **407** 289–293 (2005).

56. J. W. Emsley, in: *Encyclopedia of nuclear magnetic resonance*, edited by D. M. Grant and R. K. Harris (Wiley, Chichester, 1996), p. 2781–2787.

57. J. Courtieu, J. P. Bayle, and B. M. Fung, Variable angle sample spinning NMR in liquid crystals, *Prog. Nucl. Magn. Reson. Spectrosc.* **26** 141–169 (1994).

58. B. M. Fung, ^{13}C NMR studies of liquid crystals, *Prog. Nucl. Magn. Reson. Spectrosc.* **41** 171–186 (2002).

59. D. Goldfarb, Z. Luz, and H. Zimmermann, Deuterium magnetic resonance in the discotic columnar mesophases of hexaalkyloxytriphenylenes: the conformation of the aliphatic side chains, *J. Chem. Phys.* **78** 7065–7072 (1983).

60. X. Zhao, M. Edén, and M. H. Levitt, Recoupling of heteronuclear dipolar interactions in solid-state NMR using symmetry-based pulse sequences, *Chem. Phys. Lett.* **342** 353–361 (2001).

61. S. V. Dvinskikh, H. Zimmermann, A. Maliniak, and D. Sandström, Measurements of motionally averaged heteronuclear dipolar couplings in MAS NMR using R-type recoupling, *J. Magn. Reson.* **168** 194–201 (2004).

62. S. V. Dvinskikh, Z. Luz, H. Zimmermann, A. Maliniak, and D. Sandström, Molecular characterization of hexaoctyloxy-rufigallol in the solid and columnar phases: A local field NMR study, *J. Phys. Chem. B* **107** 1969–1976 (2003).

63. X. Zhao, J. L. Sudmeier, W. W. Bachovchin, and M. H. Levitt, Measurements of NH bond length by fast magic-angle spinning solid-state NMR spectroscopy: a new method for the quantification of hydrogen bonds, *J. Am. Chem. Soc.* **123** 11097–11098 (2001).

64. S. V. Dvinskikh, H. Zimmermann, A. Maliniak, and D. Sandström, Heteronuclear dipolar recoupling in liquid crystals and solids by PISEMA-type pulse sequences, *J. Magn. Reson.* **164** 165–170 (2003).

65. S. V. Dvinskikh, H. Zimmermann, A. Maliniak, and D. Sandström, Heteronuclear dipolar recoupling in solid-state nuclear magnetic resonance by amplitude-, phase-, and frequency-modulated Lee-Goldburg cross-polarization, *J. Chem. Phys.* **122** 044512 (2005).

66. K. Yamamoto, V. L. Ermakov, D. K. Lee, and A. Ramamoorthy, PITANSEMA-MAS, a solid-state NMR method to measure heteronuclear dipolar couplings under MAS, *Chem. Phys. Lett.* **408** 118–122 (2005).

67. S. V. Dvinskikh and V. I. Chizhik, Cross-Polarization with Radio-Frequency Field Phase and Amplitude Modulation under Magic-Angle Spinning Conditions, *J. Exp. Theor. Phys.* **102** 91–101 (2006).
68. V. Ladizhansky, E. Vinogradov, B.-J. van Rossum, H. J. M. de Groot, and S. Vega, Multiple-spin effects in fast magic angle spinning Lee-Goldburg cross-polarization experiments in uniformly labeled compounds, *J. Chem. Phys.* **118** 5547–5557 (2003).
69. A. A. Nevzorov and S. J. Opella, A "Magic Sandwich" pulse sequence with reduced offset dependence for high-resolution separated local field spectroscopy, *J. Magn. Reson.* **164** 182–186 (2003).
70. S. V. Dvinskikh, V. Castro, and D. Sandström, Probing Segmental Order in Lipid Bilayers at Variable Hydration Levels by Amplitude- and Phase-Modulated Cross-Polarization NMR, *Phys. Chem. Chem. Phys.* **7** 3255–3257 (2005).
71. M. M. Maricq and J. S. Waugh, NMR in rotating solids, *J. Chem. Phys.* **70** 3300–3316 (1979).
72. N. C. Nielsen, H. Bildsoe, H. J. Jakobsen, and M. H. Levitt, Double-quantum homonuclear rotary resonance – efficient dipolar recovery in magic-angle-spinning nuclear-magnetic-resonance, *J. Chem. Phys.* **101** 1805–1812 (1994).
73. V. Ladizhansky and S. Vega, Polarization transfer dynamics in Lee-Goldburg cross polarization nuclear magnetic resonance experiments on rotating solids, *J. Chem. Phys.* **112** 7158–7168 (2000).
74. J. D. Gross, D. E. Warschawski, and R. G. Griffin, Dipolar recoupling in MAS NMR: a probe for segmental order in lipid bilayers, *J. Am. Chem. Soc.* **119** 796–802 (1997).

CHAPTER 5

PHASE BIAXIALITY IN NEMATIC LIQUID CRYSTALS
Experimental evidence for phase biaxiality in various thermotropic liquid crystalline systems

KIRSTEN SEVERING[1,*] AND KAY SAALWÄCHTER[2]

[1]*Albert-Ludwigs-Universität Freiburg, Institut für Makromolekulare Chemie Stefan-Meier-Strasse 31, D-79104 Freiburg, Germany. E-mail: kirsten@makro.uni-freiburg.de*
[2] *Martin-Luther-Universität Halle-Wittenberg, Institut für Physik, Friedemann-Bach Platz 6, D-06108 Halle/Saale, Germany. E-mail: kay.saalwaechter@physik.uni-halle.de*

1.   INTRODUCTION

Ever since their discovery in the 1880s, a variety of different phases with increasingly complex molecular arrangements in thermotropic liquid crystalline materials has been found. Sometimes, these discoveries have occurred accidentally in the past. However, in the case of the biaxial nematic phase, the discovery was stimulated by a theoretical prediction published by M. J. Freiser [1] in 1970. Molecules with rotational symmetry are capable of forming a uniaxial nematic phase, in which they exhibit a one-dimensional long-range orientational order but no degree of long-range positional ordering. In contrast to this ubiquitous liquid crystalline phase, molecules which significantly deviate from this cylindrical shape can be expected to form a biaxial nematic phase, which is characterized by a three-dimensional orientational order. By taking into account the interactions between asymmetric molecules within the framework of the Maier-Saupe theory, Freiser could show that the first-order transition from an isotropic to a uniaxial nematic state should be followed by a second-order phase transition at lower temperatures. Sometimes, molecules with sufficient deviation from axial symmetry, which had been identified as being biaxial in optical investigations [2], actually turned out to be uniaxially nematic when studied by other techniques [3]. Thus,

---

* Corresponding Author

A. Ramamoorthy (ed.), Thermotropic Liquid Crystals, 141–170.
© 2007 *Springer.*

Freiser's prediction stimulated not only a synthetic effort in the design of molecules with the desired shape to form a biaxial nematic phase but also a search for an experimental method to unambiguously prove the existence of this phase biaxiality.

In 1980, the first justified claim for the discovery of a biaxial nematic phase was made for a lyotropic liquid crystal comprised of the ternary system of potassium laurate-1-decanol-D_2O [4]. In addition to a uniaxial phase (micelles of a bilayer structure), there were two further nematic phases. One of the phases was found to be uniaxial as well, probably corresponding to a phase with cylindrical micelles. Existing in a temperature range in between these two uniaxial nematic phases was a third phase which was found to be biaxial. The phases were classified by microscopic studies as well as deuterium NMR measurements. Three years later, Galerne and Marcerou studied the same system by conoscopy, leading to a complete determination of the ordering tensor in all three nematic phases [5].

An accepted experimental proof for the existence of a biaxial nematic phase in a thermotropic liquid crystal remained missing for a very long time. However, in recent years, biaxial nematic phases have been found in liquid crystalline polymers as well as in liquid crystals made of rod-disc mesogens, banana-shaped (bent-core) molecules, and organo-siloxane tetrapodes. Here, some characteristics of these systems and the corresponding experimental procedure for the investigation of phase biaxiality will be introduced. Further details for the individual systems can be found in the cited literature.

2.   A LONG-DISCUSSED CANDIDATE

The first thermotropic compound for which a biaxial nematic phase was claimed is the laterally broadened mesogen shown in Figure 5-1. The assignment of the phases was based on the observation of the optical textures and the X-ray diffraction pattern [2].

Roughly ten years later, this compound was investigated by means of deuterium NMR spectroscopy [3]. For this purpose, the mesogen was specifically deuterated in the three oxymethylene groups linking the three phenyl rings to a single one in the rod-like part of the molecule. The spectra obtained from this compound

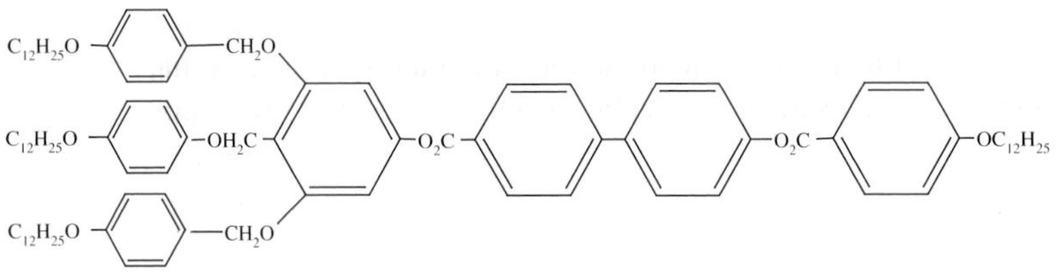

Figure 5-1. Chemical structure of the laterally broadened mesogen. Figure adapted from [3]

contained two different quadrupolar doublets, one originating from the deuterons in the outer methylene groups (*3,5 deuterons*) and another from the deuterons in the inner methylene group (*4 deuterons*). Furthermore, Luckhurst and coworkers investigated the protonated mesogen with the aid of the deuterated spin probe hexamethylbenzene, HMB-d_{18}.

To measure the biaxiality parameter

$$(5\text{-}1) \quad \tilde{\eta} = (\tilde{q}_{xx} - \tilde{q}_{yy})/\tilde{q}_{zz},$$

it is, in principle, sufficient to determine the principal components $\tilde{q}_{xx}$, $\tilde{q}_{yy}$ and $\tilde{q}_{zz}$ of the motion-averaged quadrupolar tensor $\tilde{\mathbf{q}}$ by aligning a monodomain sample of the liquid crystal with the magnetic field along each of the principal axes in turn. However, since this procedure is extremely difficult, another approach was followed here. The uniform alignment of the major director parallel to the magnetic field was intentionally removed by spinning the sample above some critical speed about an axis perpendicular to the magnetic field. This results in a uniform distribution of the director in the plane orthogonal to the spinning axis [6]. In the case of a uniaxial nematic phase, the resulting deuterium NMR spectrum is a so-called two-dimensional powder pattern. This spectrum is dominated by two quadrupolar splittings in the ratio 2:1, each respectively associated with the components of $\tilde{\mathbf{q}}$ parallel ($\tilde{q}_{zz}$) and perpendicular ($\tilde{q}_{xx} = \tilde{q}_{yy}$) to the director.

The situation in a biaxial nematic phase is obviously more complex, as the three components $\tilde{q}_{zz}$, $\tilde{q}_{yy}$ and $\tilde{q}_{xx}$ of $\tilde{\mathbf{q}}$ all take different values. In this case, it is expected that to minimize the magnetic free energy, the least negative component of the partially averaged susceptibility tensor, which is associated with the major director, and the intermediate component will be randomly distributed in a plane orthogonal to the spinning axis. Consequently, the most negative component will align along the spinning axis. The observed deuterium NMR spectrum then corresponds to a two-dimensional powder pattern, where the ratio of the two components $\tilde{q}_{zz}$ and $\tilde{q}_{xx}$ is no longer 2:1. Studying this kind of spectra therefore provides unambiguous evidence for the biaxiality of the investigated phase.

For the deuterated mesogen, ^{2}H NMR spectra of (a) a static sample and (b) a sample spinning at 3.7 Hz in the nematic phase at a temperature of 87° C are shown in Figure 5-2. The larger splitting in the static spectrum can be assigned to the 4-deuterons, while the small splitting originates from the 3,5-deuterons. The outer pair of broad lines found in the spinning spectrum is the same as the large splitting observed in the static spectrum and thus corresponds to the 4-deuterons with the director aligned parallel to the magnetic field. The shoulders on the central peak exhibit a splitting of approximately one half of the outer one, which suggests that the nematic phase is uniaxial. Due to significant overlap of the individual features, no further information can be extracted from the signals of the 3,5-deuterons.

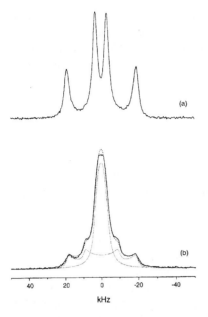

Figure 5-2. Deuterium NMR spectrum for (a) a static sample of the deuterated mesogen and (b) a sample spinning at 3.7 Hz in the nematic phase (87° C). The simulated spectrum, shown as the dashed line was obtained for $\tilde{\eta} = 0$; the dotted lines indicate the contributions of the two sets of equivalent deuterons. Figure reprinted with permission from J. R. Hughes, G. Kothe, G. R. Luckhurst, J. Malthête, M. E. Neubert, I. Shenouda, B. A. Timimi, and M. Tittelbach, *J. Chem. Phys.*, **107**, 9252 (1997). Copyright 1997, American Institute of Physics

The two-dimensional powder pattern is a sum of different Lorentzian lines originating from all orientations of the major and secondary directors with respect to the magnetic field [7],

$$(5\text{-}2) \quad L(\nu) = \pi^{-1} \int_0^{2\pi} T_2 / \{1 + 4\pi^2 T_2^2 [\nu - \tilde{\nu}(\beta)]^2\} d\beta,$$

where T_2 is the inverse of the line width in the static spectrum and the central frequency $\tilde{\nu}(\beta)$ varies with the angle β between the major director and the magnetic field according to

$$(5\text{-}3) \quad \tilde{\nu}(\beta) = \nu_0 \pm \frac{3}{4}\tilde{q}_{zz}[\frac{1}{2}(3\cos^2\beta - 1 + \tilde{\eta}\sin^2\beta)].$$

Using information from the static spectrum ($\tilde{q}_{zz}$, the quadrupolar splitting, and T_2), two-dimensional powder spectra were simulated according to (5-2) and (5-3) by choosing different values for the biaxiality parameter.

Figure 5-3 shows such spectra obtained with four different biaxiality parameters. For comparison, the experimental spectrum was superimposed on each of the

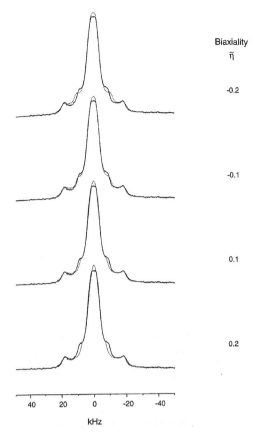

Biaxiality
$\tilde{\eta}$

-0.2

-0.1

0.1

0.2

Figure 5-3. Simulated two-dimensional powder patterns obtained with non-zero values for the biaxiality parameter (dotted lines). The experimental spectrum is superimposed on each of the simulated ones (solid lines). Figure reprinted with permission from J. R. Hughes, G. Kothe, G. R. Luckhurst, J. Malthête, M. E. Neubert, I. Shenouda, B. A. Timimi, and M. Tittelbach, *J. Chem. Phys.*, **107**, 9252 (1997). Copyright 1997, American Institute of Physics

simulations. The best agreement between the experimental and simulated spectra was observed for a biaxiality parameter of 0.0 (dashed line in Figure 5-2), supporting the suggestion of a uniaxial nematic phase.

The same investigations were carried out with the non-deuterated mesogen. The 18 equivalent deuterons in the spin probe HMB-d_{18} resulted in a single quadrupolar splitting (Figure 5-4 (a)). The sample spun at 3.7 Hz yielded a two-dimensional powder pattern with well-resolved features (Figure 5-4 (b)). For the non-deuterated mesogen the best agreement between experiment and theory was again achieved for a biaxiality parameter of 0.0. Additional results obtained at lower temperatures did not give any indication for the presence of a biaxial nematic phase either. All these experiments suggest that the phase was, in contrast to the results gained from other experimental methods, indeed uniaxial.

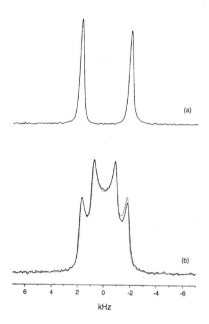

Figure 5-4. Deuterium NMR spectrum of HMB dissolved in the mesogen for (a) a static sample and (b) a sample spun at 3.7 Hz in the nematic phase (87° C). The dashed line shows the simulated spectrum for the uniaxial case. Figure reprinted with permission from J. R. Hughes, G. Kothe, G. R. Luckhurst, J. Malthête, M. E. Neubert, I. Shenouda, B. A. Timimi, and M. Tittelbach, *J. Chem. Phys.*, **107**, 9252 (1997). Copyright 1997, American Institute of Physics

3.   PHASE BIAXIALITY IN NEMATIC LIQUID-CRYSTALLINE POLYMERS

In principle, by lowering the temperature, the orientational correlation between neighboring molecules grows, which in turn is expected to support the formation of a biaxial nematic phase. Because of the onset of crystallization or the formation of smectic phases, a biaxial nematic phase in a low molar mass system could not be achieved experimentally for a long time. These problems paved the way for the concept of a biaxial nematic polymer.

Weissflog and Demus [8] succeeded in synthesizing a new type of liquid crystalline material, in which the rotation of the molecules about their long molecular axis was hindered by the introduction of laterally attached substituents. However, experiments conducted on these liquid crystals proved the existence of conventional uniaxial nematic phases only. Beyond that, a further restriction of the rotation of the mesogens by connecting them via terminal spacers to a polymer backbone (*end-on polymers*) resulted in uniaxial nematic and smectic phases. Consequently, the next step was the synthesis of a new class of liquid crystalline side-chain polymers, in which the mesogenic moiety was laterally attached to the polymer backbone. The idea was that this *side-on* connection should be successful in hindering the rotation

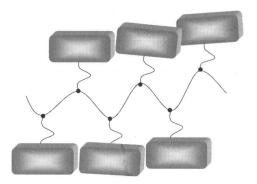

Figure 5-5. Schematic representation of a side-on side-chain polymer

of the mesogen and therefore support the formation of a biaxial nematic phase. A schematic sketch of a liquid crystalline polymer is given in Figure 5-5.

3.1 Conoscopic Investigations

Conoscopy allows the quantitative determination of the biaxiality in terms of $2V_x$, which is the angle between the two optical axes. In order to do so, a biaxially oriented monodomain has to be prepared, where both, the major director parallel, and the minor directors perpendicular to the mesogenic long axis, have to be aligned. In 1991, Leube et al. [9] published a conoscopy study conducted on a liquid crystalline side-on side-chain polymer, in which the polymer backbone consisted of a methacrylate chain. In this system a naphthalene unit served as the central core of the mesogenic unit. This naphthalene unit produced an effective broadening of the mesogen, which is expected to enhance the stability of the biaxial nematic phase [10]. The mesogen also exhibited a well pronounced anisotropy of polarizability, which is crucial to allow a reliable detection of a biaxial nematic phase with optical methods. The chemical structure of the investigated polymer is shown in Figure 5-6.

Due to the high rotational viscosity of the polymer even directly below the nematic-isotropic transition temperature $T_{n,i}$, the preparation of a homogeneously oriented sample using conventional alignment techniques was not successful.

Since conoscopic experiments require macroscopic uniformly oriented samples, mixtures of the polymer with 30%, 40% and 50 % of a cyanophenylester, a low molar mass liquid crystal, were investigated. The addition of a low molar mass liquid crystal serves to lower the glass transition temperature T_g and consequently the rotational viscosity. These mixtures could therefore be oriented using a procedure that is depicted in Figure 5-7.

Leube et al. achieved a homeotropic alignment by using an appropriate surface treatment and confirmed it by conoscopy. Furthermore, conoscopic observations, namely a "Maltese cross" that did not change upon rotation, proved that a macroscopic alignment of the minor directors was not present. Afterwards, they obtained

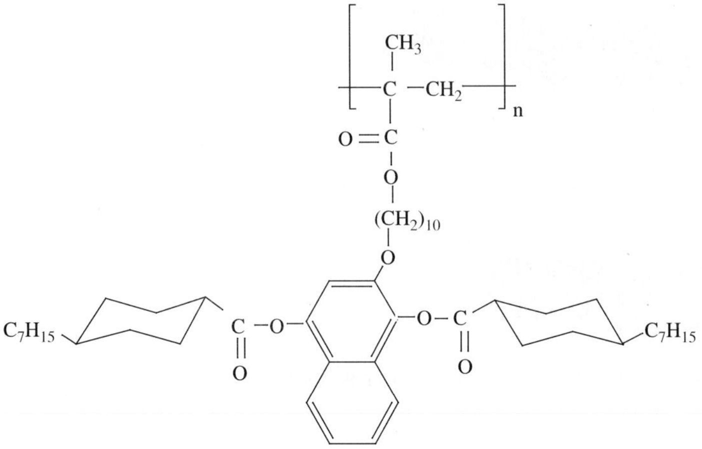

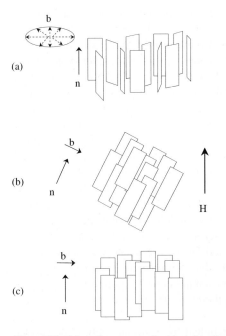

$$\left[\begin{array}{c} CH_3 \\ | \\ C\!-\!CH_2 \\ | \\ O\!=\!C \\ | \\ O \\ | \\ (CH_2)_{10} \\ | \\ O \end{array} \right]_n$$

C_7H_{15} ... $C\!-\!O$ ... $O\!-\!C$ ... C_7H_{15}
$\parallel$ $\parallel$
O O

Figure 5-6. Chemical structure of the investigated liquid crystalline side-on side-chain polymer

(a)
b
n

(b)
b
n
H

(c)
b
n

Figure 5-7. Orientation process for biaxially aligned samples, with **n** being the main director and **b** the secondary director. (a) Homeotropically oriented sample, polydomain with respect to **b**; (b) tilted structure in the presence of the magnetic field **H**, biaxially aligned; c) relaxed biaxial structure without the magnetic field. Figure adapted from [9]

a uniform orientation of the minor director by the application of a magnetic field **H** parallel to the main director. This procedure resulted in a tilted orientation of the phase, which was again observed by conoscopy. This was the first indication for the presence of a biaxial nematic phase, since a uniaxial nematic phase would have aligned along the magnetic field and would therefore not have changed by applying the field in this way. After annealing the sample outside the magnetic field, a reorientation of the main director occurred. This time, upon rotation of the sample, Leube et al. observed a biaxial conoscopic interference pattern. A quantitative value of $2V_x$ was obtained from the distance between the turning points of the two hyperbolas in this interference pattern.

Temperature dependent measurements of $2V_x$ showed phase biaxiality for mixtures containing more than 50% polymer. As seen in Figure 5-8, higher degrees of biaxiality, i.e., greater values of $2V_x$, were found for those mixtures with higher polymer contents, while a temperature dependence for the individual mixtures was not observed within their respective nematic and smectic phases.

3.2 Deuterium NMR Investigations

Conoscopy is known to be prone to artifacts because of its sensitivity to the symmetry of the refractive index, which might be tampered due to surface effects and flow phenomena. By using deuterium NMR spectroscopy, Severing et al. [11] confirmed phase biaxiality in a polymeric liquid crystal similar to that studied earlier by Leube. To evaluate different parameters that bias the formation of a biaxial nematic phase and gain a more general picture of the phase biaxiality in nematic liquid crystalline polymers, the investigations were expanded to side-chain polymers of different chemical constitutions as well as to mixtures of polymers and low molar mass liquid crystals [12].

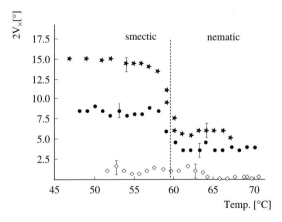

Figure 5-8. Temperature dependence of the angle between the two optical axes ($2V_x$) in mixtures with 50% (diamonds), 60% (circles) and 70% (asterisks) polymer; the dashed line indicates the approximate location of the phase transition from the nematic to the smectic phase. Data replotted from [9]

The investigated samples included two different classes of side-chain polymers consisting of a polysiloxane[1] backbone [14] with the mesogens either terminally attached [15] via a flexible spacer of four or six carbon atoms (called *end-on 4* and *end-on 6*, respectively) or laterally attached [16] via spacers of six or eleven carbon atoms (called *side-on 11* and *side-on 6*, respectively). Due to the relatively high glass transition temperature, the side-on 6 sample was only investigated in a mixture with a low molar mass liquid crystal (called *side-on 6 mix*). In addition to the pure polymer, the side-on 11 sample was also investigated in a mixture with a different low molar mass liquid crystal (called *side-on 11 mix*).

The chemical structures of the two different polymer types are given in Figure 5-9.The side-on 6 mix sample was prepared by dissolving the polymer in benzene in the presence of 10% 4-octyloxyl-4'-hexyloxybenzoate, followed by the evaporation of the solvent under high vacuum conditions. For the side-on 11 mixture 2% of 8CB (4-n-octyl-4'-cyanobiphenyl) was added as described above. In all samples, 2% of perdeuterated hexamethylbenzene (HMB-d_{18}) was dissolved as a nuclear spin probe. Due to the restricted diffusive motion of the probe within the liquid crystal, the probe reveals the symmetry of the particular liquid crystalline phase. The use of a spin probe rather than a deuterated mesogen itself offers the

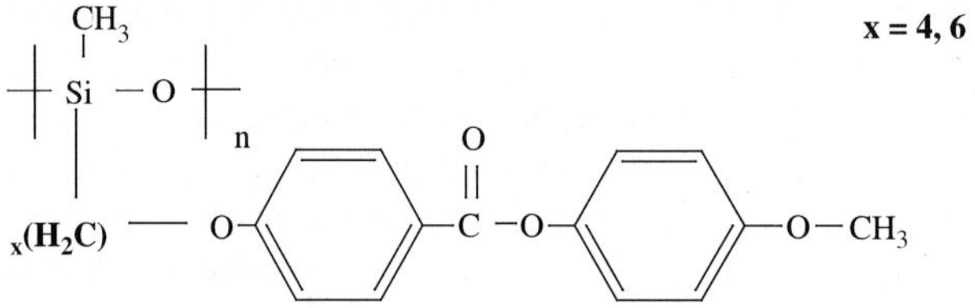

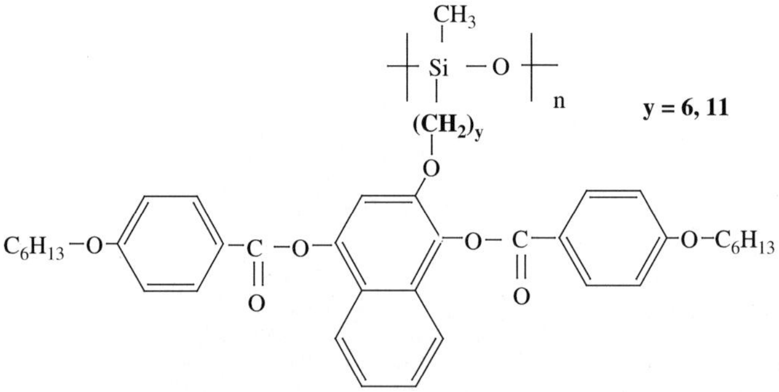

Figure 5-9. Chemical structure of the different types of investigated polymers: end-on polymer (top), side-on polymer (bottom). Figure adapted from [13]

[1] The advantage of siloxanes as compared to methacrylates is a much lower glass transition temperature, which enables (in most cases) the formation of a mondomain even for the bulk polymer.

advantage of observing narrower spectral lines, as was seen in Section 1. However, one has to keep in mind that the volume over which the probe averages depends on the diffusion coefficient and therefore decreases somewhat with decreasing temperature. The samples were prepared in standard NMR glass tubes between silicon spacers coated with Teflon on the inner side facing the sample. The NMR probe used for measurements under various angles was a modified BRUKER goniometer double resonance probe, which allowed for fast sample rotations (90° in less than 150 ms) by using a rotary magnet controlled by trigger pulses from the NMR console. Figure 5-10 shows some pictures of this modified NMR probe.

Due to the relatively high viscosity of the polymer systems as compared to the sample studied by Luckhurst (see Section 1), the technique introduced by Yu and Saupe [4] was used. Therein, the biaxiality parameter of different samples is derived from measurements of the quadrupolar splitting $\Delta\nu_q(\beta)$ at different angles β between the main director of an annealed monodomain and the static magnetic field. At a particular angle β, the quadrupolar splitting is[2]

$$(5\text{-}4) \qquad \nu_q(\beta) = \nu_0 \pm \frac{3}{4} q_{zz} \frac{1}{2} (3\cos^2\beta - 1 - \eta\sin^2\beta),$$

Figure 5-10. The modified NMR Probe (left), the sample holder (upper right), and the angular adjustment scale (lower right)

---

[2] This equation was introduced in (5-3), differing in the sign of η only.

where ν_0 is the deuterium resonance frequency, q_{zz} is the z principal component of the quadrupole tensor $\mathbf{q}$, and η is the biaxiality parameter defined as[3]

$$(5\text{-}5) \qquad \eta = \frac{q_{yy} - q_{xx}}{q_{zz}}.$$

The components q_{zz} and q_{yy} can be measured by respectively orienting the main director along and perpendicular to the magnetic field. Because $\mathbf{q}$ is traceless $(q_{xx} + q_{yy} + q_{zz} = 0)$, (5-5) can be rewritten as

$$(5\text{-}6) \qquad \eta = 1 + \frac{2q_{yy}}{q_{zz}}.$$

In the case of a uniaxial nematic phase ($\eta = 0$), the negatively signed 90°-splitting (q_{yy}) is therefore exactly half of the 0°-splitting (q_{zz}), while the ratio is different from 1:2 in the case of a biaxial nematic phase ($\eta \neq 0$). The determination of the biaxiality parameter according to this procedure is called the *tilt experiment*.

Under the assumption of exponential relaxation, the spectra were fitted with two Lorentzians separated by $\Delta\nu_q$. For Yu and Saupe, the determination of the quadrupolar splitting of a tilted sample was achieved by simply rotating the sample to the desired angle and acquiring the signal. In this system however, inaccuracies in the determination of the 90°-splitting might occur due to the director relaxing during the measurement for two reasons. First, because here the viscosity is considerably lower than in the system of Yu and Saupe the director relaxation is much faster. Secondly, due to the low concentration of the deuterated spin probe, the director can relax during the comparatively long time needed for the acquisition of a spectrum with a sufficient signal-to-noise ratio. Therefore, the technique introduced by Zhou and Frydman [17] was used, whereby the sample is positioned along the desired angle β for the actual pulse sequence and acquisition only. For the considerably longer recycle delay, which is needed for the spins to return to their thermal equilibrium, the sample is then rotated back to its equilibrium position along the magnetic field.

For the creation of a three-dimensional monodomain, the sample should in principle undergo repeated rotation followed by the relaxation of the major director (q_{zz}), so as to force the minor directors (q_{xx} and q_{yy}) into a minimal energy configuration. In our case it was not possible to experimentally verify a planar polydomain of q_{xx} and q_{yy} when simply tilting the sample about an angle of 90°. This means that a tilting-relaxation procedure to produce a monodomain was not necessary. From this observation we drew the important conclusion that the alignment of the secondary director is probably rather fast, i.e. on the order of tens of ms. It could either occur via a magnetic torque in the q_{xx}-q_{yy} plane combined with a low local viscosity for rotations around q_{zz}, or due to accumulating weak director

---

[3] Note that (5-5) and (5-1) differ in the sign of the biaxiality parameter η because of the different conventions used in the respective publications.

relaxation/viscuous friction effects during the first of the few hundred sample flips that make up the total number of recorded transients.

The main results of the NMR experiments are summarized below. The tilt experiment for the determination of the biaxiality parameter was conducted on all different polymer systems for several temperatures within the liquid crystalline phase. Multiple measurements were done for each temperature and both angles (0° and 90°). Figure 5-11 shows the result for a tilt experiment conducted on the side-on 11 polymer at 316 K. The 90°-splitting is notably smaller than half of the 0°-splitting, as would have been expected for the case of a uniaxial nematic phase. From the ratio of the quadrupolar splittings, the biaxiality parameter was calculated according to (5-6) and for this temperature was $\eta \sim 0.05$.

Figure 5-12 (a) displays the average biaxiality parameter as a function of the reduced temperature $T/T_{n,i}$ for the different investigated systems. For the sake of clarity, the error intervals based on the maximum error of multiple measurements, which vary between 0.005 and 0.015, are omitted in this graph. Two-dimensional (2D) ^{2}H-correlation experiments were conducted for sample orientations of 0° and 90° to determine the degree of heterogeneous broadening present in the different systems. Such an underlying heterogenous contribution can be caused by a director distribution that is stable on the time scale of the NMR experiment and might interfere with a precise measurement of the quadrupolar splitting. A comparison between the line width of a peak along the diagonal and the line width of that same peak parallel to the anti-diagonal provides a measure for the degree of the imhomogeneous contribution to the line width [11, 13]. The difference between these line widths is taken as the interval in which an exact determination of the

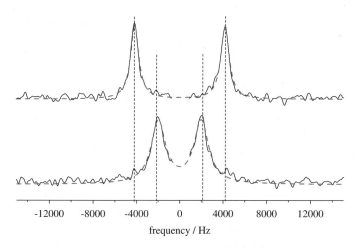

Figure 5-11. Result for a tilt experiment conducted on the side-on 11 polymer at 316 K: 0°-spectrum (top) and 90°-spectrum (bottom); the dashed lines are the Lorentzian fits. The outer and inner pairs of dashed lines respectively indicate the 0°-splitting and half of this splitting. See [12] for more spectra, fits, and residuals

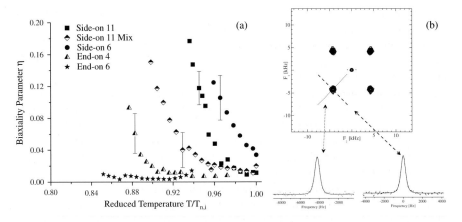

Figure 5-12. (a) Temperature dependence of the biaxiality parameter for the different investigated polymer systems, (b) exemplary 2D spectrum for the side-on 11 polymer at 316 K. Adapted from [13]

quadrupolar splitting is possible. Figure 5-12 (b) shows one exemplary 2D spectrum for the side-on 11 polymer at 316 K along with the parallel and anti-parallel projections. For each polymer, an error interval based on the 2D spectra for one exemplary temperature is included in the graph in Figure 5-12 (a).

From the results of the investigations of the end-on polymers, where the mesogenic units are terminally connected to the polymer backbone, it could be seen that a long spacer enables a decoupling of the liquid crystal and the polymer. The end-on 6 polymer displayed uniaxial behavior over the entire region of the nematic phase and in this sense resembled the behavior of a low molar mass liquid crystal. In contrast to this, the end-on 4 polymer exhibited a biaxial nematic phase at low temperatures. This proved the important influence of the spacer length between the mesogenic unit and the polymer backbone, as has been found in earlier studies conducted on cholesteric liquid crystals [18]. The formation of a biaxial nematic phase was observed, although the biaxiality was clearly less pronounced in the end-on mesogen when compared with the side-on mesogen. A reduction of the spacer length from 11 to 6 carbon atoms for the side-on polymer resulted in an increase of the biaxiality parameter at comparably reduced temperatures. For the side-on 6 polymer, the existence of a uniaxial nematic phase could not be proven unambiguously. The addition of a low molar mass liquid crystal to the side-on 11 polymer, which induced a downward shift of the glass transition temperature, resulted in a decreased temperature dependence of the biaxiality parameter. The onset of increasing biaxiality parameters was shifted towards lower reduced temperatures.

The importance of polymer dynamics for the formation of a biaxial nematic phase is shown in Figure 5-13, where the biaxiality parameter for the different systems is plotted as a function of T/T_g. It can be seen that the temperature dependent behavior was very similar for all investigated systems except for the side-on 11 mix

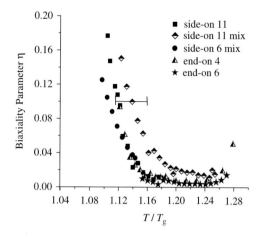

Figure 5-13. The biaxiality parameter for the different systems as a function of T/T_g. Data replotted from [13]

system.[4] The formation of a biaxial nematic phase occured upon approaching the glass transition temperature at $T/T_g \sim 1.16$. This clearly demonstrates the surprising role of the glass transition and thus the importance of polymer dynamics for the stabilization of the biaxial nematic phase, as discussed in more detail in [12].

We note that a potential artifact which might mimic phase biaxiality is the occurrence of a dynamic frequency shift of the quadrupolar splitting due to director fluctuations on intermediate timescales. This has been discussed by Frezzato, Moro, and Kothe [19]. However, this frequency shift is predicted to decrease with increasing correlation times (and thus with decreasing temperature) and therefore cannot serve as an explanation for the observation of an increasing biaxiality parameter at lower temperatures.

3.3   Effects of Phase Biaxiality on Cholesteric Phase Structure

The effect of potential phase biaxiality on the structure of cholesteric liquid crystals was investigated in detailed optical studies by Ogawa and Finkelmann [16]. In cholesteric liquid crystals, a characteristic fingerprint texture is obtained when the helical axis is oriented parallel to the surface using homeotropic boundary conditions. In the case of a uniaxial cholesteric liquid crystal, black pseudo-isotropic lines separated by a distance of half of the pitch length are observed under crossed polarizers in the regions of local alignment of the director perpendicular to the surface. This is schematically depicted in Figure 5-14 (a). If the fingerprint texture is formed by a biaxial cholesteric polymer, one characteristic change is expected, as can be seen in Figure 5-14 (b), where the locations of homeotropic alignment of the director are no longer pseudo-isotropic but birefringent. This has indeed been

---

[4] The rather wide error interval for the side-on 11 mix sample originates from the difficulty to assign the glass transition temperature T_g, which was rather smeared out in the DSC scans for this sample.

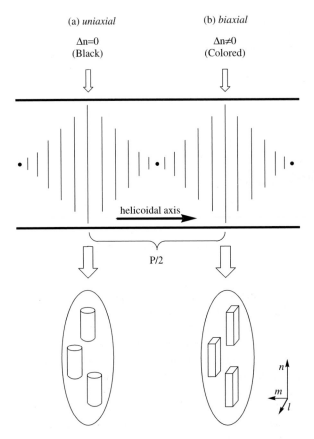

Figure 5-14. Schematic representation of (a) a uniaxial cholesteric phase and (b) a biaxial cholesteric phase. Figure reprinted from H. Ogawa, E. Stibal-Fischer, and H. Finkelmann, *Macromol. Chem. Phys.* **205**, 593 (2004) with permission from Wiley-VCH

found by Ogawa and Finkelmann. In a mixture of a low-molar mass liquid crystal (4-cyanophenyl-4-*trans*-pentylcyclohexane) with a biaxial cholesteric copolymer[5] with laterally attached mesogenic units, periodic lines still existed. These lines, however, were no longer black but colored. The microscope picture on the left side of Figure 5-15 shows a conventional fingerprint texture obtained by a mixture with 10% cholesteric polymer, while the one on the right side depicts a distorted fingerprint texture from a mixture with 80% cholesteric polymer.

An increasing phase biaxiality obviously disturbed the smooth optical period-icity of the cholesteric phase structure. Whether these defects were due to a modification of the continuous twist of the main director or to the biaxial optics

---

[5] For the synthesis of the biaxial cholesteric copolymer, the side-on 6 polymer from the NMR study (see above) and cholesterylcarbonate in a concentration range from 1 to 10% were used.

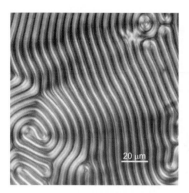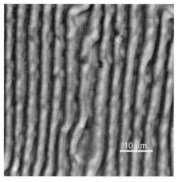

Figure 5-15. Fingerprint texture of a low molar mass liquid crystal with 10% (left) and 80% (right) biaxial cholesteric polymer. Figure reprinted from H. Ogawa, E. Stibal-Fischer, and H. Finkelmann, *Macromol. Chem. Phys.* **205**, 593 (2004) with permission from Wiley-VCH

and a non-uniform orientation of the minor directors was not answered on the basis of these experiments.

4. PHASE BIAXIALITY IN COMBINED DISC-ROD MESOGENS

Rod- and disc-shaped mesogens are likewise capable of forming uniaxial nematic phases N_u. In the early 1970s, Alben [20] published a theoretical calculation based on a mean-field lattice model in which he predicted the appearance of a biaxial nematic phase N_B for mixtures containing certain ratios between discs and rods. The phase diagram corresponding to his prediction is schematically shown in Figure 5-16.

However, thermodynamics requires the system to phase separate into two N_u phases. This important point was ignored by Alben and was first noted very much later [21, 22]. The long search to find a miscibility between the two types of mesogens was not successful for a long time [23, 24].

To avoid phase separation between the two moieties, a vast effort was devoted towards the synthesis of chemically linked disc-rod molecules [25–27]. Beyond that, mixtures of prolate and oblate mesogens have stimulated theorists to perform intensive computer simulations. Simulations and molecular field theories predict the biaxial nematic phase to occur around the minimum of the transition temperature $T_{n,i}$ from the nematic to the isotropic phase [22]. Furthermore, a strong decrease of the transition enthalpy is expected upon approaching the tricritical point.

In 2003 Mehl and Kouwer [28] published a miscibility study of a novel mesogen bearing both a calamitic and a discotic part. The synthetic concept as well as the chemical structure of this mesogen is shown in Figure 5-17. Surprisingly, this combined rod-disc compound **1** was separately miscible with either of the pure mesogens, i.e., the disc **2** and the rod **3**, over the entire composition range. The mixing behavior was determined by studying contact samples using optical microscopy, which gave no indication for a miscibility gap and revealed the presence of a nematic phase by a typical schlieren or marble texture. Polarization microscopy

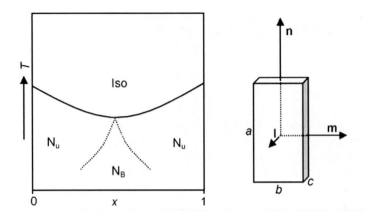

Figure 5-16. Schematic phase diagram containing the N_B phase. For board-shaped molecules, x scales with the ratio b/a; for mixtures of discs and rods, x represents the fraction of discs. The exact shape of the phase transitions between the N_B and the N_U phase depends on the model used. Figure reprinted with permission from P. H. J. Kouwer and G. H. Mehl, *J. Am. Chem. Soc. Commun.*, **125**, 11172. Copyright 2003 American Chemical Society

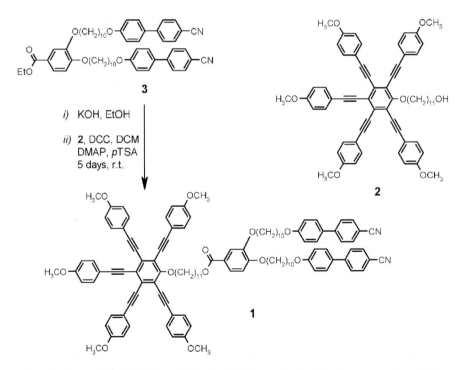

Figure 5-17. Synthesis of the disc-rod mesogen. Details can be found in the corresponding literature. Figure reprinted with permission from P. H. J. Kouwer and G. H. Mehl, *J. Am. Chem. Soc. Commun.*, **125**, 11172. Copyright 2003 American Chemical Society

as well as differential scanning calorimetry experiments were conducted on discrete mixtures consisting of either the combined mesogen with the rods or the combined mesogen with the discs, and were used for the construction of the phase diagrams, one of which is given in Figure 5-18.

A minimum in the clearing temperature was observed when the fraction of the disc mesogen was approximately 0.56, as can be seen in Figure 5-18 (a). As predicted in theoretical and simulation studies for the occurrence of a biaxial nematic phase, the latent heat of the phase transition strongly decreased upon approaching this composition range, reaching a minimum also at a fraction of 0.56 (Figure 5-18 (b)). A reliable determination of the exact value for the transition enthalpies in this composition range proved difficult due to the limit of the sensitivity of the instrument used. However, another strong indication for the presence of a biaxial nematic phase was the fact that a second phase transition between the glass transition and the clearing temperature could be observed in the DSC traces of these mixtures (Figure 5-18 (c)). These transitions were not accompanied by a change of the textures, as observed by optical microscopy during prolonged annealing, and X-ray investigations did not indicate the presence of any positional ordering.

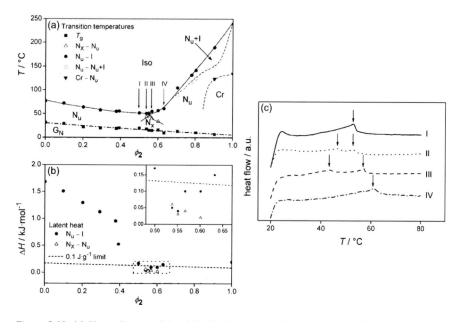

Figure 5-18. (a) Phase diagram of **1** and **2**. The lines are a guide to the eye. DSC traces of mixtures I-IV are shown in diagram (c). (b) Latent heat values for the N_u-I and N_x-N_u transitions of mixtures of **1** and **2**. The dotted line represents the maximum resolution of the DSC for integrating phase transitions unambiguously. The inset shows a magnification from the data in the box. (c) Normalized DSC traces of **1** and **2** (in the N_x phase region) in the following ratios: (I) 0.50:0.50; (II) 0.46:0.54; (III) 0.43:0.57; and (IV) 0.37:0.63. Figure reprinted with permission from P. H. J. Kouwer and G. H. Mehl, *J. Am. Chem. Soc. Commun.*, **125**, 11172. Copyright 2003 American Chemical Society

All these observations are in good agreement with the theoretical prediction of a biaxial nematic phase. This conclusion will have to be confirmed using further experimental techniques, by performing for example, solid state NMR experiments on specifically deuterated samples.

Only very recently have individual disc and rod-shaped mesogens been synthesized and found to be completely miscible in the nematic phase. The miscibility behavior has been investigated and confirmed by polarization microscopy, DSC and X-ray diffraction studies [29].

5.   BIAXIAL NEMATIC PHASES IN LIQUID CRYSTALS CONSISTING OF BANANA MESOGENS

Another class of mesogen, which has proven to be a successful candidate for the formation of a biaxial nematic phase, is made up of the so-called bent-core, or banana mesogens. In this kind of mesogen, two rod-like molecules are chemically linked together via a central unit. The molecular structure of such a banana mesogen is schematically given in Figure 5-19.

The shape of these mesogens is clearly biaxial. However, the stability of a biaxial nematic phase strongly depends on the inter-arm angle θ_v. A theoretical model based purely on repulsive molecular interactions [30] as well as another model including attractive and repulsive anisotropic interactions [31] both predict the Landau point, i.e., the point where a direct transition from the isotropic state to the biaxial nematic state occurs, to be achieved when the inter-arm angle adopts the tetrahedral value of 109.47°. On the other hand, the stability of the biaxial nematic phase is limited by the crystallization of the system. Therefore, the range of angles for which a biaxial nematic phase can be expected is extremely small, on the order of $\pm 2°$ from the tetrahedral value.

Within the last few years, two different experimental techniques could successfully prove the existence of a biaxial nematic phase in liquid crystals based on bent-core mesogens. Madsen et al. [32] found indications for a phase biaxiality with polarization microscopy and conoscopy and also obtained additional quantitative evidence by means of deuterium NMR. Acharya and co-workers [33] used X-ray

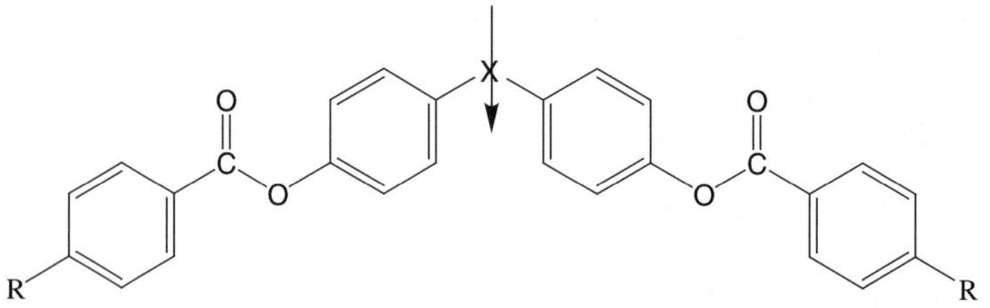

Figure 5-19. Molecular structure of a bent-core mesogen. R is an aliphatic chain; X represents the central core, which is an oxadiazole for the examples discussed in the following. The arrow indicates the direction of the transverse dipole moment. Figure adapted from [33]

diffraction measurements as the experimental method by which they successfully confirmed phase biaxiality in the same system.

5.1 Deuterium NMR Investigations

Samulski and co-workers [32] investigated two different bent-core mesogens, where the core consisted of an oxadiazole and the aliphatic chain R in the structure shown in Figure 5-19 was either an OC_{12}- or a C_7-unit. Figure 5-20 shows polarization microscopy pictures as well as conoscopic results for the OC_{12} sample.

Figure 5-20 (a) shows a birefringent schlieren texture in a cell with a homeotropic surface treatment. Since this alone is not definitive evidence for a biaxial nematic phase,[6] additional optical experiments were conducted. In a homeotropic surface-treated wedge cell "dark" states were observed (Figure 5-20 (b)). Conoscopic investigations of these dark states revealed the transformation of an apparently uniaxial structure to an increasingly biaxial structure upon lowering the temperature, as can be seen in Figure 5-20 (c). These optical results strongly indicated phase biaxiality, however, surface-induced biaxiality can never be ruled out completely. Therefore, deuterium NMR spectroscopy was used to gain unambiguous and quantitative results.

As in the study of Luckhurst and colleagues discussed in Section 1 and in the experiments conducted on liquid crystalline polymers introduced in Section 2.2, the solute "probe" molecule used for the NMR investigations was HMB-d_{18}. The rotation technique discussed in Section 1 was used to isotropically distribute the respective z and y components of the susceptibility tensor in a plane perpendicular

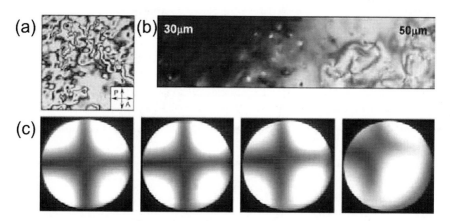

Figure 5-20. Microscopy in the nematic phase of OC_{12}. a) Homeotropic $4\,\mu m$ cell; b) wedge-shape cell with homeotropic surface treatment; c) opening of the isogyres on cooling the dark nematic domain in b), sample temperature from left to right of 204, 202, 201, and 200° C. Figure reprinted with permission from L. A. Madsen, T. J. Dingemans, M. Nakata, and E. T. Samulski, *Phys. Rev. Lett.*, **92**, 145505 (2004). Copyright 2004 American Physical Society

---

[6] A nonuniform director configuration would also appear birefringent.

to the rotation axis. The experimental spectrum was compared to calculated ones assuming different biaxiality parameters. Due to a much lower viscosity, considerably higher rotation frequencies (hundreds of Hz) were necessary to generate a planar director distribution. The use of a theory derived by Photinos *et al.* [34] to include rotational averaging on the experimental timescale was required due to these high rotation rates.

Figure 5-21 shows different calculated spectra assuming biaxiality parameters of 0, 0.11, and 0.22. The experimental spectrum for the C_7-mesogen at 174° C (within the nematic phase) rotating at 230 Hz is superimposed on each of the simulations. The best agreement between experiment and simulation can be observed for a biaxiality parameter of 0.11 (Figure 5-21 (b)), which demonstrates the biaxial nature of the nematic phase. A biaxiality parameter of 0.08 has been measured at 184° C in the same way. One should keep in mind that because of the indirect nature of spin probe measurements, the value of the biaxiality parameter obtained in this way most probably is a lower bound. However, the biaxiality in this system as well as in the polymeric system discussed in section 2.2 is still far smaller than that of

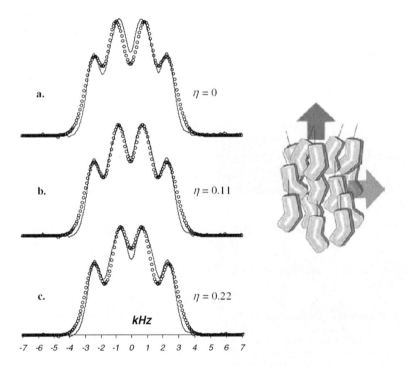

Figure 5-21. Experimental spectrum for C_7 at 174° C at a rotation frequency of 230 Hz (solid line) along with simulations (circles). The biaxiality parameter η used for the calculation is given on the right side of the spectra; schematic representation of a biaxial nematic phase made of bent-core mesogens (far right). Figure reprinted with permission from L. A. Madsen, T. J. Dingemans, M. Nakata, and E. T. Samulski, *Phys. Rev. Lett.,* **92**, 145505 (2004). Copyright 2004 American Physical Society

the lyotropic liquid crystal [4], where D_2O acts as a spin probe. Yet, in the latter case diffusion averaging occurs over much larger length scales between mesoscopic aggregates of variable shape. Thus, the systems are hardly comparable.

5.2 X-Ray Diffraction Measurements

Three different types of bent-core molecules, the C_7 and OC_{12} introduced in Section 4.1 and an additional one with a heterocyclic core, were investigated by means of X-ray diffraction (XRD) experiments. In general, the X-ray pattern of a conventional uniaxial nematic phase exhibits one pair of diffusive spots at small angles and a second pair at large angles along orthogonal directions, which originate from a short-range positional correlation associated with the molecular length and width. However, preliminary results [35] obtained from X-ray studies conducted on these mesogens suggested a biaxiality in the nematic phase. This is because a splitting of the small angle reflection into two pairs was observed.

Such a pattern is shown in Figure 5-22, where the nematic director **n** was oriented using a strong magnetic field (5 T) in a capillary sample. The occurrence of the four small angle peaks cannot be due to the presence of cybotactic groups, i.e., pre-transitional fluctuations pertaining to the underlying smectic C phase, for several reasons. These include the fact that the four peaks persisted over the entire range of the nematic phase, while cybotactic clusters would only be observable in a limited temperature range.

To really demonstrate the biaxiality of this nematic phase, additional XRD measurements were carried out using another type of sample cell. This cell was

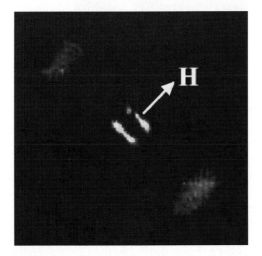

Figure 5-22. Two-dimensional XRD pattern in the presence of a magnetic field **H** within the nematic phase of C_7. Figure reprinted with permission from Bharat R. Acharya, Andrew Primak, and Satyendra Kumar, *Phys. Rev. Lett.*, **92**, 145506 (2004). Copyright 2004 American Physical Society

prepared using insulated beryllium plates coated with polyimide film, between which a homogeneous director orientation was achieved by rubbing in antiparallel directions. This setup was used to orient both the nematic director **n** (z axis) and the minor director **m** (x axis, defined as the mean orientation of the molecular apex bisector) in the presence of an electric field. Exemplary results from these electric field experiments are shown in Figure 5-23. In the absence of an electric field

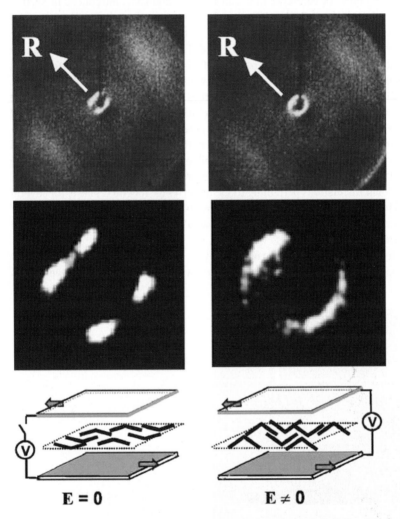

Figure 5-23. Two-dimensional XRD pattern of OC_{12} at zero electric field (left) and in the presence of an electric field applied perpendicular to the substrate (right). The corresponding small angle patterns are magnified in the middle panels. The schematics at the bottom represent the respective experimental setups and molecular orientations; the arrows indicate the rubbing direction R. Figure reprinted with permission from Bharat R. Acharya, Andrew Primak, and Satyendra Kumar, *Phys. Rev. Lett.*, **92**, 145506 (2004). Copyright 2004 American Physical Society

(left) four diffusive peaks were observed in the small angle region, as can be seen in the magnification in the middle panels. When an electric field was applied, no changes occurred until the electric field exceeded a threshold value of 6×10^6 V/m (Fredericks transition). At that point, **m** reoriented perpendicular to the Be-plates and the two pairs of reflections changed to one pair of arcs along the rubbing direction. The wide angle reflections remained essentially unchanged, confirming that **n** was still parallel to the rubbing direction. The fact that two distinct diffraction patterns were observed in the the absence and the presence of an electric field ruled out the possibility of a uniaxial nematic phase, since in that case any direction perpendicular to the main director would be equivalent.

To verify the conclusions drawn above, different XRD patterns were calculated using a simple model: the observed XRD pattern was taken to be the product of the molecular form factor $f(\mathbf{q})$ (i.e., the Fourier transform of the molecular electron density) that explicitly includes the bent-core shape, and the structure factor $S(\mathbf{q})$ (i.e. the density-density correlation). Figure 5-24 shows the calculated small angle patterns by applying the model and using geometrical values like the molecular length and the apex angle θ_v for the bent-core mesogen. The two patterns in the figure correspond to: a) the xz-plane in the case of zero electric field and b) the yz-plane when the electric field is switched on and perfect biaxial order is assumed.

The main features of the two experimental interference patterns in Figure 5-23 – two pairs of diffuse peaks with an azimuthal angular separation of $\sim 80°$ in the absence and one pair of reflections in the presence of the electric field – were found in the calculated patterns as well. This unambiguously proves the biaxial nature of the investigated nematic phase. Latest polarization microscopy and X-ray experiments conducted on azo-substituted achiral bent-core mesogens revealed the existence of a uniaxial and a biaxial nematic phase as well as three smectic phases [36].

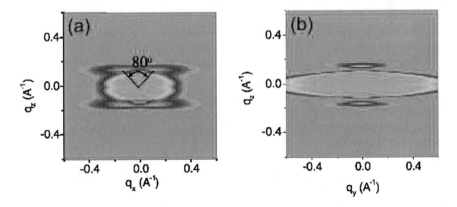

Figure 5-24. Calculated small angle XRD patterns, $f(\mathbf{q}) \times S(\mathbf{q})$, in the a) xz-plane and b) the yz-plane for perfect biaxial order. Figure reprinted with permission from Bharat R. Acharya, Andrew Primak, and Satyendra Kumar, *Phys. Rev. Lett.*, **92**, 145506 (2004). Copyright 2004 American Physical Society

The fact that the angle θ_v of 140° between the two mesogenic arms in the mesogens investigated above is far outside the range for which a biaxial nematic phase has been predicted, indicates the important role of the large transverse electrical dipole for the stabilization of the biaxial nematic phase.

6.   ORGANO-SILOXANE TETRAPODES

In a 2004 study [37], polarized IR spectroscopy, conoscopy, and optical textures suggested the phase biaxiality in organo-siloxane tetrapodes. In these tetrapodes, the mesogens are connected to the siloxane core through four siloxane spacers. This supermolecular system forms quasi-flat platelets, while at the same time, it avoids problems associated with polydispersity and coil structure found in linear polymers. The two different types of tetrapodes investigated are shown in Figure 5-25 (a) and (b); the local structure of the liquid crystalline tetrapodes, which has been confirmed in X-ray diffraction measurements, is drawn in Figure 5-25 (c).

The three spatial components of the IR absorbance, one along (A_z) and two perpendicular (A_x, A_y) to the nematic director were measured using planar homogeneously and homeotropically oriented samples. These different sample geometries were obtained using suitable surface treatment techniques. In Figure 5-26, the temperature dependence of the intensity of the absorbance components for the phenyl stretching band $(1160\,\mathrm{cm}^{-1})$ as well as for the carbonyl stretching band $(1738\,\mathrm{cm}^{-1})$ in tetrapode A is shown. The phenyl stretching band was found to be an excellent indicator of both the nematic and the biaxial ordering. In the nematic phase, two regions can clearly be distinguished: In the first, the two transverse components $(A_x$ and $A_y)$ are equal, corresponding to a uniaxial nematic phase, N_U. Then, in the second region, these transverse components start to diverge. This region consequently corresponds to a biaxial nematic phase N_B.

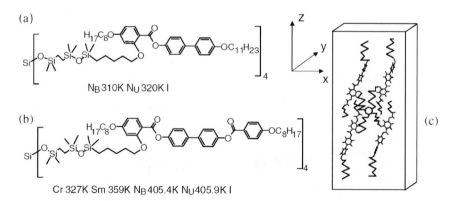

Figure 5-25. Molecular structures of the two tetrapodes: (a) *A* with asymmetric mesogens and (b) *B* with symmetric mesogens. (c) Platelet formed by the tetrapode *A*. Figure reprinted with permission from K. Merkel, A. Kocot, J. K. Vij, R. Korlacki, G. H. Mehl, and T. Meyer, *Phys. Rev. Lett.*, **93**, 237801. Copyright 2004 American Physical Society

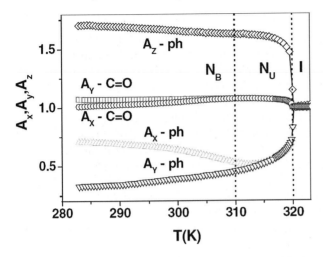

Figure 5-26. Absorbance components normalized with those of the isotropic phase for the tetrapode *A*: A_x (triangle), A_y (nabla), A_z (diamond) values for the phenyl ring stretching band; A_x (circle) and A_y (square) values for the carbonyl stretching band. Figure reprinted with permission from K. Merkel, A. Kocot, J. K. Vij, R. Korlacki, G. H. Mehl, and T. Meyer, *Phys. Rev. Lett.*, **93**, 237801. Copyright 2004 American Physical Society

Using the three components of absorbance, the four scalar order parameters S, D, C, and P proposed by Straley [10] can be calculated [38, 39]. The observed sequence of the phase transitions is shown in Figure 5-27. At 320 K the uniaxial-isotropic phase transition is observed. In the isotropic phase (I) all order parameters are zero,

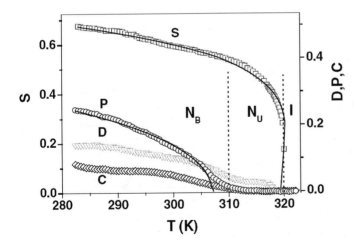

Figure 5-27. Order parameters for tetrapode **A**: S (square), P (circle), D (nabla), and C (diamond). Figure reprinted with permission from K. Merkel, A. Kocot, J. K. Vij, R. Korlacki, G. H. Mehl, and T. Meyer, *Phys. Rev. Lett.*, **93**, 237801. Copyright 2004 American Physical Society

while in the uniaxial nematic phase N_U, S, corresponding to the nematic order parameter, and D, associated with the molecular biaxiality for a uniaxial nematic phase, are different from zero. Then, as the temperature is lowered to $310\,K$, a second transition to a biaxial nematic phase N_B occurs. Here, both C, the molecular biaxiality in a biaxial phase and P, the phase biaxiality in the system, are also nonzero. Observations of texture and conoscopic experiments support the phase sequence found in the infrared experiments. The phase sequence for tetrapode **B** was determined in the same way. In addition to a smectic phase, this tetrapode also exhibits a uniaxial N_U and a biaxial nematic phase N_B found in tetrapode **A**.

In a recent publication, these findings have been nicely confirmed by deuterium NMR investigations of a spin probe dissolved in the organo-siloxane tetrapodes [40], where the same sample-flip technique was applied as in [4, 11]. In their study, the authors point out that the mode of stabilization of the biaxial nematic phase in tetrapodes and liquid-crystalline side-chain polymers, i.e., the lateral fixation of the mesogenic group, may in fact be similar.

7.   SUMMARY

In this chapter, various liquid crystalline systems for which phase biaxiality has recently been shown are discussed. Furthermore, the different experimental methods applied to these systems are introduced.

Conoscopy provides an extremely sensitive method with which to determine the degree of biaxiality. By the early 1990's, conoscopic measurements had already indicated the presence of phase biaxiality in a nematic side-on liquid crystalline side-chain polymer [9]. However, the method's sensitivity is also its weak point because surface effects may induce optical biaxiality in an actual uniaxial system. For this reason, deuterium NMR was used to confirm phase biaxiality in a liquid crystalline polymer system similar to the one investigated with conoscopy by Leube [11–13]. Due to the fairly high viscosity of the polymeric samples, the tilt experiment, employed by Yu and Saupe to show phase biaxiality in a lyotropic liquid crystal [4], was used. The results obtained in this way are in good agreement with observations of optical textures in a biaxial cholesteric copolymer [16], where phase biaxiality disturbs the smooth optical periodicity of the cholesteric phase structure.

Mixtures of discs and rods have been predicted to form a biaxial nematic phase [20]. Avoiding the problem of phase separation between the two moieties, chemically linked disc-rod molecules have been studied. Optical textures, misci-bility studies, and DSC experiments all resulted in the conclusion that, for a mixture with a certain ratio between the combined mesogen and the pure disc mesogen, a biaxial nematic phase had indeed been found [28]. However, this conclusion will have to be confirmed using additional experimental techniques, as, for example, deuterium NMR spectroscopy.

Using different experimental methods, two research groups recently indepen-dently confirmed phase biaxiality in liquid crystals consisting of banana mesogens. The first group conducted deuterium NMR experiments to unambiguously verify the biaxial nature of the nematic phase [32]. The relatively low viscosity necessitated the use of the rotation method for the investigation of the biaxiality parameter. The biaxiality of the nematic phase was further supported by the observation of optical textures and conoscopic investigations. Another group used X-ray diffraction experiments [33, 35] to successfully demonstrate phase biaxiality in the same system. To obtain a biaxially oriented sample, surface effects were exploited and magnetic or electric fields were applied during the X-ray measurements.

Finally, polarized IR spectroscopy is another sensitive method for the study of phase symmetry in liquid crystals. It was recently applied to prove phase biaxiality in organo-siloxane tetrapodes [37] – a supermolecular system forming quasi-flat platelets. Planar homogenously and homeotropically oriented samples were studied in order to derive all relevant order parameters from the three components of the IR absorbance. The presence of both a uniaxial and a biaxial nematic phase was detected, and again, optical textures and conoscopic observations supported these findings.

8.   REFERENCES

1. M. J. Freiser, *Phys. Rev. Lett.*, **24**, 1041 (1970)
2. J. Malthête, H. T. Nguyen, and A. M. Levelut, *J. Chem. Soc. Chem. Commun.*, 1548, (1986); J. Malthête, L. Liébert, A. M. Levelut, and Y. Galerne, *Compt. Rend. Acad. Sci. Paris*, **303**, 1073 (1986)
3. G. R. Luckhurst, J. R. Hughes, G. Kothe, J. Malthête, M. E. Neubert, I. Shenouda, B. A. Timimi, and M. Tittelbach, *J. Chem. Phys.*, **107**, 9252 (1997)
4. L. Yu and A. Saupe, *Phys. Rev. Lett.*, **45**, 1000 (1980)
5. Y. Galerne and J. P. Marcerou, *Phys. Rev. Lett.*, **51**, 2109 (1983)
6. S. G. Carr, G. R. Luckhurst, R. Poupko, and H. J. Smith, *Chem. Phys.*, **7**, 278 (1975)
7. S. M. Fan et al. *Chem. Phys. Lett.*, **204**, 517 (1993)
8. W. Weissflog and D. Demus, *Cryst. Res. Technol.*, **19**, 55 (1984)
9. H. F. Leube and H. Finkelmann, *Makromol. Chem.*, **192**, 1317 (1991)
10. J. P. Straley, *Phys. Rev. A*, **10**, 1881 (1974)
11. K. Severing and K. Saalwächter, *Phys. Rev. Lett.*, **92**, 125501 (2004)
12. K. Severing, A. Hasenhindl, E. Stibal-Fischer, H. Finkelmann, and K. Saalwächter, *J. Phys. Chem. B*, **110**, 15680 (2006)
13. K. Severing, Dissertation, Albert-Ludwigs-Universität Freiburg, Germany (2005)
14. H. Finkelmann and G. Rehage, *Makromol. Chem. Rapid Commun.*, **1**, 31 (1980)
15. S. Kniesel, Dissertation, Albert-Ludwigs-Universität Freiburg, Germany (2005)
16. H. Ogawa, E. Stibal-Fischer, and H. Finkelmann, *Macromol. Chem. Phys.* **205**, 593 (2004)
17. M. Zhou, V. Frydman, and L. Frydman, *J. Am. Chem. Soc.* **120**, 2178 (1998)
18. H. Finkelmann, *Phil. Trans. R. Soc. Lond.* A309, 105 (1983)
19. D. Frezzato, G. J. Moro, and G. Kothe *J. Chem. Phys.*, **119**, 6931 (2003)
20. R. Alben, J. Chem. Phys., **59**, 4299 (1973)
21. G. R. Luckhurst *Mol. Phys.*, **53**, 1535 (1984)
22. P. Palffy-Muhoray, J. R. de Bruyn, and D. A. Dunmur, *J. Chem. Phys.*, **82**, 5294 (1985)
23. R. Pratibha and N. V. Madhusudana, *Mol. Cryst. Liq. Cryst. Lett.*, **1**, 111 (1985)

24. R. Hashim, G. R. Luckhurst, F. Prata, and S. Romano, *Liq. Cryst.*, **15**, 283 (1993)
25. I. D. Fletcher and G. R. Luckhurst, *Liq. Cryst.*, **18**, 175 (1995)
26. J. J. Hunt, R. W. Date, B. A. Timimi, G. R. Luckhurst, and D. W. Bruce, *J. Am. Chem. Soc.*, **123**, 10115 (2001)
27. R. W. Date and D. W. Bruce, *J. Am. Chem. Soc.*, **125**, 9012 (2003)
28. P. H. J. Kouwer and G. H. Mehl, *J. Am. Chem. Soc. Commun.*, **125**, 11172 (2003)
29. D. Apreutesei and G. H. Mehl, *Chem. Commun.*, 609 (2006)
30. P. I. C. Teixeira, A. J. Masters, and B. M. Mulder, *Mol. Cryst. Liq. Cryst.*, **323**, 167 (1998)
31. G. R. Luckhurst, *Thin Solid Films*, **393**, 40 (2001)
32. L. A. Madsen, T. J. Dingemans, M. Nakata, and E.T. Samulski, *Phys. Rev. Lett.,* **92**, 145505 (2004)
33. B. R. Acharya, A. Primak, and S. Kumar, *Pramana* **61**, (2003); reprinted in *Phys. Rev. Lett.*, **92**, 145506 (2004)
34. D. J. Photinos and J. W. Doane, *Mol. Cryst. Liq. Cryst.*, **76**, 159 (1981)
35. B. R. Acharya and S. Kumar, *Bull. Am. Phys. Soc.* **45**, 101 (2000)
36. V. Prasad, S. Kang, K. A. Suresh, L. Joshi, Q. Wang, and S. Kumar, *J. Am. Chem. Soc.* **127**, 17224 (2005)
37. K. Merkel, A. Kocot, J. K. Vij, R. Korlacki, G. H. Mehl, and T. Meyer, *Phys. Rev. Lett.,* **93**, 237801 (2004)
38. D. Dunmur and K. Toriyama, in *Handbook of Liquid Crystals*, edited by D. Demus *et al.*, 189 (Wiley-VCH, Weinheim 2001)
39. K. Merkel, A. Kocot, J. K. Vij, G. H. Mehl, and T. Meyer, *J. Chem. Phys.*, **121**, 5012 (2004)
40. J. L. Figueirinhas, C. Cruz, D. Filip, G. Feio, A. C. Ribeiro, Y. Frère, T. Meyer, and G. H. Mehl. *Phys. Rev. Lett.*, **4**, 107802 (2005)

CHAPTER 6

NMR STUDY OF SELF-DIFFUSION

ANATOLY KHITRIN

Department of Chemistry, Kent State University, Kent, OH 44242-0001

1.   INTRODUCTION

Self-diffusion, translational motion of molecules in liquid crystal, provides important information about internal viscosity, anisotropic intermolecular interactions, and ordering in mesophase. Although it has been extensively studied, there is a lack of reliable data for the diffusion coefficients, due to experimental limitations of the existing techniques, and the results presented in different works are often inconsistent. Coefficients of self-diffusion can be measured with various methods, such as quasielastic neutron scattering (QENS) [1], forced Rayleigh scattering (FRS) [2, 3], or NMR [4]. Most of the previously reported data for liquid crystals suffer from insufficient accuracy and may differ from one another by almost an order of magnitude. A comparison of results, obtained with different techniques for 4-pentyl-4'-cyanobiphenyl (5CB), and a review of various methods is presented in [5]. The most accurate values of coefficients of self-diffusion in thermotropic liquid crystals have probably been reported in recent papers [5, 6], where an NMR technique using gradient slice selection, stimulated echo, and multi-pulse homonuclear decoupling has been developed. However, it is known that the radio-frequency power required for homo- or heteronuclear proton decoupling creates considerable heating of the sample, even in thermotropic liquid crystals [7]. The resulting gradients of temperature affect the order parameter, create convective flows, and reduce the accuracy of measurements.

For liquids, a pulsed-gradient NMR technique is the most direct method of measuring the diffusion coefficients. It monitors spatial displacement of molecules in the direction of the applied magnetic field gradients and does not require any assumptions about the long-time behavior of the velocity autocorrelation function, as QENS, or chemical modification of molecules, as FRS. The method is routinely used for liquids, where anisotropic spin interactions are averaged by molecular motions and NMR peaks are sharp, usually on the order of few Hz.

A. Ramamoorthy (ed.), Thermotropic Liquid Crystals, 171–178.

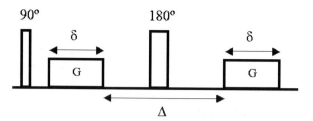

Figure 6-1. A scheme of pulsed-gradient spin-echo NMR sequence for measuring coeffiecients of self-diffusion

The spin-echo pulse sequence [8] for measuring coefficients of self-diffusion is shown in Figure 6-1. Transverse magnetization created by the initial 90° pulse is dephased by the first gradient pulse. A combination of the 180° pulse and the second gradient reverses this evolution and refocuses the magnetization. Such refocusing is perfect only if a nucleus producing the signal does not change its spatial position, otherwise the reversal of the evolution is incomplete. An additional decay of a spin echo amplitude S, resulting from molecular diffusion, is described by the Stejskal-Tanner relation [8]:

$$(6\text{-}1) \qquad S \propto exp\{-(\gamma G)^2 \delta^2 (\Delta - \delta/3) D\},$$

where γ is the gyromagnetic ratio of the nuclei, G is the amplitude of the two gradient pulses, δ is their duration, Δ is the interval between the gradient pulses, and D is the diffusion coefficient. Eq. (6-1) shows that the effect of diffusion depends on the cube of the time available for applying the magnetic field gradients.

In liquid crystals, intra-molecular dipole-dipole interactions are not averaged out by molecular motions. As a result, conventional proton NMR spectra are very broad, with linewidths reaching tens kHz, and the free induction signals decay about four orders of magnitude faster than in liquids. There is not enough time even for strong magnetic field gradients to produce a noticeable effect. In other words, line-broadening decreases an accuracy of gradient labeling of a spatial position and makes conventional pulsed-gradient NMR techniques inefficient. In substances with short T_2, like liquid crystals, it is impossible to measure diffusion without applying some line-narrowing technique.

2.   COLLECTIVE COHERENT RESPONSE SIGNALS

Recently, we have found that long and weak radio-frequency pulses can excite very sharp proton NMR response signals in liquid crystals [9]–[11]. For 5CB, as an example, the linewidth of the coherent long-lived collective response signal can be as narrow as 12 Hz.

The phase of the signal excited by a "soft" pulse is opposite to the phase of a signal excited by a "hard" 90° pulse. The process of excitation is non-linear, and experiments with two consecutive weak pulses, the second one having a 90° phase shift, show a very unusual behavior [10]. During a lifetime of the signal, individual spins can perform thousands of almost random rotations in their fluctuating local fields. Long lifetime of the signal is a result of correlated motion of a large number of spins. Larger intensities of the signals are found in cases when the dipole-dipole couplings are partially averaged by molecular motions [12]. Such collective response signals are excited at the generator, rather then the Larmor frequency and, therefore, carry no spectroscopic information. But they are suitable for encoding spatial position in the presence of gradients and can be used in imaging [13, 14] or diffusion measurements [15]. In fact, excitation at the generator frequency has an extra advantage of being insensitive to a static field inhomogeneity caused by non-perfect shims or magnetic susceptibility variations.

The example in Figure 6-2 demonstrates the line narrowing which can be achieved by using the new type of excitation of NMR signals. Figure 6-2A shows the

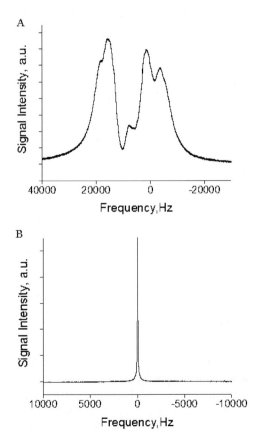

Figure 6-2. (A) Conventional ^{1}H NMR spectra of 5CB excited 5.5 μs 90°-pulse; (B) ^{1}H NMR spectrum of 5CB excited with 100 ms pulse of RF amplitude $\gamma B_1/2\pi = 55$ Hz, acquisition delay is 200 μs, the linewidth is 37 Hz

conventional ^{1}H NMR spectrum of 5CB (4'-*n*-pentyl-4-biphenylcarbonitrile). Its width, about 35 kHz, is determined by residual dipolar couplings between proton spins. With soft excitation, the linewidth is decreased by more than 1,000 times (Figure 6-2B). Please note that the frequency scales in Figures 6-2A and 6-2B are different. Actual phase of the signal excited by the soft pulse is opposite to that of the conventional spectra. For convenience, the spectrum in Figure 6-2B is shown with 180° phase shifts.

At present, there is no good understanding of physical origin of the coherent response signals. A theoretical consideration [16] suggests that such response to weak excitation may be a general phenomenon, although it gives no estimate for the signal intensity. Discussion [17] relates this unusual type of response to non-equilibrium saturation. Simulations for small (ten spins) static clusters [16] reproduce most of the experimental features of the signals. For systems with unrestricted network of dipolar couplings, it was found that molecular and segmental motions greatly enhance the signals intensity, although no explicit dependence on T_1 or T_2 has been found. Experimental study of various organic solids [12] highlighted the important role of fast molecular motions. We have found that soft excitation produces intense signals in soft polymers (polybutadiene, polyurethane, PDMS, collagen), and were able to acquire high-quality 3D images of such systems. For all tested liquid crystals, soft excitation pulses produce intense coherent response signals suitable for NMR imaging or diffusion measurements. It is possible that there exist more efficient methods of exciting long-lived signals in systems with dipolar-broadened spectra. However, insufficient theoretical understanding of these signals makes it difficult today to identify promising schemes.

Proton NMR peaks produced in liquid crystals by soft excitation pulses are considerably narrower than what can be achieved with multi-pulse decoupling, and such signals can be used for very accurate measurement of coefficients of self-diffusion. With this type of excitation, the radio-frequency power is at least six orders of magnitude smaller than an average power needed for homo- or hetero-nuclear proton decoupling. Therefore, the radio-frequency heating of the sample does not affect the measurements.

3.   GRADIENT-ECHO WITH SOFT EXCITATION PULSE

Here we present the results of measuring the coefficients of parallel self-diffusion for two thermotropic nematic liquid crystals, 5CB (4-pentyl-4'-cyanobiphenyl) and EBBA (N-(4-ethoxybenzylidene)-4-butylaniline), which have been studied previously with various techniques. Basically, the pulse sequence is a standard sequence for diffusion measurement [8] shown in Figure 6-1. The only differences are a long and weak excitation pulse, instead of a hard $\pi/2$ pulse, and use of gradients of opposite sign, instead of using a π-pulse in the middle between the gradient pulses. The π-pulse is not used because, as we have found, it destroys the coherent response signal. It is a consequence of the fact that the Hamiltonians representing

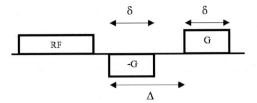

Figure 6-3. A gradient-echo pulse sequence with soft excitation pulse for measuring coefficient of self-diffusion in liquid crystals

the dipolar interactions and the chemical shifts interactions do not commute. The pulse sequence is presented in Figure 6-3.

We used the liquid crystals 5CB and EBBA, as purchased from Aldrich, without further purification. The experiments have been performed with a Varian Unity/Inova 500 MHz NMR spectrometer equipped with a standard z-gradient probe. The accuracy of the temperature stabilization was 0.1°C. The samples had the 5 mm height in 5 mm flat-bottom NMR tubes. The small height of the samples

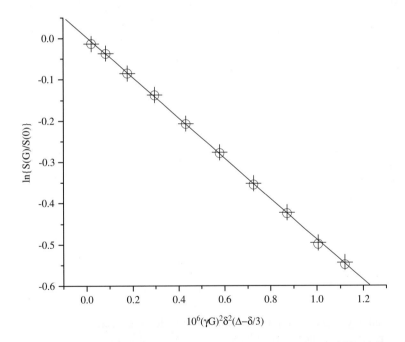

Figure 6-4. Amplitudes of the echo signal in 5CB at 25°C for different widths of the gradient pulses. The maximum width of the gradient pulse δ is 15.6 ms; $\Delta = 40$ ms; the delay between the RF and the first gradient pulses is 1 ms; the gradient amplitude is 20 G/cm. The excitation pulse has the duration of 0.1 s and the amplitude of radio-frequency field $\gamma B_1/2\pi = 25$ Hz. The amplitudes of the phased (O) and Absolute value(+) signals are shown. The diffusion coefficient obtained from a linear fit is $D = 4.90 \cdot 10^{-7}$ cm^2/s

was chosen to reduce the possibility of convective flows. We were able to use short samples because of comparatively strong signals, and because the peaks are excited at the generator, rather than at the Larmor, frequency. Therefore, good shimming was not necessary. The gradients have been calibrated with a special model sample, a 5 mm flat-bottom NMR tube filled with 5CB and a Teflon insert of known geometry.

The amplitudes of the gradient-echo signal for 5CB at 25°C are shown in Figure 6-4 for different durations δ of the gradient pulses. Two sets of points are presented for the phased and absolute-value signals. Coincidence of the two sets and an excellent linear fit according to Eq. (6-1) demonstrate that there are no phase or shape distortions and that the measurement is accurate. The obtained value of the coefficient of parallel self-diffusion in 5CB at 25° C was found to be $(4.9 \pm 0.1) \cdot 10^{-7} \, cm^2/s$.

The temperature dependences of the coefficients of self-diffusion for 5CB and EBBA are shown in Figure 6-5 and Figure 6-6, respectively. The corresponding previously reported data [6] are shown for comparison. One can see that our data for EBBA in Figure 6-6 for the temperature range not including the phase transition points demonstrate practically ideal Arrhenius temperature dependence of the diffusion coefficient. For 5CB in Figure 6-5, there is a significant deviation as the temperature increases and approaches the temperature of transition to isotropic

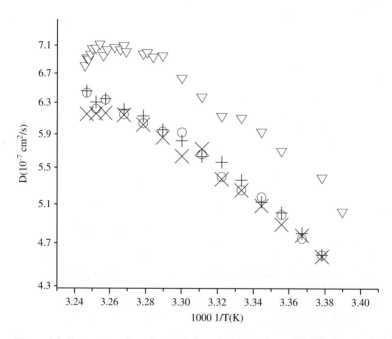

Figure 6-5. Temperature dependence of the coefficient of parallel diffusion for 5CB in the range 35-23°C; data sets (+) and (×) are obtained for decreasing, and (O) for increasing temperature, (∇) is the data [6]

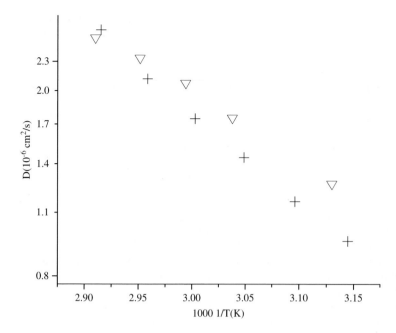

Figure 6-6. Temperature dependence of the coefficient of parallel diffusion for EBBA in the range 70-45°C (+), (∇) is the data. [6] The experiment is performed upon cooling

state. Besides that, we have found that the values of the diffusion coefficients in a rather broad interval near the transition temperature depend on a history of heating/cooling even when very slow changes in temperature are performed. Differences between the values obtained for the same sample at the same temperature sometimes considerably exceed the accuracy of our measurements. Typical scattering of data in different experimental runs can be seen in Figure 6-5 where two sets of points are shown for decreasing temperature (starting from the isotropic state) and one for increasing temperature. These history-dependent variations of the coefficients of self-diffusion are especially large near the point of phase transition to isotropic state.

In conclusion, we would like to mention some of the advantages of the described technique for studying self-diffusion in liquid crystals. (a) The linewidth of NMR peaks excited with "soft" pulses is much smaller than what can be achieved with a radio-frequency decoupling. As a result, weaker gradients can be used. In the presented examples we used 20 G/cm compared to 160 G/cm used in [5, 6]. Weaker gradients mean reduced heating of the probe, mechanical forces, and line distortions. Therefore, higher accuracy of the diffusion measurements can be achieved. (b) No RF pulses are applied simultaneously with the gradient pulses. There is no distortion from the interference between the gradient and RF pulses. (c) Radio-frequency heating is negligible in our case. The radio-frequency power we use for the signal excitation is at least six orders of magnitude smaller than the average power of the decoupling sequences. We believe that the technique can be used to study diffusion in biologically interesting lyotropic systems where the problem of radio-frequency heating is especially severe.

4. REFERENCES

1. A. J. Leadbetter, F. P. Temme, A. Heidemann, and W. S. Howells, Self-diffusion tensor for two nematic liquid crystals from incoherent quasielastic neutron scattering at low momentum transfer, *Chem. Phys. Lett.* **34**, 363–368 (1975).
2. D. R. Spiegel, A.L. Thompson, and W.C. Campbell, Forced Rayleigh scattering studies of tracer diffusion in a nematic liquid crystal: The relevance of complementary gratings, *J. Chem. Phys.* **114**, 3842–3847 (2001).
3. W.Urbach, H. Hervet, and F. Rondelez, On the application of forced Rayleigh light scattering to mass diffusion measurements, *J. Chem. Phys.* **83**, 1877–1887 (1985).
4. J.E. Tanner, Use of the stimulated echo in NMR diffusion studies, *J. Chem. Phys.* **52**, 2523–2526 (1970).
5. S. V. Dvinskikh and I. Furó, Anisotropic self-diffusion in the nematic phase of a thermotropic liquid crystal by ^{1}H-spin-echo nuclear magnetic resonance, *J. Chem. Phys.* **115**, 1946–1950 (2001).
6. S. V. Dvinskikh, I. Furó, H. Zimmermann, and A. Maliniak, Anisotropic self-diffusion in thermotropic liquid crystals studied by ^{1}H and ^{2}H pulsed-field-gradient spin-echo NMR, *Phys. Rev. E* **65**, 061701-1/9 (2002).
7. B. M. Fung, D. S. L. Mui, I. R. Bonnell, and E. L. Enwall, Evaluation of broad-band decoupling sequences applied to liquid-crystal solutions, *J. Magn. Reson.* **58**, 254–260 (1984).
8. E. O. Stejskal and J. E. Tanner, Spin diffusion measurements: spin echoes in the presence of time-dependent field gradient, *J. Chem. Phys.* **42**, 288–292 (1965).
9. A. K. Khitrin, V. L. Ermakov, and B. M. Fung, Information storage using a cluster of dipolar-coupled spins, *Chem. Phys. Lett.* **360**, 161–166 (2002).
10. A. K. Khitrin, V. L. Ermakov, and B. M. Fung, Nuclear magnetic resonance molecular photography, *J. Chem. Phys.* **117**, 6903–6906 (2002).
11. A. K. Khitrin, V. L. Ermakov, and B. M. Fung, NMR implementation of a parallel search algorithm, *Phys. Rev. Lett.* **89**, 277902-1/4 (2002).
12. B. M. Fung and V. L. Ermakov, The presence of long-lived spin states in organic solids with rapid molecular motions. *J. Magn. Reson.* **169**, 351–359 (2004).
13. V. Antochshuk, M.-J. Kim and A. K. Khitrin, High-resolution NMR imaging of objects with dipolar-broadened spectra, *J. Magn. Reson.* **167**, 133–137 (2004).
14. M. J. Kim and A. K. Khitrin, Magnetic resonance imaging of objects with dipolar-broadened spectra using soft excitation pulses, *Magn. Reson. Imaging* **23**, 865–869 (2005).
15. M. J. Kim, K. Cardwell, and A. K. Khitrin, Nuclear magnetic resonance study of self-diffusion in liquid crystals, *J. Chem. Phys.* **120**, 11327–11329 (2004).
16. A. K. Khitrin, V. L. Ermakov, and B. M. Fung Coherent response signals of dipolar-coupled spin systems, *Z. Naturforsch* **59a**, 209–216 (2004).
17. A. K. Khitrin and V. L. Ermakov, Spin processor, http://arxiv.org/abs/quant-ph/0205040 (2002).

CHAPTER 7

STRUCTURE AND DYNAMICS OF THERMOTROPIC LIQUID CRYSTALLINE POLYMERS BY NMR SPECTROSCOPY

ISAO ANDO[1] AND TAKESHI YAMANOBE[2]

[1] *Department of Chemistry and Materials Science, Tokyo Institute of Technology, Ookayama, Meguro-ku, Tokyo, Japan (At present: Professor Emeritus of Tokyo Institute of Technology; Visiting Professor of Department of Applied Chemistry, Keio University, Yagami, Kohoku-ku, Yokohama, Japan.)*
[2] *Department of Chemistry, Gunma University, Tenjin-cho, Kiryu, Gunma, Japan*

1.   INTRODUCTION

Thermotropic liquid-crystalline (LC) polymers are one of important soft materials such as polymer liquid crystals, polymer gels, vulcanized elastomers, *etc.*, in which molecular motion is much higher compared with solid polymers [1, 2]. LC polymers have both positional and orientational orders in the LC phase. The amount of order in a liquid crystal is quite small compared with a crystal. The polymer chain is rapidly rotating around the long chain axis and diffusing in the directions parallel and perpendicular to the long chain axis. Such anisotropic structures induce various characteristic properties to be used as materials. In order to develop new polymer materials, LC polymer design has been performed on the basis of advanced polymer science and technology. The properties of LC polymers are closely related to their structures and dynamics. From such situations, the establishment of methods for elucidating the structures and dynamics is important for making reliable LC polymer design and developing new advanced polymer gels. A considerable amount of efforts directed toward the rigid-rod polyesters and polyesters with long flexible side chains as a consequence of their ability to form thermotropic LC phases [3–20]. Also, polysiloxanes with *n*-alkyl side chain form the thermotropic LC phase. Prior to and concurrent with these studies it was shown that rigid-rod polymers with long alkyl side chains such as poly(γ-octadecyl L-glutamate) (POLG, that is PG-18) can form the thermotropic cholesteric, smectic and columnar phases [19, 20]. It has been demonstrated that the rigidity of the main chain and the flexibility of long olefinic side chains take an important role for formation of the thermotropic liquid crystals.

179

A. Ramamoorthy (ed.), Thermotropic Liquid Crystals, 179–233.
© 2007 *Springer.*

From such a background, the structure and dynamics of the main chain and side chains of LC polymers have been sophisticatedly characterized by solid-state NMR and pulse field-gradient spin-echo (PFGSE) NMR to understand an important role for formation of the thermotropic liquid crystals [21–25]. Details of these works will be introduced.

2.   STRUCTURE AND DYNAMICS OF CRYSTALLINE AND LIQUID CRYSTALLINE POLYESTERS WITH FLEXIBLE SIDE CHAINS

2.1   Rigid-rod Polyesters with 1, 4-dialkyl Esters of Pyromellitic Acid and 4, 4'-biphenol

B-Cn polyesters(n: the carbon number of the n-alkyl side chain) such as B-C6 indicated in Figure 7-2 have been used [13, 26].

The B-Cn polyesters form two layered crystals (K1 and K2) and two kinds of LC phases (LC-1 and LC-2) as the phase behavior shown in Figure 7-1. In the layered K1 and K2 crystalline phases, the aromatic main chains are in fully extended conformation with a repeat length of 1.66 nm and are regularly packed within a layer as determined from X-ray diffraction. The LC-2 phase, the lower temperature mesophase, has a layered segregated structure similar to that of the crystalline phase although its fundamental structure is remarkably altered in several aspects from the crystal structure. The main chains are still in an elongated conformation (a repeat length of 1.66 nm) as the crystalline phase, but they are packed into a layer having positional order only along the chain axis but not in the lateral direction. The side chains placed between the layers are in a molten state, which gives rise to the LC fluidity of the phase. This type of LC phase appears only for the B-Cn polyesters in which the alkyl side chains are longer than $n = 14$. On the other hand, the LC-1 phase, the higher temperature mesophase observed for all specimens displays

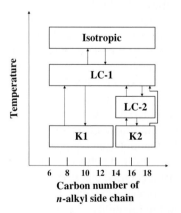

Figure 7-1. Schematic illustration of thermotropic phase behavior in B-Cn polyesters

a nematic-like optical texture, but a classic nematic phase cannot be postulated because it still exhibits a lateral packing spacing as large as that in the LC-2 phase. A biaxial nematic LC phase has been tentatively proposed such that the layers are retained, but there are frequently irregularities in their packing. ^{13}C TOSS CP-MAS

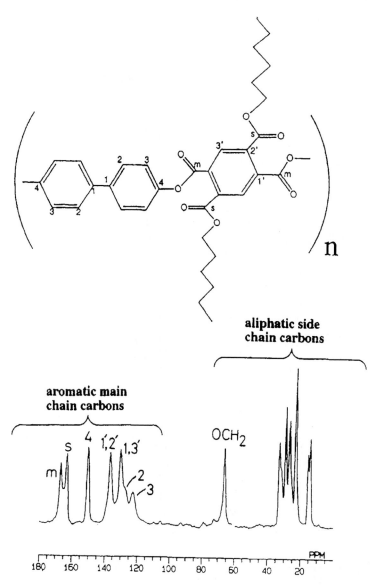

Figure 7-2. ^{13}C TOSS CP-MAS NMR spectrum as observed for the crystalline phase of B-C6 with peak assignments

(total suppression of sidebands cross polarization-magic angle spinning) spectrum of B-C6 in the crystalline phase is shown together with the chemical structure and peak assignment in Figure 7-2 [26]. Further, ^{13}C TOSS CP-MAS spectra of B-C6, B-C12, B-C14, B-C16 and B-C18 in the crystalline state at 25° C are shown in Figure 7-3. Here, peaks in the region of 100-180 ppm can be assigned to the aromatic main-chain carbons and the peaks in the region of 10-80 ppm to the

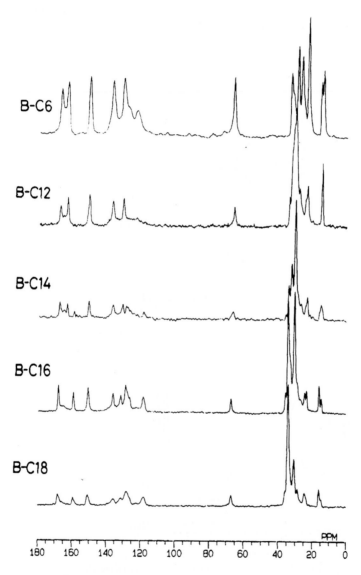

Figure 7-3. ^{13}C TOSS CP-MAS NMR spectra of B-C*n* polyesters in the crystalline phase

aliphatic side-chain carbons. The signal of the inner CH$_2$ carbons in *n*-alkyl side chain with the carbon number of 16 splits into two peaks at 34.3 and 30.5 ppm and that in *n*-alkyl side chain with the carbon number of 12 shows a single peak at ca. 30 ppm [21, 26–28]. Peak at 34.3 ppm comes from the inner CH$_2$ carbons in the *all-trans zigzag* form and peak at 30.5 ppm comes from the inner CH$_2$ carbons undergoing fast exchange between *trans* and *gauche* conformations. Therefore, it can be said that the fraction of the inner CH$_2$ carbons in the *all-trans zigzag* form is increased with an increase in the CH$_2$ carbon number of the side chain.

The conformational behavior of the side chains in the LC phase is employed. The expanded aliphatic region in the observed ^{13}C MAS NMR spectra of B-C16 in the LC-2 phase at 96° C and isotropic phase at 148° C are shown in Figure 7-4.

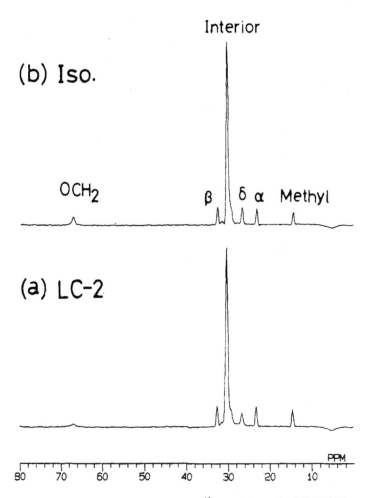

Figure 7-4. Expanded aliphatic region of the ^{13}C gated decoupling-MAS NMR spectra for (a) LC-2 and (b) isotropic phases of B-C16

Peaks of the inner CH_2 carbons in the LC-2 and isotropic phases appear at 30.5 and 30.1 ppm, respectively. This means that the inner CH_2 carbons are undergoing fast exchange between *trans* and *gauche* conformations and thus are in the molten state and fluid-like mobile state. Also, the similar result is obtained as the isotropic phase and LC-1 phase.

The expanded aromatic region in the ^{13}C MAS NMR spectra of B-C12 in the LC-1 phase, and B-C16 in the LC-2 phase and isotropic phase is shown in Figure 7-5 [26]. The chemical shifts of the aromatic carbons are listed in Table 7-1. The peaks in

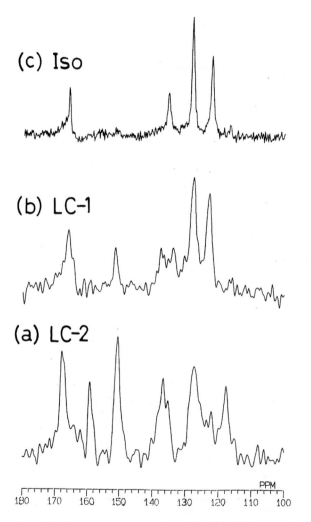

Figure 7-5. Expanded aromatic region of the ^{13}C MAS spectra for (a) the LC-2 phase of B-C16 as observed by the CP method and (b) the LC-1 phase of B-C12 and (c) the isotropic phase of B-C16 as observed by the gated decoupling-MAS method

Table 7-1. Observed ^{13}C NMR chemical shifts for the aromatic carbons of the B-Cn polyesters in the liquid crystalline and isotropic phases

	^{13}C chemical shift /ppm								
	C = O(m)[a]	C = O(s)[b]	C4	C2'	C1'	C3'	C1	C2	C3
iso(B-C16)	165.7	164.2	150.8	134.8	134.8	130.3	130.3	128.2	121.9
LC-1(B-C12)	165.7	165.0	150.7	137.0	134.0	130.3	130.3	128.2	122.6
LC-2(B-C16)	168.1	158.2	150.7	136.6	131.0	130.0	128.6	128.6	126.0/118.0

[a] (m): in the main chain.
[b] (s): in the side chains.

the LC-1 and LC-2 phases are much broader than those in the crystalline phase. The chemical shift values are similar to those in the crystalline phase. This is especially evident in the spectra of the LC and crystalline phases which are identical, indicating that in the LC phase all the aromatic groups are on a coplane, with the pyromellitic ester functionalities situated asymmetrically about the axis drawn between the C1' and C2' carbons. In the LC phase, the twisted conformation as the crystalline phase can also be assumed according to rough correspondence of the chemical shift values although a significant difference in the chemical shifts of the carbonyl carbons is observed as seen from Table 7-1. This difference may result from a greater degree of molecular motion of the main chain and side chains in the mesophase.

2.2 Structure and Dynamics of Rod like Poly(p-biphenylene terephthalate) with Long n-alkyl Side Chains

Rod like poly(p-biphenylene terephathalates) having long n-alkyl side chains with different lengths exhibit well-defined thermotropic LC phase behavior as elucidated by DSC and X-ray diffraction [29, 30]. These polymers form the nematic LC phase at high temperatures and two ordered mesophases at low temperatures. The structure of the mesophases in the low temperature range depends on the carbon number of the n-alkyl side chain, n. One of the mesophases is the hexagonal columnar phase in the polymers with n of 8 to 12. Another is the layered phase in the polymer with n = 18. In the polymers with n of 13 to 16, the hexagonal columnar phase and layered phase coexist.

It has been shown that the hexagonal columnar phase forms honeycombed network and has the microcavity with diameter of ca. 3 nm (Figure 7-6) [30]. It is the first type of aromatic polyester with flexible side chains. Further, it is expected that existence of the microcavity in the hexagonal columnar phase leads to possibility to be used as soft materials with anisotropic field and so may be applied to smart membranes. For this, structural and dynamic behavior of the polymers must be characterized with high precision.

The conformational and dynamical behavior of the main chain and long n-alkyl side chains of the polymers in the solid state and LC phase over a wide range

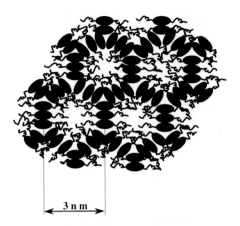

Figure 7-6. Schematic diagram for the space in the long channel cavity in the columnar phase of PBpT-O12 viewed from the top. Closed ellipsoids: the main chain, and curved strings linked with the closed ellipsoid: *n*-alkyl side chains

of temperatures has been characterized by ^{13}C CP-MAS NMR, and the relation between the structural and dynamic behavior and the phase transition behavior has been elucidated [30]. Poly(*p*-biphenylene terephthalate) with *n*-dodecyl side chains (PBpT-O12) synthesized by polycondensation of 4,4'-dihydroxybiphenyl with the 2,5-bis(dodecyloxy)terephthaloyl chloride in dry pyridine at 60° C is employed.

The observed ^{13}C CP-MAS NMR spectrum of PBpT-O12 in the solid state at room temperature is shown in Figure 7-7 together with the chemical structure. The assignments of peaks are made by using reference data for poly(L-glutamate) with *n*-octadecyl side chains as reported previously [21, 31, 32], and for rod aromatic B-C*n* polyester as reported by Sone, et al. [26]. The methyl and methylene carbons of *n*-alkyl side chains appear in the range of $0 \sim 80$ ppm, and the main chain carbons appear in the range of $100 \sim 180$ ppm. These peaks in the range of $0 \sim 80$ ppm are straightforwardly assigned to the CH_3, α-CH_2, β-CH_2, int-CH_2, γ-CH_2, and OCH_2 carbons from high field, respectively. The eight carbons for the main chain are assigned to the C4, C7, C6, C2, C8, C3, C5 and C1 carbons from high field by reference data reported previously [21, 31, 32].

Figure 7-8 shows expanded ^{13}C CP/MAS NMR spectra of PBpT-O12 in the range of $0 \sim 80$ ppm in a wide range of temperatures from room temperature to 160° C in order to clarify temperature change of the side chain conformation [31]. In the signal region derived from the interior CH_2(int-CH_2) carbons, two peaks appear at temperatures from room temperature to 80° C. As above-mentioned, at low temperatures the long *n*-alkyl side chains crystallize in the *all-trans zigzag* conformation (the immobile state), but above any given temperature the side chain crystallites melt and then the side chain carbons are undergoing rapid exchange between the *trans* and *gauche* conformations (the mobile state). The CH_2 carbons in the immobile state appear at about 33 ppm, and in the mobile state appear at about 30 ppm. This shows that the ^{13}C chemical shift value becomes a measure to

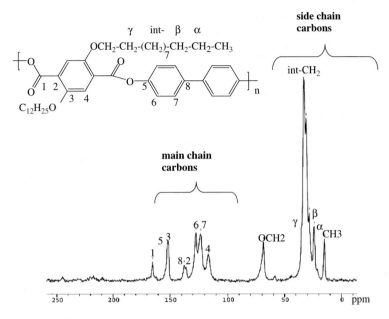

Figure 7-7. ^{13}C CP-MAS NMR spectrum of PBpT-O12 in the solid state at room temperature

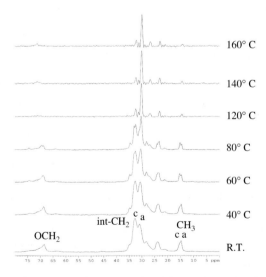

Figure 7-8. Expanded ^{13}C CP-MAS NMR spectra of the *n*-dodecyl side chain carbons of PBpT-O12 as a function of temperature

determine whether the long *n*-alkyl side chains take in the immobile state or mobile state. In the temperature range from room temperatures to 80° C the downfield peak (as indicated by c) appears at about 33 ppm and on the other hand the upfield peak (as indicated by a) appears at about 30 ppm. The downfield-peak intensity is decreased with an increase in temperature and, on the other hand, the upfield-peak intensity is increased with an increase in temperature. Above 120° C, the downfield peak completely disappears and only the upfield peak appears. Small peak in the vicinity of 33 ppm comes from the γ-CH$_2$ carbon.

The methyl signal consists of two peaks. The intensity of the downfield peak (indicated by c) at about 15 ppm is decreased with an increase in temperature in the temperature range from room temperature to 80° C and disappears above 120° C. On the other hand, the intensity of the upfield peak (as indicated by a) at about 14.5 ppm is increased with an increase in temperature. This behavior is very similar to that of the int-CH$_2$ carbons. Thus, it can be assigned that the side-chain methyl carbons in the immobile state appear at about 15 ppm and in the mobile state appear at about 14.5 ppm.

From the above experimental results, it can be said that in the temperature range from room temperature to 80° C the long *n*-alkyl side chains of PBpT-O12 do not take only the *all-trans zigzag* conformation in the immobile state, but the mobile state. At temperatures above 120° C, the *all-trans zigzag* form completely disappears. Such a behavior occurs also in the case of PG-18 with long *n*-octadecyl side chains, in which the main chain takes the α-helical conformation [21, 31–33].

The ^{13}C chemical shift values for the side chain carbons of PBpT-O12 as a function of temperature are assigned with reference data for *n*-paraffin, *n*-C$_{19}$H$_{40}$. If we look at the ^{13}C chemical shift value of the int-CH$_2$ carbons in the mobile state, in which are undergoing rapid exchange between the *trans* and *gauche* conformations by varying temperature, the ^{13}C chemical shift values move to upfield from 30.9 to 30.0 ppm as the temperature is increased from room temperature to 160° C. According to the γ-effect concept, it can be said that such an upfield shift comes from an increase of the *gauche* fraction by temperature elevation [73–76]. Such ^{13}C chemical shift behavior is very similar to that of *n*-C$_{19}$H$_{40}$ in the liquid state. On the other hand, the ^{13}C chemical shift values for the *all-trans zigzag* conformation of the int-CH$_2$ carbons in the immobile state are in the range of 32.5 to 32.8 ppm and then the temperature dependence is very small.

In Figure 7-9 are shown expanded ^{13}C CP-MAS NMR spectra of PBpT-O12 in the range of 80 to 180 ppm as a function of temperature. Peaks appeared in the range of 100 ~ 180 ppm are assigned to the main chain carbons numbered by the chemical structure (as shown in Figure 7-7). Thus, the eight peaks are assigned to the C4, C7, C6, C2, C8, C3, C5 and C1 carbons from upfield. These main chain carbon signals change drastically at temperatures between 80 and 120° C. Peaks of the C4, C2 and C3 carbons which come from the terephthalate moiety in the main chain disappear completely at temperatures more than 120° C. This means that the terephthalate moiety in the thermotropic LC phase initiates to undergo rotational motion at a

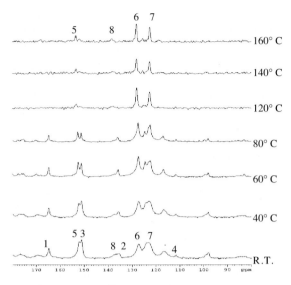

Figure 7-9. Expanded ^{13}C CP-MAS NMR spectra of the main chain carbons in PBpT-O12 as a function of temperature

frequency of about 60 kHz corresponding to the frequency of the ^{1}H decoupling frequency in these experiments in the temperature region more than 120° C.

From the above results, it is found that in the temperature range from room temperature to 80° C the long *n*-alkyl side chains of PBpT-O12 does not melt completely. There is a mixture of the immobile and mobile states. The former is that the long *n*-alkyl side chains crystallize in the *all-trans zigzag* conformation, and the latter is that the long *n*-alkyl side chain carbons are undergoing fast exchange between the *trans* and *gauche* conformations. At temperatures above 120° C the

Table 7-2. Observed ^{13}C NMR chemical shifts for the side chain carbons of PBp-12 as a function of temperature

Temperature/°C	^{13}C chemical shifts / ppm								
	OCH$_2$	γ-CH$_2$	int-CH$_2$		β-CH$_2$		α-CH$_2$		CH$_3$
-30	67.8	–	32.5	–	–	–	24.2	–	15.1	–
rt	68.5	35.0	32.8	30.9	28.3	–	24.3	23.9	15.3	14.8
40	68.7	34.9	32.6	30.7	28.3	–	24.3	23.8	15.3	14.7
60	68.8	34.7	32.6	30.5	28.1	26.6	24.0	23.4	15.3	14.7
80	68.8	34.8	32.5	30.3	28.2	26.9	24.1	23.6	15.2	14.6
120	–	34.4	–	30.2	–	26.8	–	23.1	–	14.4
140	–	32.3	–	30.1	–	26.7	–	22.0	–	14.4
160	–	32.2	–	30.0	–	26.7	–	22.9	–	14.2
n-C$_{19}$H$_{40}$(liquid)		32.6	30.4	23.3	14.4					
(crystal)		34.6	32.9				24.9		15.0	

all-trans zigzag form completely disappears, and the terephthalate moiety initiates to undergo rotational motion at a frequency of about 60 kHz corresponding to the frequency of the ^{1}H decoupling frequency in the present experiments at temperatures above 120° C. Then, the phase transition from the columnar phase to the nematic LC phase occurs.

PBpT-O12 in the nematic LC phase at 160° C is mechanically stretched to obtain highly-oriented fibers by cooling at room temperature. The nature of the channels with a diameter of about 3 nm in polyester fibers may be expected as one of materials with separation function. The nature of the inside of the fibers has been clarified through the observation of the diffusion coefficients of probe small-size molecules, methane and ethane, in direction parallel and perpendicular to the long channel by means of pulse field-gradient spin-echo(PFGSE) ^{1}H NMR spectroscopy [34, 35].

2.3   Poly(diethylsiloxane)

Certain organic-inorganic hybrid polymers such as poly(di-*n*-alkylsiloxanes) [36–56], poly(di-*n*-alkylsilanes) [57–59], and polyphosphazenes [60] that form the thermotropic LC phase have inorganic backbones and symmetrically di-substituted organic side chains, but lack typical mesogenic groups. The high molecular weight poly(diethylsiloxane) (PDES) considered in this work is one type of such polymers. At room temperature the polymer is a two-phase system consisting of the LC and isotropic regions. The properties of the LC poly(di-*n*-alkylsiloxane) family have been extensively studied using differential scanning calorimetry (DSC)[40, 42, 43, 45, 46, 48–53], optical polarizing micrograph (OPM) [40, 43–45, 49, 50], wide angle X-ray scattering (WAXS) [41, 51, 52], small angle X-ray scattering (SAXS) [51], Raman spectroscopy [48], pulse[1]H NMR [54, 55, 61], and solid state[2]H, ^{13}C and ^{29}Si NMR [41–46, 55, 60, 71]. Further, the structure of the polymer in the solid phase, the liquid crystalline phase and the isotropic phase has been characterized by static solid-state ^{29}Si NMR and ^{29}Si CP-MAS NMR [39, 40, 61, 71] and solid state ^{17}O NMR [71].

Static solid-state ^{29}Si NMR and ^{1}H solid echo NMR spectral analyses. It is known that the transition from the crystalline phase to the biphasic phase consisting of the isotropic region and LC region occurs at 17° C [40]. The LC region in the biphasic phase changes to the isotropic phase over ca. 50° C. This transition temperature depends on molecular weight [40]. Such a behavior can be clarified by using static solid-state ^{29}Si NMR at temperatures from -20 to 50° C as shown in Figure 7-10, where the sample is not rotated in the NMR probe [61]. At -20° C, the observed static solid-state ^{29}Si NMR spectrum in the crystalline phase shows a typical powder pattern and chemical shift anisotropy is very large as reported already. On the other hand, the observed static ^{29}Si NMR spectra at 20, 30 and 40° C appear as superposition of a asymmetrical and sharp powder pattern and a sharp peak which come from the LC and isotropic regions in the biphasic phase, respectively [61]. The chemical shift anisotropy of the asymmetrical and sharp powder pattern becomes very small because the two chemical shift components perpendicular to the long

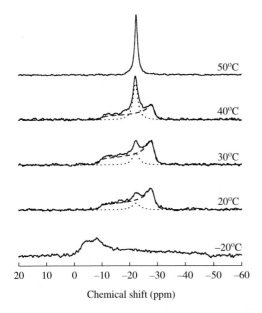

Figure 7-10. Observed static solid-state ^{29}Si NMR spectra of PDES at temperatures from -20 to 50°C by single 90° pulse with high power ^{1}H decoupling. The solid, dashed and dotted lines indicate the observed spectra, the simulated LC component and the simulated isotropic component, respectively

axis are averaged out by fast rotation of PDES chain around the long chain axis. The fractions of the LC region at 20, 30 and 40° C are approximately estimated to be 0.86, 0.83 and 0.55, respectively, by computer fitting. The fraction of the LC region in the biphasic phase decreases with an increase in temperature. Then, the signal corresponding to the LC region disappears completely at 50° C. This shows that PDES is in the isotropic phase.

The ^{1}H spin-spin relaxation time (T_2) values of the LC and isotropic regions in the biphasic phase are determined together with their fractions by ^{1}H solid echo (Figure 7-11) and Hahn spin echo measurements [61, 72, 90]. The ^{1}H free induction decay (FID) measured by solid echo method show the presence of the three T_2 components. The appearance of the three T_2 components can be explained as follows. In this temperature range, PDES is in the biphasic phase consisting of the LC and isotropic region as seen from static solid-state ^{29}Si NMR spectra (Figure 7-10) [61]. The fractions of the three components corresponding to the shorter, intermediate and longer ^{1}H T_2 components at 20° C are 0.50, 0.31 and 0.19, respectively, and those at 30° C are 0.42, 0.27 and 0.31, respectively. In the LC region, the polymer chains are strongly interacting with each other by forming the ordered structure. In such a situation it is thought that the ^{1}H T_2 values of the methyl and methylene groups are different from each other by their different dipolar-dipolar interactions. It may be supported from the experimental finding that the fraction ratio of the first ^{1}H T_2 component to the second ^{1}H T_2 component is nearly 3:2

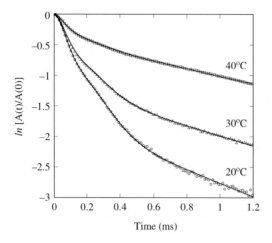

Figure 7-11. Temperature dependence of ^{1}H FID signal of PDES at 20(○), 30 (Δ) and 40°C(□) as measured by solid echo method

and this corresponds to the ratio of their proton numbers. On the other hand, the fraction ratio of the third ^{1}H T_2 component to the sum of the first and second ^{1}H T_2 components agrees nearly with that of the LC component to the isotropic component as estimated from static solid-state ^{29}Si NMR. Therefore, the three ^{1}H T_2 components can be reasonably explained as mentioned above. The fractions of the LC component in the biphasic phase at 20, 30 and 40° C are estimated to be 0.81, 0.69 and 0.32, respectively. It is shown that the fraction of the LC region is decrease with an increase in temperature as agreed with the static solid-state ^{29}Si NMR results. In the diffusion experiments, the decay signal of the second ^{1}H T_2 component corresponding to the LC region is used. The ^{1}H T_2 values for the liquid crystalline region and the isotropic region are about 0.2 and 7.4 ms, respectively. The ^{1}H T_2 value of PDES in the LC region is much shorter than that of the isotropic region. Thus, it is thought that it is very difficult to determine the diffusion coefficient of PDES in the LC region by PFGSE ^{1}H NMR because of extremely short ^{1}H T_2. On the other hand, it is possible to the diffusion coefficient of PDES in the isotropic region by PFGSE ^{1}H method because of sufficiently long ^{1}H T_2.

In addition to the use of ^{1}H, ^{13}C, and ^{15}N NMR chemical shifts, it has been shown that observation of solid-state ^{17}O chemical shifts and quadrupolar couplings adds important information on the higher-ordered and hydrogen-bonded structures of polypeptides and peptides in the solid state [62–65]. Solid-state ^{17}O NMR has been shown to be useful, moreover, for determining the local structures of alkali silicates [66–70]. Therefore, it is expected that solid-state ^{17}O NMR may reveal further dimensions of the structural and dynamic behavior of PDES in the crystalline, biphasic, and isotropic phases, in addition to the ^{1}H, ^{13}C, and ^{29}Si NMR results reported previously. Due to the low natural abundance of ^{17}O nuclei (0.037%), the NMR analysis requires ^{17}O labeling. The structure and dynamics of ^{17}O-enriched

PDES with a high molecular weight in its crystalline phases in the biphasic region consisting of the LC and isotropic phases have been characterized through the observation of ^{17}O NMR chemical shifts and quadrupolar couplings determined by solid-state ^{17}O NMR spectra in combination with *ab initio* MO calculations.

^{17}O NMR spectral analysis and structural characterization. Figure 7-12 shows 67.8 MHz static solid-state ^{17}O NMR spectra of [17]O-enriched PDES at −80° C (a), −70° C (b), −60° C (c), −40° C (d), −20° C (e), 0° C (f), 10° C (g), 25° C (h), 40° C (i), 60° C (j), and 80° C (k). The spectra were recorded using single pulse excitation. According to the DSC diagram, spectra (a) and (b) represent PDES in the low-temperature β_2 crystalline phase, spectra (c)-(f) in the high-temperature β_2 crystalline phase, spectra (g) and (h) in the biphasic system, and spectra (i)-(k) in the isotropic phase [71].

Spectrum (a) at −80° C shows a typical quadrupole powder pattern which is a consequence of the combination of chemical shift and quadrupolar interactions. Spectrum (b) at −70° C is almost the same as spectrum (a), suggesting that PDES has a magnetically equivalent oxygen site in the β_1 crystalline phase at −80 and −70° C, and that the molecular motion is frozen. As expected, spectra (c)-(f) of PDES at −60, −40, −20, and 0° C are slightly different from spectra (a) and (b) due to difference in the crystalline structures, but still represent a quadrupolar powder pattern characteristic of a rather rigid system.

The quadrupole powder patterns observed in the β_1 and β_2 crystalline phases suddenly disappear between 0 and 10° C, as the polymer forms the biphasic system. This shows that PDES chains are undergoing anisotropic fast molecular rotation,

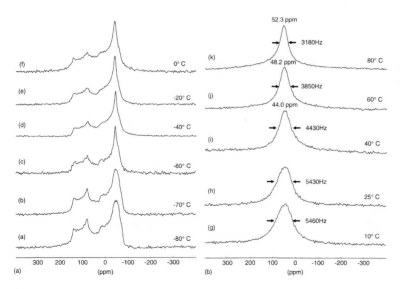

Figure 7-12. 67.8 MHz static solid-state ^{17}O NMR spectra of [17]O-enriched PDES at -80°C (a), -70°C (b), -60°C (c), -40°C (d), -20°C (e), 0°C (f), 10°C (g), 25°C (h), 40°C (i), 60°C (j), and 80°C (k), acquired using single pulse excitation

with a partial averaging of the chemical shift anisotropy and the quadrupolar inter-action. Indeed, the fractions of the LC region are 83% and 78% at 10 and 25° C, respectively, as determined from ^{29}Si static NMR spectra. The linewidth in spectrum (g) at 10° C (5460 Hz) is almost the same as that in spectrum (h) at 25° C (5430 Hz) in the biphasic system. Although the rotation at 25° C should be more rapid than that at 10° C in the LC region, the residual chemical shift anisotropy and the quadrupolar interaction are no longer affected by anisotropic motion. It might therefore be assumed that the anisotropic motion in the LC region is uniaxial free rotation. The static solid-state ^{29}Si NMR results [47, 61, 77] show that the polymer chain rotates about the long chain axis in the LC region, too.

At 4430 Hz, the linewidth of spectrum (i) at 40° C in the isotropic phase is largely decreased compared to linewidth in the LC region, thus implying that the molecular motion of PDES becomes isotropic in the isotropic phase. As temperature is further increased, the linewidth continues to decrease, becoming 3180 Hz at 80° C. On the other hand, the ^{17}O chemical shift moves higher frequency at temperatures above 40° C, perhaps, due to a change of the conformational distribution of the PDES main chain.

Similar spectral changes are observed at 40.7 MHz (Figure 7-13). In this case, spectra (a)-(f) in the β_1 and β_2 crystalline phases were recorded using the cross-polarization (CP) method from ^{1}H nuclei to ^{17}O nuclei, which resulted in increased sensitivity relative to single pulse ^{17}O excitation. However, the CP efficiency decreases with an increase in temperature in the β_2 crystalline phase, which is

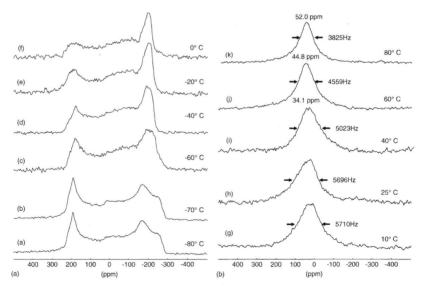

Figure 7-13. 40.7 MHz static solid-state ^{17}O NMR spectra of ^{17}O-enriched PDES at -80°C (a), -70°C (b), -60°C (c), -40°C (d), -20°C (e), 0°C (f), 10°C (g), 25°C (h), 40°C (i), 60°C (j), and 80°C (k). Spectra (a)–(f) were acquired using ^{1}H-^{17}O CP, spectra (g)–(k) were acquired using single pulse excitation

consistent with increased mobility. The biphasic and isotropic phases could only be observed using the single-pulse method, as fast molecular motion interfered with the CP process.

It is also noted that the half-height linewidth (5460 Hz) in spectrum (g) observed at 67.8 MHz (Figure 7-12) is smaller than that (5710 Hz) in spectrum (g) observed at 40.7 MHz (Figure 7-13). Such dependence of the linewidth on observed frequency shows that the second-order quadrupolar interaction remains operable in the LC region of the biphasic phase. This phenomenon supports the notion that the molecular motion of PDES in the LC region of the biphasic phase is not isotropic, but rather that PDES chains in the LC region have some preferred orientation. Similarly, the linewidth in the isotropic phase observed at 40.7 MHz is larger than that observed at 67.8 MHz, indicating that the quadrupolar interaction and short-range order are still present. This may come from the existence of polymer chain entanglements, as reported previously [61].

Computer simulations of selected spectra observed at 40.7 and 67.8 MHz in the crystalline phases are shown in Figure 7-14. The simulations included the effects of the chemical shift anisotropy (CSA) as well as the quadrupolar inter-action. The ^{17}O chemical shifts and quadrupolar parameters of PDES determined are shown together with those of ^{17}O-enriched cyclic tetramer D_3Et_6 as the model compound of PDES in Table 7-3. These include the quadrupolar coupling constant (C_Q), the quadrupolar asymmetry parameter (η_Q), the isotropic chemical shift (δ_{iso}), $\xi_{CS} = \delta_{33} - \delta_{iso}$, $|\delta_{33} - \delta_{iso}| > |\delta_{11} - \delta_{iso}| > |\delta_{22} - \delta_{iso}|$, the chemical shift asymmetry parameter ($\eta_{CS} = (\delta_{22} - \delta_{11}/\delta_{33} - \delta_{iso})$) and the Euler angles describing

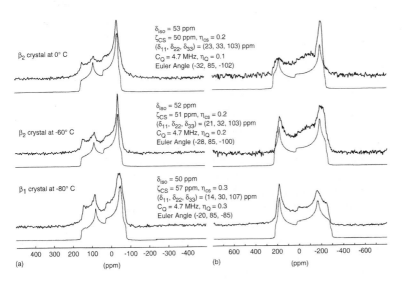

β_2 crystal at 0° C

δ_{iso} = 53 ppm
ξ_{CS} = 50 ppm, η_{cs} = 0.2
(δ_{11}, δ_{22}, δ_{33}) = (23, 33, 103) ppm
C_Q = 4.7 MHz, η_Q = 0.1
Euler Angle (-32, 85, -102)

β_2 crystal at -60° C

δ_{iso} = 52 ppm
ξ_{CS} = 51 ppm, η_{cs} = 0.2
(δ_{11}, δ_{22}, δ_{33}) = (21, 32, 103) ppm
C_Q = 4.7 MHz, η_Q = 0.2
Euler Angle (-28, 85, -100)

β_1 crystal at -80° C

δ_{iso} = 50 ppm
ξ_{CS} = 57 ppm, η_{cs} = 0.3
(δ_{11}, δ_{22}, δ_{33}) = (14, 30, 107) ppm
C_Q = 4.7 MHz, η_Q = 0.3
Euler Angle (-20, 85, -85)

| 400 | 300 | 200 | 100 | 0 | -100 | -200 | -300 | -400 |

(a)   (ppm)

| 600 | 400 | 200 | 0 | -200 | -400 | -600 |

(b)   (ppm)

Figure 7-14. 40.7 (a) and 67.8 (b)MHz static solid-state ^{17}O NMR spectra of PDES in the crystalline phase together with theoretically simulated spectra

Table 7-3. Observed [17]O chemical shifts and electric field parameters of PDES in the liquid crystalline, crystalline and isotropic phases phases

Sample/phase	C_Q/MHz[a]	η_Q[b]	δ_{iso}/ppm[c]	ζ_{CS}/ppm[d]	η_{CS}[b]	Euler angles[f]
PDES/β_1 crystalline phase	4.7	0.3	50	57	0.3	$(-20°, 85°, -85°)$
PDES/β_2 crystalline phase PDEA/at $-60°C$	4.7	0.2	52	51	0.2	$(-28°, 85°, -100°)$
PDES/β_2 crystalline phase PDEA/at $0°C$	4.7	0.1	53	50	0.2	$(-32°, 85°, -102°)$
PDES/liquid crystal	1.7	0.0	53	21	0.0	$(0°, 0°, 0°)$

[a] Quadrupolar coupling constant.
[b] Quadrupolar asymmetry parameter.
[c] Isotropic chemical shift $\delta_{iso} = 1/3(\delta_{11} + \delta_{22} + \delta_{33}).(|\delta_{33} - \delta_{iso}| > |\delta_{11} - \delta_{iso}| > |\delta_{22} - \delta_{iso}|)$.
[d] Chemical shift anisotropy $\zeta_{CS} = \delta_{33} - \delta_{iso}$.
[e] Asymmetry parameter of shielding $\eta_{CS} = (\delta_{22} - \delta_{11})/(\delta_{33} - \delta_{iso})$.
[f] Euler angles between the principal axes of the chemical shift tensor and the quadrupolar tensor.

the relative orientations from the quadrupolar to chemical shift tensors defined previously [64].

The difference in spectral lineshape between the β_1 crystalline phase, which does not exhibit molecular motion, and β_2 crystalline phase, is caused by a decrease in η_Q, ξ_{CS} and η_{cs}. Clearly, the oxygen sites in both phases are similarly coordinated. The small decrease in [17]O η_Q, ξ_{CS}, and η_{CS} values between -60 to $0°$C in the β_2 crystalline phase may indicate the onset of the slow libration motion of PDES chains about a specific axis. From the static solid-state ^{29}Si NMR result [47, 61, 71], due to the onset of oscillations around the molecular chain axis and small conformational transition in the β_2 crystalline phase, the ^{29}Si NMR chemical shift anisotropy is reduced by 25% compared with the β_1 crystalline phase. The static solid-state ^{17}O NMR result supports the solid-state ^{29}Si NMR result.

In the LC phase, the simulations could be best performed using the complete [17]O spectrum, which included inner satellite signals [$(-3/2, -1/2)$ and $(1/2, 3/2)$]. Figure 7-15 shows 54.2 MHz spectrum of the biphasic system taken at 25°C with 1 MHz spectral width (a) and its theoretical simulated spectrum (b). As mentioned above, the fraction of the LC region at 25°C is about 78% in the biphasic phase; thus, the major signal comes from the LC region. The central transition and inner satellite signals are clearly observed in the spectrum, which shows that quadrupolar interactions still affect the lineshape in the LC phase. From this spectrum, the C_Q value of 1.7 MHz and the η_Q value of 0 have been determined as shown in Table 7-3. These NMR parameters have been discussed as they relate to anisotropic molecular motion of PDES in the LC region.

Molecular motion of PDES in the liquid crystalline phase. In the observed [17]O spectra of PDES in the biphasic (mainly LC) phase, the typical quadrupole pattern of the [17]O signal disappears as a result of fast uniaxial molecular rotation

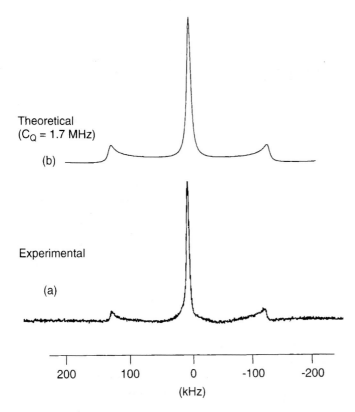

Theoretical
(C_Q = 1.7 MHz)

(b)

Experimental

(a)

200   100   0   -100   -200
(kHz)

Figure 7-15. 54.2 MHz static solid-state ^{17}O NMR experimental spectrum of PDES in the biphasic phase at 25° C with 1 MHz spectral width (a), and its simulated spectrum with C_Q of 1.7 MHz (b). Note that the horizontal scale is in kHz, rather than ppm

(Figure 7-16 and 7-17) [71]. This rotation averages both of the chemical shift and quadrupolar interactions, such that the NMR powder pattern becomes axially symmetrical ($\eta_Q = \eta_{CS} = 0$), the principal axes of the chemical shift tensor and quadrupolar tensor coincide, and the C_Q value decreases from 4.7 MHz in the crystalline β_2 phase to 1.7 MHz in the liquid crystalline region (Figure 7-15 and Table 7-3).

It is possible to determine the orientation of the rotation axis in the molecular frame that corresponds to the observed changes by considering the following relationships between the components of the electric field gradient tensor (V) and chemical shift tensor(δ) [69, 70].

$$(7\text{-}1) \quad V_{zz(LC)} = 1/2(3\cos^2\beta_V - 1)[V_{zz(\beta2cryst)} - 1/2(V_{xx(\beta2cryst)} + V_{yy(\beta2cryst)})]$$
$$+ 3/4(V_{xx(\beta2cryst)} - V_{yy(\beta2cryst)})\sin^2\beta_V \cos 2\alpha_V$$

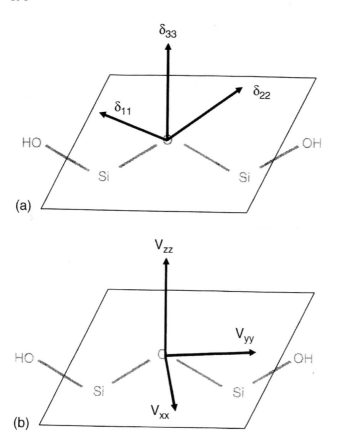

Figure 7-16. Directions of the principal axes of the ^{17}O shieldings (a); directions and quadrupolar tensor (b) of model molecule of PDES with the *trans-trans* conformation

$$(7\text{-}2) \quad \delta - \delta_{(LC)} = 1/2(3\cos^2\beta_{CS} - 1)[\delta_{33(\beta2cryst)} - 1/2(\delta_{11(\beta2cryst)} + \delta_{22(\beta2cryst)})]$$
$$+ 3/4(\delta_{11(\beta2cryst)} - \delta_{22(\beta2cryst)})\sin^2\beta_{CS}\cos 2\alpha_{CS}$$

where the determined NMR parameters are $V_{zz(LC)} = C_Q = 1.7\,\text{MHz}$, $V_{zz(\beta2cryst)} = 4.7\,\text{MHz}$, $V_{xx(\beta2cryst)} = -2.7\,\text{MHz}$, $V_{yy(\beta2cryst)} = -2.0\,\text{MHz}$, $\delta - \delta_{(LC)} = 74 - 41 = 33\,\text{ppm}$, $\delta_{33(\beta2cryst)} = 103$ ppm, $\delta_{11(\beta2cryst)} = 23\,\text{ppm}$, and $\delta_{22(\beta2cryst)} = 33\,\text{ppm}$ from simulated spectrum shown as Figure 7-17. The Euler angles $(\alpha_{cs}, \beta_{cs})$ and (α_V, β_V) are defined for the transformation from the principal axis system of the chemical shift tensor and quadrupolar tensor to the molecular rotating frame z-axis. The $V_{zz(\beta2cryst)}$ and $\delta_{33(\beta2cryst)}$ parameters are aligned in a direction perpendicular to the Si-O-Si plane from Figure 7-18 so that $\beta_{CS} = \beta_V$. Thus, the Euler angles obtained are $\beta_{CS} = \beta_V = 44 \sim 46$, $\alpha_V = 0 \sim 71$ and $\alpha_{CS} = 39° \sim 90°$, respectively. Thus, it is thought that PDES molecules in the LC region rotate rapidly around the molecular axis, with the tilted angle of $44° \sim 46°$ for the normal vector of

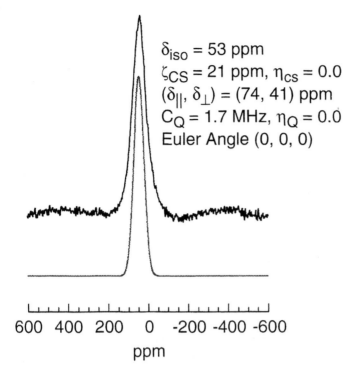

$\delta_{iso} = 53$ ppm
$\zeta_{CS} = 21$ ppm, $\eta_{CS} = 0.0$
$(\delta_{\parallel}, \delta_{\perp}) = (74, 41)$ ppm
$C_Q = 1.7$ MHz, $\eta_Q = 0.0$
Euler Angle (0, 0, 0)

600 400 200   0   -200 -400 -600

ppm

Figure 7-17. 67.8 MHz ^{17}O static solid-state NMR spectra of PDES in the biphasic phases together with theoretically simulated spectra67.8 MHz ^{17}O static solid-state NMR spectra of PDES in the biphasic phases together with theoretically simulated spectra

the Si-O-Si plane shown as Figure 7-18. The rotational axis about the molecular frame of the PDES molecules in the LC region can be determined from the static solid-state ^{17}O NMR result [71], because there exist two interaction tensors such as the chemical shift and quadrupolar in static ^{17}O solid-state NMR spectrum.

2.4 Poly(L-glutamate)s with Long *n-alkyl* Side Chains

It has been shown that a series of α-helical poly(L-glutamate) with *n*-alkyl side chains of various lengths(PALG) form a crystalline phase composed of paraffin-like crystallites together with the α-helical main-chain packing into a characteristic structure if the number of methylene carbon in the side chains is longer than 10 [19, 20, 31, 32, 73]. The polymers form thermotropic liquid crystals by melting of the side-chain crystallites above ca. 50° C. The long *n*-alkyl side chains play a role of solvent to form liquid crystals. In order to obtain detailed information about the structure and dynamics of these liquid crystals, it is very essential to study the structures and motion of the main chains and side chains at various temperatures.

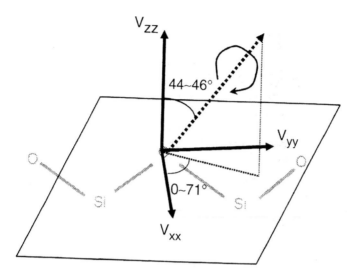

Figure 7-18. Molecular rotation axis for PDES molecular frame and principal axis system of quadrupolar tensor

In this section, the relationship between the LC behavior and NMR parameters of PALGs are clarified based on the solid state NMR results. For convenience, all of polymer samples are designated by the letter PG followed by the number of carbon atoms in the *n*-alkyl side chain.

Structure and dynamics of the main chain and side chains of PG-ns at room temperature. Figure 7-19 shows the ^{13}C CP-MAS NMR spectra of a series of PG-ns in the solid state [32]. As seen from these spectra, the ^{13}C chemical shift and the peak intensity vary depending on the side chain length, which reflect the structure and dynamics of PG-ns. The peak assignments are shown in Figure 7-19 [32, 73].

It has shown that ^{13}C chemical shifts of the C_α and CO(amide) carbons in polypeptides are displaced depending on the main chain conformation such as the α-helix and β-sheet forms [74]. In particular, the ^{13}C chemical shift of the CO(amide) carbon is not strongly affected by the amino-acid residues and thus may be used to identify the main-chain conformation. The ^{13}C chemical shifts of the C_α and CO(amide) carbons range from 57.2 – 58.1 ppm and 175.9 – 176.7 ppm, respectively, within experimental errors, and are close to those of poly(γ-benzyl L-glutamate) (PBLG) that takes the α-helix conformation. These chemical shift values are almost independent of the side-chain length, which implies that the main-chain conformation of a series of PG-ns takes the α-helix conformation irrespective of the *n*-alkyl side-chain length. Similar behavior can be applied to the chemical shifts of the CO(ester) carbon. The ^{13}C chemical shift of the CO(ester) carbon of PBLG with the α-helix conformation is 172 ppm. The corresponding ^{13}C chemical shifts for a series of PG-ns are very close to 172 ppm except for PG-1. Therefore, the conformation around the CO(ester) carbon in PG-ns is similar to that in PBLG.

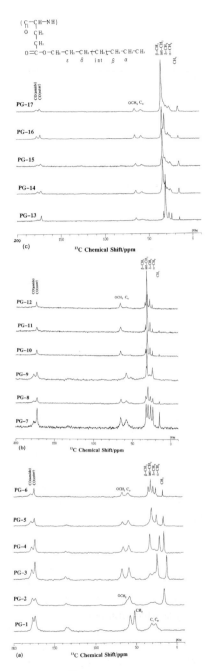

Figure 7-19. ^{13}C CP-MAS NMR spectra of PG-ns in the solid state: (a) PG-1–PG-6, (b) PG-7–PG-12 and (c) PG-13–PG-17

The conformations of the main-chain carbon and the CO(ester) carbons of the side chains are similar to those of PBLG irrespective of the side chain length.

For PG-1 – PG-5, the number of peaks for the side-chain carbons increases with the side-chain length and then their chemical shifts change. For example, the ^{13}C chemical shift values of the terminal CH$_3$ carbon for PG-1, 2, 3, 4 and 5 are 51.7, 14.8, 10.8, 14.2 and 14.6 ppm, respectively. As the CH$_3$ carbon of PG-1 is adjacent to oxygen, the ^{13}C chemical shift of the CH$_3$ carbon appears at a lower field than the others. Additional chemical shift change on going from PG-2 to PG-5 arises from the effect of the neighboring group. Similarly, the effects of the neighboring oxygen atom and the end group influence on the ^{13}C chemical shift of the side-chain CH$_2$ carbons for PG-1 – PG-5.

For PG-ns with the *n*-alkyl side chains longer than PG-6, peaks of the CH$_3$, α-CH$_2$, δ-CH$_2$, interior CH$_2$(int-CH$_2$) and β-CH$_2$ carbons appear from high field. The ^{13}C chemical shifts of the side chain carbons coincide with those of PG-6 – PG-8 within experimental errors. The ^{13}C chemical shifts of the int-CH$_2$ carbons of PG-6 – PG-8 are similar to those of liquid *n*-C$_{19}$H$_{40}$. This means that the int-CH$_2$ carbons of PG-6 – PG-8 are in the mobile state as the CH$_2$ of liquid paraffins. The ^{13}C chemical shift behavior of PG-9 is complex. The structure and mobility of PG-9 seem to be sensitive to the sample preparation condition.

In PG-10 – PG-13, the averaged ^{13}C chemical shift of the int-CH$_2$ carbons is 30.4 ppm that is very close to the ^{13}C chemical shift for the amorphous phase in polyethylene. It is known that peak of the CH$_2$ carbons in *n*-paraffins and polyethylene appears at a higher field by 4-6 ppm if any carbon atoms three bonds away are in a *gauche* rather than a *trans* conformation (γ-*gauche* effect) [75]. The int-CH$_2$ carbons for both of PG-6 – PG-8 and PG-10 – PG-13 are undergoing fast exchange between *trans* and *gauche* conformations in the mobile state [76]. The ^{13}C chemical shift difference of the int-CH$_2$ carbons between PG-6 – PG-8 and PG-10 – PG-13 shows the different *trans /gauche* fraction ratio, *i.e.* the *gauche* fraction of the int-CH$_2$ carbons for PG-6 – PG-8 is higher than that for PG-10 – PG-13. In Figure 7-20(a) are shown the dependences of ^{1}H T_2 of the side chain carbons on the side-chain length at room temperature. As seen from Figure 7-20(a), the ^{1}H T_2 increases in going from PG-3 to PG-10 and becomes almost plateau between PG-10 and PG-13. Such a behavior shows that the mobility of the side chains increases in going from PG-3 to PG-10. In this case, the side-chain terminal gets the high degree of freedom for mobility as the side-chain length increases. The ^{1}H T_2 plateau as observed between PG-10 and PG-13 indicates that the mobility of the side-chain terminal carbons is restricted in the same degree. By taking the ^{13}C chemical shift behavior of the int-CH$_2$ and the behavior of ^{1}H T_2 into consideration, it is shown that interchain interactions between the side chains become strong in PG-n (n $\geq$ 10). In other words, the structure and the mobility of PG-ns are cooperatively governed by interchain interactions for PG-n (n $\geq$ 10), while the nature of the main-chain governs the structure and the mobility for PG-n (n < 10).

The side-chain part of ^{13}C CPMAS spectra of PG-14 is different from that of PG-13. The ^{13}C chemical shift of the int-CH$_2$ carbons moves downfield at about

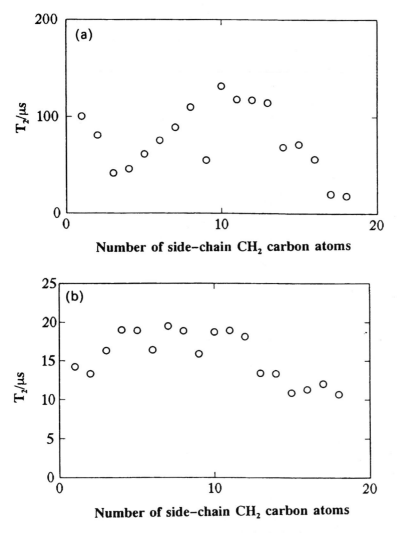

Figure 7-20. Dependence of the ¹H T_2 of (a) the side-chains and (b) the main-chain of PG-ns on the side-chain length at room temperature

33.3 ppm that is typical ¹³C chemical shift for the crystalline phase of polyethylene and *n*-paraffins. In Figure 7-21 are shown expanded ¹³C CP-MAS NMR spectra of PG-15 at room temperature and 53° C at which the side chains of PG-15 melt. The side chains of PG-15 form the crystalline phase at room temperature. At 53° C, the spectrum becomes simple like the spectra of PG-6 – PG-8, which shows that the melting of the side chains occurs. For the spectrum of PG-15 at room temperature, peaks of the CH₂ carbons in the crystalline and amorphous phases can be assigned as follows. As the peak positions for the amorphous phase (int-CH₂(A), γ-CH₂(A),

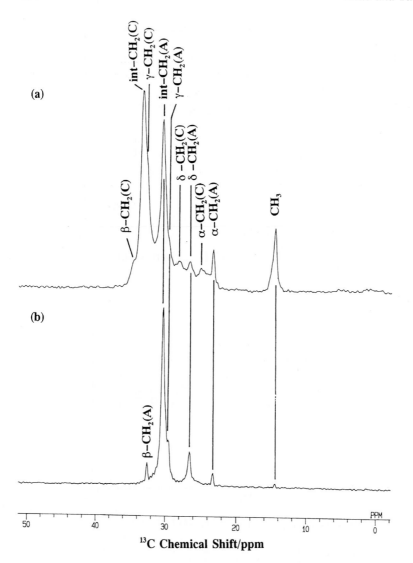

Figure 7-21. ^{13}C CP-MAS NMR spectra of PG-15 in the solid state at (a) room temperature and in the LC state at (b) 55°C

δ-CH$_2$(A) and α-CH$_2$(A)) are the same as those for Figure 7-21(b), the int-CH$_2$ carbons in the amorphous phase are in the mobile state and undergo rapid exchange between *trans* and *gauche* conformations. In the crystalline phase, the side chain carbons take the *all-trans zigzag* conformation.

It is known that the ^{13}C chemical shifts of the CH$_2$ carbons in the *n*-alkyl side chains can give useful information about not only the conformation, but also the crystallographic form. The ^{13}C chemical shift value of 33.3 ppm is similar to that of *n*-paraffins taking the *all-trans zigzag* conformation in the pseudohexagonal or orthorhombic form and polyethylene taking the *all-trans zigzag* conformation in the orthorhombic form. Therefore, PG-ns in the crystalline state may take the orthorhombic or pseudeohexagonal form rather than the triclinic or monoclinic form.

For the OCH$_2$ carbon, the ^{13}C chemical shift is independent of the side-chain length and the peak for the crystalline phase is not observed. The CH$_2$ carbons outer from δ-CH$_2$ in the side chains, at least, contribute to form the crystallites. For the terminal CH$_3$ carbon, only one peak is observed. However, the linewidth becomes broad in forming the side-chain crystallites. The mobility of the terminal CH$_3$ carbon is affected by the crystallization of the side chains. The CH$_2$ carbons between the δ-CH$_2$ and α-CH$_2$ carbons contribute to formation of the side-chain crystallites.

The peak intensity of ^{13}C CP-MAS spectra is affected by molecular mobility. As seen from Figure 7-19, the peak intensity of the C$_α$ and CO(amide) carbons changes depending on the side-chain length without change of their ^{13}C chemical shifts. The peaks for these carbons decrease with the side-chain length until PG-9 and disappear for PG-10 – PG-13. Such a behavior seems to be connected to the side-chain mobility as shown in Figure 7-21(a), *i.e.* the increment of ^{1}H T_2 of the side-chain carbons decreases with the side-chain length. The corresponding peaks of the C$_α$ and CO(amide) carbons disappear in the ^{1}H T_2 plateau region of the side-chain carbons of PG-10 – PG-13. As the ^{1}H T_2s of the main-chain become short and almost constant, the overall motion of the α-helical main chain as activated by the side-chain motion affects the peak intensity behavior. It can be explained by the line broadening caused by the reduced efficiency of high power ^{1}H decoupling. In general, the full linewidth at half height of the ^{13}C signal (Δδ) can be written as

$$(7\text{-}3) \qquad \Delta\delta = (\pi T_{20})^{-1} + (\pi T_{2C})^{-1} + (\pi T_{2m})^{-1}$$

where the first term represents the intrinsic linewidth due to the inhomogeneous static field and so on [79]. This term is independent of temperature. The second term is the contribution from the distribution of the isotropic chemical shift caused by the distribution of conformation or crystal structure. Therefore, this term is temperature-dependent because the start of molecular motion faster than an NMR timescale can average out the local distribution of conformation or crystal structure. The third term arises from the ^{13}C-^{1}H dipolar interaction. As the high-power dipolar-decoupling is usually enough strong to reduce the dipolar interaction between ^{13}C and ^{1}H, the third term is negligible. However, if the rate of molecular motion is close to the ^{1}H decoupling frequency, the applied radio-frequency (*rf*) cannot reduce the dipolar interactions effectively. Therefore, the third term has the maximum contribution to the peak width if the rate of molecular motion is close to the decoupling frequency. By taking these terms into account, it can be understood that the linewidth varies with the frequency of the molecular motion, as shown schematically in Figure 7-22.

From Figure 7-22, it is seen that the linewidth is relatively broad at slow motion (region A) because of the first and second terms. The onset of molecular motion induces a broad line width due to the third term (region B). At fast molecular motion, the peaks become very sharp (region C) as a result of diminishment of the second and third terms. Because the intensity of one peak is a constant, the peak height can be a good measure for linewidth. In addition to the variation in linewidth, peak-height behavior is also shown in Figure 7-22. Since the region B corresponds to a frequency of ^{1}H decoupling field of 60 KHz, the main-chain for PG-10 – PG-12 is undergoing reorientation at corresponding frequency.

For PGs with $n \geq 14$, the C_α and CO(amide) carbons give intense peaks at the chemical shift position for the α-helix form. Also, a large decrease in ^{1}H T_2s of the side chains and the main chain carbons occurs at $n = 14$ as seen from Figure 7-20(a) and (b), respectively. As mentioned above, in PG-n($n \geq 14$) side-chain interactions are very strong enough to restrict the molecular motion and induce the formation of the crystallites. The existence of peaks for the C_α and CO(amide) carbons and the behavior of ^{1}HT_2s of the side chain and the main chain carbons result from formation of the side-chain crystallites.

Structure and mobility of the main chain and side chains of a series of PG-ns at various temperatures [21, 32, 33]. In the thermotropic LC state for PG-ns, the side chains act as a solvent. In Figure 7-20(a), ^{1}H T_2s of the side chain carbons for PG-ns($n \geq 10$) are not so long compared with PG-ns($n < 10$) at room temperature. Figure 7-23 shows the dependence of ^{1}H T_2 of the side-chain carbons on the side-chain length at 80° C. From this figure, it is easy to find the discontinuity between $n = 9$ and 10. The side-chain mobility of PG-ns($n < 10$) is much lower than that of PG-ns with longer side chains. ^{1}H T_2s for PG-ns($n < 10$) increase slightly with an increase in temperature, while those for PG-ns($n \geq 10$) increase drastically. This critical difference arises from interchain interactions. In PG-ns($n \geq 10$), side-chain interactions become strong and then can become lubricant for main-chain interactions. In other words, at high temperature, the PG main-chain is floating

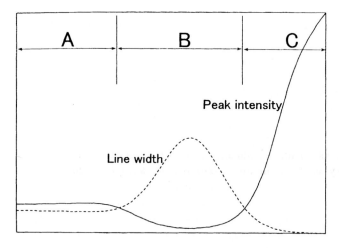

Figure 7-22. Schematic behavior of the half height width and peak intensity of the ^{13}C signal

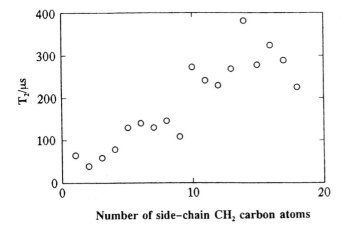

Figure 7-23. Dependence of the ^{1}H T_2 of the side-chain carbons of PG-ns on the side-chain length at 80°C

in the sea of the *n*-paraffin-like side chains. On the contrary, for PG-n (n < 10), the side-chains are too short to weaken main-chain interactions. Therefore, the side-chain mobility is restricted by interactions between the main-chains even at high temperature.

Information about the molecular mobility of all parts of PG-ns can be obtained by ^{13}C VT (temperature variable) CP-MAS NMR spectra. In Figure 7-24 are shown the ^{13}C CP-MAS NMR spectra of PG-6 as a function of temperature. As seen from this figure, it is easy to see drastic change in the spectra by temperature variation. At temperatures below −60° C, all the peaks are broad. At −40° C, the intensities of peaks for the CO(ester) and OCH$_2$ decrease and the CH$_2$ signal initiates to split. At −20° C, peaks of the CO(ester) and OCH$_2$ carbons completely disappeared. In addition, the intensities of the CH$_2$ peaks become very weak. Peaks for the CO(amide), C$_\alpha$ and methyl carbons become strong at this temperature. At 0° C, peaks of the CO(ester) and OCH$_2$ carbons still do not reappear. Peaks for the CH$_2$ carbons reappear with a narrower linewidth than those at low temperatures. Above 25° C, sharp peaks for the CO(ester) and OCH$_2$ carbons reappear. On other hand, peaks of the CO(amide) and C$_\alpha$ carbons disappear. Peaks for the CH$_2$ carbons become stronger and sharper with an increase in temperature.

Such a behavior of the ^{13}C CP-MAS NMR spectra reflects the molecular mobility of the sample. Peaks as shown in Figure 7-24 can be classified into three groups. The first group is peaks for the side chain carbons except for the terminal methyl carbon. The intensity of peaks in this group is a constant between −100 and −60° C, and the peaks become broad. This temperature range is in region A in Figure 7-22. The peak intensities of the side chain carbons start to decrease above −40° C and pass through a minimum at −20 and 0° C. This temperature range is in region B in Figure 7-22. Above 0° C, the peak intensity increases rapidly and the linewidth becomes very sharp (region C). The second group is peaks for the

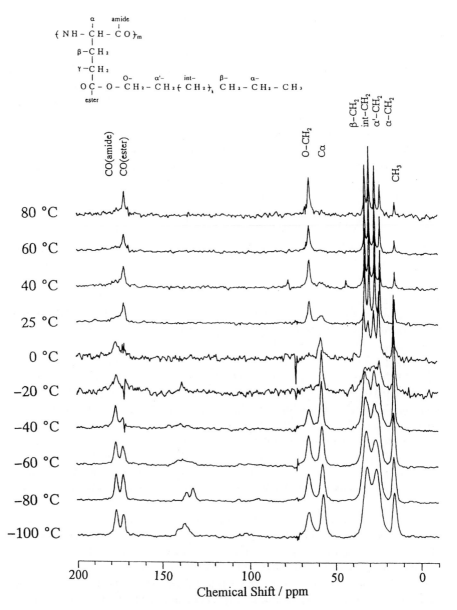

Figure 7-24. ¹³C CP-MAS NMR spectra of PG-6 as a function of temperature

main chain CO(amide) and C$_\alpha$ carbons. The peaks in this group are very strong and broad below −20° C (region A). Above 0° C, the peaks become weak and almost disappear at 80° C (region B). Peaks for the main chain carbons do not reappear in the measurement temperature range. Although the mobility of the main chain and side chain carbons may be of the same order at low temperatures, the molecular motion of the side chain carbon is much faster than that of the main chain carbons at high temperatures. The main-chain motion starts after the side chain carbons gain sufficient high mobility. The ^{13}C chemical shifts for the main chain carbons are independent of temperature, the main chain takes the right-handed α-helix conformation [32]. It is reasonable that the *n*-alkyl side chains are in the mobile state and the α-helical main chain is in the rigid state. The third group is the terminal methyl one in the side chains. As seen from Figure 7-24, the intensity of methyl peak gradually increases and passes through a maximum. A minimum is not observed. It is well-known that the methyl carbon rotates around the three-fold axis. For polypropylene, it has been reported that the methyl peak disappears at about $-160°$C [77]. Therefore, the methyl carbon already is in region C. Thus, in the side chains the terminal methyl carbon is in the mobile state, and the mobility decreases in going from the outside to the inside in the side chain.

In Figure 7-25 are shown the ^{13}C CP-MAS NMR spectra of PG-2 as a function of temperature. At temperatures below $-40°$C, all peaks are broad. This temperature range corresponds to region A. Above $-20°$C, the peaks for the side chain carbons decrease in their intensity and reach a minimum at about 20°C. In this temperature range, the mobility of the side chain carbons increases, and the rate of motion is close to the ^{1}H decoupling frequency (region B). Above 50°C, the linewidth of peaks for the side chain carbons becomes sharp and the intensity increases with temperature (region C).

For the main-chain carbons, peaks do not disappear even at 80°C. As the amide carbonyl ^{13}C chemical shift appears at about 176 ppm independent of temperature, the main chain takes the right-handed α-helix conformation. Therefore, the main chain is in the rigid state and the molecular mobility should be slower than that of the side chains. The main-chain mobility of PG-2 is much lower than that of PG-6. The longer side chains can gain higher freedom of molecular motion.

In Figure 7-26 are shown ^{13}C CP-MAS NMR spectra of PG-12 as a function of temperature. Below $-20°$C, both of peaks for the side chains and the main chain are broad, and the mobility for them is in region A. At 0°C, the ^{13}C NMR chemical shift of the int-CH$_2$ carbons moves upfield and the intensity of the OCH$_2$ carbon becomes very weak. The side chains are in region B at this temperature. The ^{13}C NMR chemical shift of the int-CH$_2$ carbons of PG-12 moves upfield by about 3 ppm as the temperature is raised from $-20°$C to 0°C. Above 0°C, the chemical shift value of ca. 30 ppm indicates that the side chains are in the amorphous phase. Below 0°C, the ^{13}C NMR chemical shift is about 32.9 ppm. Therefore, it shows that the int-CH$_2$ carbons take the *all-trans zigzag* conformation in the immobile state. Therefore, the drastic change in the side chains between -20 and 0°C shows the melting of the side-chain crystallite.

Katoh, et al. have found that the *n*-alkyl side chains of PG-18 are in the 'rotator' phase like long *n*-paraffins in the narrow temperature range below the melting point of the side-chain crystallite [33]. In the 'rotator' phase, the *n*-alkyl side chains are undergoing a rapid rotation similar to liberations around the *all-trans zigzag* chain

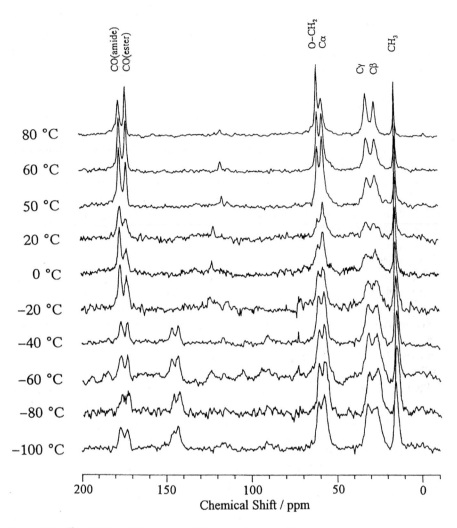

Figure 7-25. ^{13}C CP-MAS NMR spectra of PG-2 as a function of temperature

axis. As seen from Figure 7-26, at temperatures above 25° C the main-chain carbon peaks almost disappear, while peaks of the side chains are clearly present. This means that the mobility of the side chains and the main chain are in regions C and B, respectively. Compared PG-12 with PG-ns (n < 10), the transitions from regions A to C for the side chains and from regions A to B for the main chain take place in the narrower temperature range in PG-12.

In Figure 7-27, the schematic phase diagram of PG-ns based on the molecular mobility is shown. In this figure, the solid and dotted lines indicate the temperature at which the side and main chain peaks disappear, respectively. The mobility of

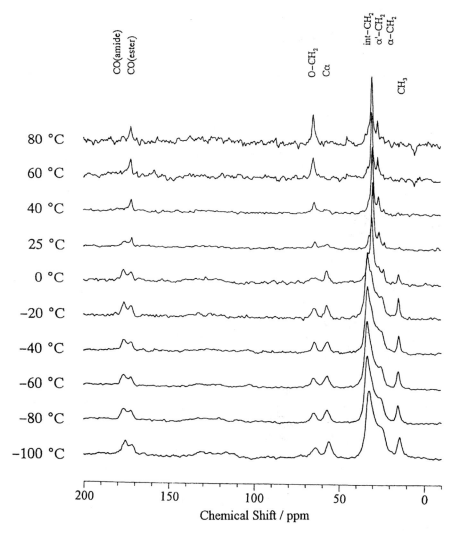

Figure 7-26. ^{13}C CP-MAS NMR spectra of PG-12 as a function of temperature

the side and main chains are in region B around these lines. As seen from this figure, the solid line goes down with a increase in the carbon number from PG-1 to PG-9. Therefore, the motion of the side-chain for PG-n with longer side-chain length is easily activated at lower temperature. At room temperature, the mobility of the side-chain increases as the side-chain length increases. The side chains are in the immobile glassy state below the solid line indicated in Figure 7-27. The main-chain mobility increases after the side-chain motion is activated. In PG-n (n ≥ 10), the temperature difference between the solid and dotted lines decreases. The temperature range of the transition from region A to C for the side chains

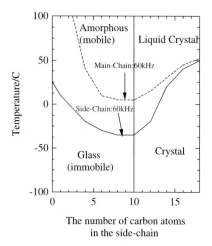

Figure 7-27. Phase diagram of PG-ns

becomes narrow as the side-chain length increases. In these PG-ns, side-chain interactions become strong and change in the side-chain mobility affects the overall dynamics. As soon as the side chains melts and then the side chains act as a solvent or a lubricant for the main-chain, the overall main-chain motion is activated.

Information about molecular motion of the side chains can be obtained from ^{13}C VT CP-MAS NNR. When peaks for the main-chain carbons disappear at any specified temperatures, it is impossible to gain information about the mobility of the main-chain in corresponding temperature range. ^{2}H NMR can give it if the amide protons of the main-chain are partially substituted by ^{2}H as shown below [80].

In Figure 7-28(a) is shown the theoretical ^{2}H NMR spectrum in which the asymmetrical parameter(η) is 0.5. Peak separations Δv_1, Δv_2 and Δv_3 have different values from each other. If the symmetrical motion takes place, the molecular motion is reflected in the observed spectrum as an axially symmetrical powder pattern. Figure 7-28(b) shows the observed ^{2}H NMR spectrum of PG-10-N-D at 40° C, where N-D means that the main-chain amide proton is substituted by [2]H. The observed spectrum shows a typical powder pattern composed of the inner peaks(Δv_1), the shoulders(Δv_2) and the outermost wings(Δv_3). Therefore, this means that molecular motion of PG-10-N-D is restricted at 40° C.

In Figure 7-28(c) is shown the observed ^{2}H NMR spectrum of PG-10-N-D in 10% chloroform solution. In this figure, two sharp ^{2}H peaks are observed. This indicates that all the N-D directions to the magnetic field in the sample are unique. In the magnetic field, it is known that PG-ns in the LC state tend to orient to the magnetic field. The long axis of the mesogenic α-helical chain becomes parallel to the magnetic field. This can be confirmed from the observation of the ^{13}C NMR chemical shift tensor components [78]. The sample forms the nematic LC phase in the magnetic field of NMR magnet. The splitting between the two peaks is about

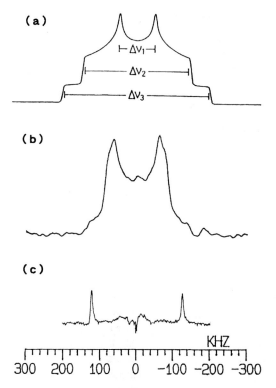

Figure 7-28. (a) Theoretical ^{2}H NMR spectrum with $\eta = 0.5$, (b) ^{2}H NMR of PG-10-N-D at 40°C and (c) ^{2}H NMR of 10% chloroform solution of PG-10-N-D

250 KHz. This splitting corresponds to Δv_3. In the α-helix form, the hydrogen bonded N-D direction is almost parallel to the α-helix axis within 5°. Therefore, the outermost wings in the ^{2}H NMR spectrum correspond to the direction of the α-helix axis. The inner peaks(Δv_1) and its shoulders(Δv_y) can be attributed to the axes perpendicular to the α-helix axis.

In Figure 7-29 are shown the temperature dependencies of Δv_1 and Δv_2 for PG-4-N-D. As seen from this figure, Δv_1 and Δv_2 are almost a constant in the measurement temperature range. The typical powder pattern with $\eta \neq 0$ and an invariance of Δv_1 and Δv_2 shows that the exchanges between the principal axes do not occur, in other words, the molecular motion is strongly restricted. From the ^{13}C CP-MAS NMR spectra, peaks of the main chain carbons decrease in the intensity, but did not completely disappear within the measurement temperature range. In PG-4, although molecular motion in PG-4-N-D takes place, the rate is much slower than several tens of KHz.

In PG-12-N-D, the existence of the outermost wings is confirmed even at 100° C. Similar to PG-4-N-D, this principal axis does not exchange with the other axes in this sample. In Figure 7-30 are shown the temperature dependencies of Δv_1 and Δv_2.

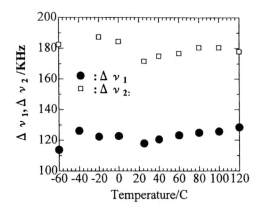

Figure 7-29. Temperature dependences of Δv_1 and Δv_2 for PG-4-N-D

As the temperature increases, Δv_1 and Δv_2 gradually increase and decrease, respectively. These slow changes show that the molecular motion increases step by step. At temperatures above 80° C, Δv_2 suddenly decreases, and then Δv_1 and Δv_2 become equal due to the disappearance of shoulder peaks. As the wings are recognized in the spectra, molecular motion in PG-12-N-D is the rotation or vibration around the α-helix axis. Peaks of the main-chain carbons as obtained by ^{13}C CP-MAS NMR completely disappeared at high temperature. This shows that the rate of molecular motion is close to the ^{1}H decoupling frequency of 60 KHz [21, 77]. The frequency difference between Δv_1 and Δv_2 is about 60 KHz. The rate of molecular motion is estimated to be more than this frequency difference. As mentioned above, the rate of molecular motion for the main-chain is close to the ^{1}H decoupling frequency of 60 MHz from the CP-MAS NMR spectra above 80° C [21, 77]. The rate of the main-chain motion obtained by ^{13}C CP-MAS NMR and ^{2}H NMR is in good agreement with each other. If the rate is much faster than 60 KHz, peaks in

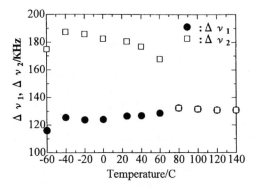

Figure 7-30. Temperature dependences of Δv_1 and Δv_2 for PG-12-N-D

the ^{13}C CP-MAS NMR spectra reappear. By taking such a situation into account, the rate of molecular motion seems to be close or a little faster than 60 KHz.

In Figure 7-31, the thermal change in the structure and mobility of PG-ns is summarized. In PG-ns(n<10), as the side-chain length is short, the occupied volume of the side chains is reduced and the main chains are close to each other. Even though the temperature is increased and the side-chain mobility is increased, there is not enough space for the main chain to move fast. Therefore, only the fluctuation of the main chain is increased with retaining the rigid rod α-helix form. In PG-ns(n ≥ 10), at low temperatures (Figure 7-31(b)) the side-chain crystallite is formed. The main-chain takes the α-helical conformation and is ordered in the side-chain crystallite. The side-chain crystallite transforms into the rotator phase below the melting point (Figure 7-31(c)). Above the melting point of the side chain crystallite, the *n*-alkyl side chains behave like liquid *n*-paraffins(Figure 7-31(d)). Even in the LC state, the main chain takes the α-helix conformation with rotation or vibration around the α-helix axis at a frequency of about 60 KHz.

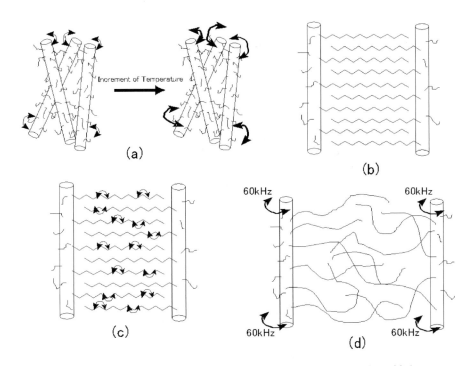

Figure 7-31. Schematic representation of the structure and molecular motion of PG-ns: (a) the temperature change for PG-ns(n < 10), (b) PG-ns(n ≥ 10) at low temperature (the side-chain crystallite), (c) PG-ns(n ≥ 10) at temperature just below the melting point (the rotator phase of the side chains) and (d) PG-ns(n ≥ 10) at high temperature (the LC state)

2.5   Poly(L-glutamate) with Unsaturated Olely Side Chains

The melting temperature of unsaturated side chains of poly(γ-oleyl L-glutamate) (POLLG) is much lower than that of saturated side chains of PG-18. Obviously, a double bond placed in the central part of the oleyl group interrupts the crystallization of the side chains [81]. Thus, the LC nature of α-helical rods may be maintained all the way to $-40°$ C. In Figure 7-32 is shown ^{13}C CP-MAS NMR spectrum of POLLG in the LC state at room temperature [81]. The assignment of each peak can be made from reference data of PG-18 and *n*-paraffins. The interior CH$_2$ peak appears at 30.7 ppm. The ^{13}C chemical shifts of the amide CO and C$_\alpha$ carbons are 176.6 and 57.5 ppm, respectively. This shows that the side chains are melting at room temperature and the main chain is in the α-helix form. The amide CO peak disappears at $-20°$ C. It can be said that from analogy of PG-18, the main chain in the LC state is undergoing fast reorientation at a frequency of 60 KHz at $-20°$ C. The ^{13}C NMR behavior of POLLG over a wide range of temperatures is very similar to the case of PG-18 in the LC state.

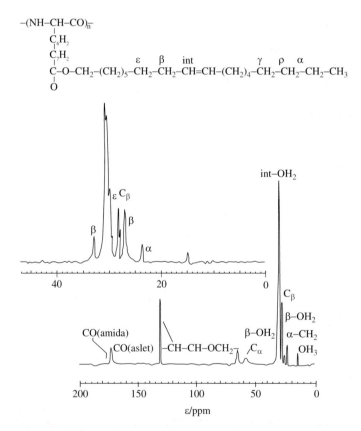

$-(\mathrm{NH-CH-CO})_{\overline{n}}$

$\mathrm{\underset{|}{C_6H_2}}$

$\mathrm{\underset{|}{C_7H_2}}$

$\underset{\mathrm{O}}{\overset{|}{\mathrm{C}}}-\mathrm{O-CH_2-(CH_2)_5-\overset{\varepsilon}{CH_2}-\overset{\beta}{CH_2}-\overset{int}{CH}=CH-(CH_2)_4-\overset{\gamma}{CH_2}-\overset{\rho}{CH_2}-\overset{\alpha}{CH_2}-CH_3}$

Figure 7-32. ^{13}C CP-MAS NMR spectrum of POLLG in the LC state at room temperature

Mohanty, et al. have studied the nature of thermotropic LC behavior exhibited by these two polypeptides, dynamics of PG-18 and POLLG by ^{1}H NMR relaxation times such as ^{1}H spin-lattice relaxation time(T_1) and spin-lattice relaxation time in the rotating frame($T_{1\rho}$) over a wide range of temperatures [82]. The ^{1}H T_1 and ^{1}H $T_{1\rho}$ are plotted against temperature in the temperature range from -120 to $120°$ C in Figures 7-33 and 7-34, where ^{1}H resonance frequency is 90 MHz and the locking field H_1 is 1 mT. According to the BPP theory [83], T_1 passes through a minimum and increases again when the correlation time τ_c for molecular motion increases further. Elevation of temperature leads to an increase in molecular motion of polymers in the solid state (the decrease of τ_c), and so T_1 and $T_{1\rho}$ decrease. A minimum in T_1 is reached at $\omega_0\tau_c = 2\pi\nu_c\tau_c = 1$, where ω_0 and ν_c are the resonance frequencies in radian per second and in Hz, respectively, and T_1 again increases. From the T_1 minimum, the correlation time τ_c for molecular motion at MHz frequencies can be obtained. On the other hand, $T_{1\rho}$ shows T_1-like behavior against temperature, but a minimum in $T_{1\rho}$ is reached at $\omega_1\tau_c = 1$, where $\omega_1/2\pi = \gamma H_1/2\pi = 42.6$ KHz [83]. From the $T_{1\rho}$ minimum, the correlation time τ_c for molecular motion at KHz frequencies can be obtained. Thus, $T_{1\rho}$ behaves similarly to T_2 and is more sensitive to lower frequency motions. As seen from Figure 7-34, T_1 of PG-18 decreases from 700 to 350 ms as the temperature is increased from -100 to $-10°$ C. This means that the molecular motion is in the slow molecular-motion region; *i.e.*, $\omega\tau_c \gg 1$. Above $-10°$ C, T_1 increases from 350 to 550 ms as the temperature is increased from -10 to $45°$ C. This means that molecular motion is nearly in the extremely narrowing region ($\omega\tau_c \ll 1$). The relaxation arises from the side chain motion, which corresponds to the γ-relaxation observed in viscoelastic measurements. This can be justified from the Arrhenius plot as shown in Figure 7-35.

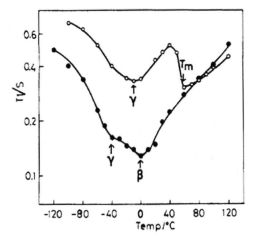

Figure 7-33. Temperature dependence of 1H T1 for PG-18(○) and POLLG(●) β: β-relaxation; γ, γ-relaxation; T$_m$, the melting point of side chain crystallite

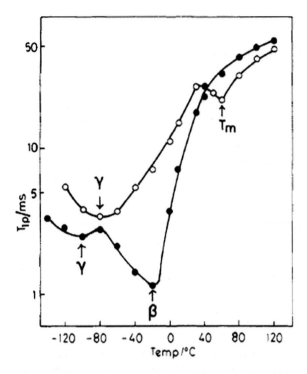

Figure 7-34. Temperature dependence of ^1H $T_{1\rho}$ for PG-18(○) and POLLG(●): β: β-relaxation; γ, γ-relaxation; T_m, the melting point of side chain crystallite

However, T_1 decreases from 550 to 327 ms as the temperature is increased from 45 to 60°C and again increases from 327 to 403 ms as the temperature is further increased from 60 to 120°C. The minimum at lower temperature depends on the observing frequency, but that at higher temperature does not. The first minimum comes from relaxation and the second one comes from the first-order melting transition.

The $T_{1\rho}$ values are plotted against temperature in Figure 7-34. Two distinct minima are observed. The $T_{1\rho}$ decreases from 6.0 to 3.0 ms as the temperature is increased from −120 to −80°C and increases from 3.0 and 29 ms through the first minimum as the temperature is further increased. The $T_{1\rho}$ decreases from 29 to 21 ms again as the temperature is increased from 30 to 60°C and increases 21 to 41 ms through the second minimum as the temperature is further increased from 60 to 120°C. Further, it can be said that below −80°C the γ-relaxation comes from the molecular motion corresponding to the rotation of methyl groups in the side chains at a frequency below ca. 40 KHz seen from the $T_{1\rho}$ minimum. The activation energy, ΔE, for the γ-relaxation can be determined by using $\tau_c = \tau_0 \exp(-\Delta E/kT)$, where τ_0 is the prefactor, k is the Boltzmann constant and T is the absolute temperature. The correlation time τ_c can be estimated using $\tau_c = 1/2\pi\nu_c$. Therefore, the activation

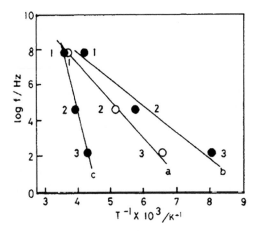

Figure 7-35. Arrhenius plots of log *f* in PG-18(a) and POLLG(b and c) against the inverse of the absolute temperature. *f* is the frequency in Hz. (a and b) from γ-relaxation. (c) from β-relaxation. 1 and 2 were obtained from ^{1}H T_1 and $T_{1\rho}$ and 3 was obtained from the viscoelastic data

energy Δ E is determined from the plots of log *f* (the frequency in Hz) against 1/T as shown in Figure 7-35, where the values of *f* for the T_1 and $T_{1\rho}$ minima are 90 MHz and 42.6 KHz, respectively. The activation energy of PG-18 is 10 Kcal/mol. This is a reasonable value for the γ-relaxation and agrees with the value of 11 Kcal/mol obtained from the mechanical relaxation by viscoelastic measurements [19].

3.   DIFFUSION OF ROD LIKE POLYMERS IN THE THERMOTROPIC LIQUID CRYSTALLINE PHASE BY FIELD-GRADIENT NMR

3.1   Rod like Polypeptides

As mentioned above, poly(γ-glutamate) with long *n*-alkyl side chains forms thermotropic LC phase by melting of the side chain crystallites [20]. From high-resolution solid-state ^{13}C NMR experiments on poly(γ-*n*-alkyl L-glutamates) [21, 32, 73, 81, 82], it has been shown that the main chain of polypeptides takes a right-handed α-helical conformation and the *n*-alkyl side chains take an all-*trans* zigzag conformation in the crystallites at temperatures below the melting point, and that at temperatures above the melting point of the side chain crystallites the side chains are undergoing fast *trans-gauche* exchange, and then the main chain undergoes fast molecular motion at a frequency of about 60 KHz [20]. Poly(*n*-dodecyl L-glutamate) (PDLG, that is PG-12) forms a typical cholesteric LC phase at temperatures above 50° C. This shows that by melting of the side chain crystallites, the rate of reorientation of the side chains is transitionally increased. It has been reported that α-helical poly(γ-glutamate)s with *n*-octadecyl side chains in the thermotropic LC state are diffusing as demonstrated by PFGSE ^{1}H NMR method, and further the diffusion coefficients in directions parallel $(D_{\parallel})$ and perpendicular $(D_{\perp})$ to the α-helical chain axis are determined and the diffusion is an anisotropic [84–86].

Further, it is well known that poly(glutamate) systems such as PBLG, poly(γ-*n*-alkyl L-glutamate), *etc.* in solvent form the isotropic, biphasic and LC phases depending on the polypeptide concentration [78,87–89]. However, in these phases the diffusional behavior of the polypeptides is not clarified.

Here, the preparation of highly-oriented PDLG films is described as a function of the main-chain length and the isotropic and anisotropic diffusion coefficients of the polypeptides as a function of temperature within the temperature range from 50 to 80° C by means of PFGSE ^{1}H NMR method [90–99], in order to elucidate how the diffusion is affected by changes of the main-chain length [84–86]. The diffusion behavior in thermotropic LC phase is analyzed by the translational diffusion equation on the basis of Kirkwood theory [100] of diffusion process for rod like polymers derived by Doi and Edwards [101]. Further, we discuss the diffusional behavior of α-helical PG-18 and chloroform as solvent in the isotropic, biphasic and LC phases by using high field-gradient ^{1}H NMR and diffusion ^{1}H NMR imaging.

The self-diffusion coefficient measurements to be introduced are carried out by using a standard PFGSE pulse sequence: (the Hahn echo sequence: $\pi/2$ pulse-τ-π pulse) [90] with field-gradient pulse in between the $\pi/2$ and π pulses, and in between the π pulse and spin echo [91]. For small diffusion coefficient measurements, high field-gradient strength must be used (for example, more than 10 Tm^{-1} (1000 G/cm) [93, 99]. The relationship between the echo signal intensity and pulse field-gradient parameters is given by

$$(7\text{-}4) \qquad A(G \text{ or } \delta)/A(0) = \exp[-\gamma^2 G^2 D \delta^2 (\Delta - \delta/3)]$$

where $A(G \text{ or } \delta)$ and $A(0)$ are echo signal intensities at $t = 2\tau$ with and without the magnetic field-gradient pulse, respectively. The field-gradient pulse width is δ. τ is the pulse interval, γ the gyromagnetic ratio of proton, G the field-gradient strength, D the self-diffusional coefficient, and Δ the field-gradient pulse interval. The echo signal intensity is measured as a function of δ. The plot of $ln[A(\delta)/A(0)]$ against $\gamma^2 G^2 \delta^2 (\Delta - \delta/3)$ gives a straight line with a slope of $-D$. In the small diffusion coefficient measurements, for example, the τ, Δ and δ values employed are 4, 4 and 0.001–0.4 ms, respectively. The diffusion coefficient D of water of 2.5×10^{-5} cm^2/s at 303 K is used as the calibration of the field-gradient strength as well-known. The experimental error for the D value is estimated to be within 5%.

As probe molecules in the biphasic phase have two-diffusion components in diffusion on the measurement time scale, the total echo attenuation is given by a superposition of contributions from the individual components as expressed by

$$(7\text{-}5) \qquad A(\delta)/A(0) = f_1 \exp[-\gamma^2 G^2 D_1 \delta^2 (\Delta - \delta/3)]$$
$$+ f_2 \exp[-\gamma^2 G^2 D_2 \delta^2 (\Delta - \delta/3)]$$

where D_i is the self-diffusion coefficient of the *i*th component, and f_i is the fraction of the *i*th component and thus $f_1 + f_2 = 1$. The fraction for the fast and slow diffusion components can be determined from the intercept of the least-squares fitted straight line.

3.2 Diffusion of Poly(n-alkyl L-glutamate)s in the Thermotropic Liquid Crystalline Phase

The self-diffusion coefficients (D) of rod like poly(*n*-alkyl L-glutamate)s having *n*-dodecyl side chains in the thermotropic LC state have been measured as a function of the main-chain length (molecular weight (M_w) of 7000, 30000 and 130000), which are corresponding to the main chain lengths (L) of ca.30, 200 and 890Å, respectively, within the temperature range from 50 to 80° C by means of PFGSE ^{1}H NMR method as shown in Table 7-4, in order to elucidate the diffusional behavior of the polypeptides in the thermotropic LC state [85]. It is found that at temperatures above the melting point of side-chain crystallites in poly(*n*-alkyl L-glutamate) the polypeptide forms the thermotropic LC phase, and then the isotropic diffusion coefficients (D_{iso}) of the rod like polypeptides are decreased with an increase in the main-chain length. From the table, it is seen that the $D_\parallel$ value is larger than the $D_\perp$ value. Their values are decreased with an increase in temperature. These agree with the results of *n*-alkanes in the rotator phase [102]. The diffusion process is analyzed by the Kirkwood theory [100] of diffusion process for rod like polymers as shown in Figure 7-36. The diffusion coefficients of PGs in the directions parallel ($D_\parallel$) and perpendicular ($D_\perp$) to the α-helical axis are determined, and the $D_\parallel$ value is found to be larger than the $D_\perp$ value.

Table 7-4. Determined diffusion coefficients of PDLG in the liquid crystalline phase at temperatures from 50 to 80°C

$D/\times10^{-7}$		Temperature/°C						
mol wt	cm^2/s	50	55	60	65	70	75	80
7000	$D_\parallel$	17.6	18.3	19.1	19.7	20.4	21.2	21.8
	D_{iso}	16.7	17.3	18.1	18.9	19.9	20.6	21.3
	$D_\perp$	15.7	16.2	17.2	18.1	19.1	19.8	20.6
	$D_\parallel/D_\perp$	1.12	1.13	1.11	1.09	1.07	1.07	1.06
30000	$D_\parallel$	10.4	11.1	11.9	12.8	13.6	14.3	15.4
	D_{iso}	9.21	9.96	9.51	10.5	11.6	12.6	13.6
	$D_\perp$	7.94	8.72	17.2	18.1	19.1	19.8	20.6
	$D_\parallel/D_\perp$	1.31	1.27	1.25	1.22	1.17	1.13	1.13
130000	$D_\parallel$	6.33	7.33	8.44	9.31	10.6	11.4	12.1
	D_{iso}	5.67	6.44	7.56	8.43	9.63	10.5	11.2
	$D_\perp$	4.77	5.65	6.78	8.43	9.73	19.8	10.8
	$D_\parallel/D_\perp$	1.33	1.30	1.24	1.22	1.22	1.17	1.12

The main-chain of PG considered in this work takes the α-helical form like a long rod. At temperatures above the melting point of the side-chain crystallites the side chains take the liquid like phase such as liquid *n*-paraffins and are working as a solvent for the main chain. The rod like polypeptide is diffusing as reported previously. The diffusion process of the polypeptide is assumed to follow the Kirkwood theory [100] for the diffusion process of rod like polymers.

The derivation of obtaining the translational diffusion coefficient of rod like polymers derived by Doi and Edwards [101] on the basis of Kirkwood theory [100], the isotropic diffusion coefficient D_{iso} of a rod like polymer chain is followed by the following equation.

(7-6) $D_{iso} = (D_{\parallel} + 2D_{\perp})/3 = [ln(L/b)/L] kT/3\pi\eta_s$

in which

(7-7) $D_{\parallel} = [ln(L/b)/L] kT/2\pi\eta_s$

(7-8) $D_{\perp} = [ln(L/b)/L] kT/4\pi\eta_s$

where $D_{\parallel}$ and $D_{\perp}$ are the diffusion coefficients in parallel to and perpendicular to the rod like polymer chain axis, respectively, L is the rod like polymer length, *b* is the diameter of the rod like polymer, η_s is the viscosity of the solvent corresponding

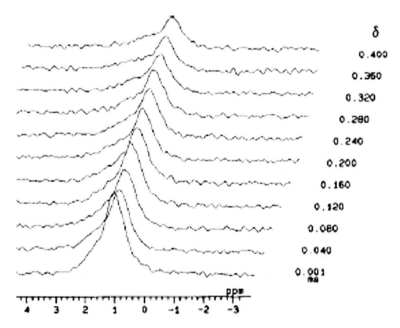

Figure 7-36. Spin echo ^{1}H NMR spectra of PG-18 in the thermotropic LC phase at 80°C as a function of field-gradient pulse length d by PFGSE NMR

to long n-alkyl side chains in the thermotropic LC state, k is the Boltzmann constant and T is the absolute temperature.

By using the standard bond lengths and bond angles determined by X-ray diffraction we can straightforwardly estimate the rod like main chain length and the diameter of α-helical polypeptides. Then, the main-chain lengths of α-helical poly(γ-n-alkyl L-glutamate)s with average molecular weights of 7000, 30000 and 130000 can be estimated to be L = ca. 30,200 and 890 Å, respectively, and the diameter of the α-helical main-chain including the ester group of side chain to b = ca. 10 Å. The diffusion process of these rod like polypeptides is expected to follow eq. (7-6). This equation shows that the plots of D_{iso} against $ln(L/b)/L$ become a straight line.

Therefore, that the plots of D_{iso} against $ln(L/b)/L$ become a straight line at any given temperature. Here, it is assumed that the viscosity of the solvent corresponding to long n-dodecyl side chains is independent of the main-chain length, which is undergoing fast exchange between the *trans* and *gauche* conformations like liquid n-alkanes. In Figure 7-37 the plots of the isotropic diffusion coefficients of PDLG in the thermotropic LC state against $ln(L/b)/L$ are shown at various temperatures. It is found that its plots become a straight line. This trend does not conflict with the theoretical prediction. Therefore, it can be said that the isotropic diffusion of rod like PDLG chains follows approximately the translational diffusion equation of rod like polymers derived by Doi and Edwards [101] on the basis of Kirkwood theory [100]. Further, it seen from Figure 7-37 that the slope of the plots of D_{iso} against $ln(L/b)/L$ is increased with an increase in temperature. This agrees with the theoretical prediction as seen from eq. (7-6) because the slope is a function of temperature.

In this polypeptide system, by melting of long n-alkyl side-chain crystallites, the side chains play a role solvent in LC system. Therefore, we may take into account the case that the viscosities η_s' and η_s'' of the solvent in parallel to and perpendicular to the rod like polymer chain axis, respectively, may be, in principle, different. In this case, η_s in eqs. (7-7) and (7-8) must be replaced by η_s' and η_s'', respectively. Then, η_s in eq. (7-6) must be replaced by $(\eta_s' + \eta_s'')/2\eta_s'\eta_s''$. When η_s' and η_s'' are equal to each other, $D_{\parallel}/D_{\perp}$ becomes 2. When η_s' and η_s'' are different from each other, $D_{\parallel}/D_{\perp}$ becomes $2\eta_s'/\eta_s''$. As predicted from this polypeptide system, we have $\eta_s' \geq \eta_s''$. Thus, we may expected $D_{\parallel}/D_{\perp} < 2$. Such a situation may be associated with the degree of the orientation of the polypeptide LC system.

Fundamental theories of diffusion for low-molecular weight liquid crystals in the nematic phase have been studied by Franklin [103–105] based on Oseen-Kirkwood hydrodynamic theory for isotropic liquids. Further theories of diffusion for low-molecular weight liquid crystals have been developed. These theories explained partially the experimental data on $D_{\parallel}$ and $D_{\perp}$. Chu and Moroi [106], and Leadbetter [107] have obtained that the anisotropy ratio of the diffusion coefficients, $D_{\parallel}/D_{\perp}$, for low-molecular weight liquid crystals are expressed by $[2\gamma(1-S)+2S+1]/[\gamma(S+2)+1-S]$, where $\gamma = \pi d/4l$ in which l is the length and d of the diameter of the

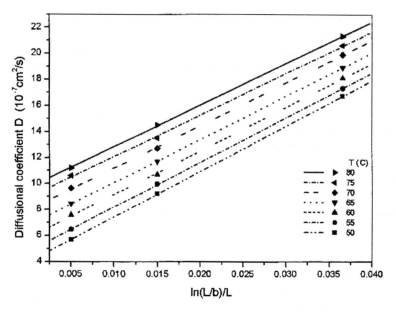

Figure 7-37. The plots of the isotropic diffusion coefficients of PG-12 in the thermotropic LC state against *ln*(L/*b*)/L are shown at various temperatures

rod like molecules. This equation shows that if S becomes less than 1, $D_{\parallel}/D_{\perp}$ is reduced.

From Table 7-4, it is seen that the $D_{\parallel}$ value is larger than the $D_{\perp}$ value. Their values are decreased with an increase in temperature. These agree with the previous results of PG-18 in the thermotropic LC state. Further, as seen from Figure 7-37, it can be said that these experimental results can be qualitatively explained by the Kirkwood theory. Also, the experimental results agree with the theoretical prediction. As above-mentioned, when η_s' and η_s'' are equal to each other, $D_{\parallel}/D_{\perp}$ becomes 2, when η_s' and η_s'' are different, $D_{\parallel}/D_{\perp}$ becomes $2\eta_s'/\eta_s''$, and as predicted from this polypeptide system, we have $\eta_s' \geq \eta_s''$ and, thus, we may expected $D_{\parallel}/D_{\perp} < 2$. As seen from Table 7-4, at 50° C the ratio of $D_{\parallel}$ to $D_{\perp}$ is 1.33 and at 80° C is 1.12. This means $\eta_s' \geq \eta_s''$. These values are close to those of low-molecular weight liquid crystals [108] such as PAA(*p*-azoxyanisole) and DMBBA (*p*-methoxybenzylidene-*p'*-*n*-butylaniline) to be 1.33 and 1.44, respectively.

3.3 Diffusion of Poly(diethylsiloxane) in the Thermotropic Liquid Crystalline Phase and Isotropic Phase

As mentioned in section 2.3, molecular motion of PDES has been studied by NMR method. Nevertheless, diffusional behavior of PDES in the LC phase has never been clarified. It can be expected that a PDES chain in the isotropic phase behaves as a random-coil because of the flexible main chain and, on the other hand, PDES chains in the LC phase behave as rigid-rod and then are diffusing.

It has been shown that pulse field-gradient spin-echo ^{1}H NMR can provide very useful information for elucidating the diffusion process of rod like polymers in the LC phase. Sometimes, it is difficult to determine the diffusion coefficient of polymers with short ^{1}H T_2(^{1}H$T_2 < 2$ ms) in the isotropic phase and the LC phase by PFGSE ^{1}H NMR because the echo signal is predominantly decayed by short T_2 before decayed by diffusion. For this reason, modified pulse field-gradient NMR techniques must be used for PDES systems with short ^{1}H T_2. Then, the diffusional behavior of PDES through determination of the diffusion coefficients of PDES in the LC phase and the isotropic phase as a function of temperature by using PFGSE ^{13}C NMR method with ^{1}H CW decoupling and PFGSE ^{1}H NMR method will be introduced [61].

3.4 Diffusion of PDES in the Isotropic Phase and in the Isotropic Region of the Biphasic Phase

Figure 7-38(a) shows the plots of $ln[A(G)/A(0)]$ for PDES in the isotropic phase at 50, 60, 70 and 80° C against $\gamma^2 G^2 \delta^2 (\Delta - \delta/3)$ by changing $G = 0 - 1160$ G/cm at $\Delta = 500$ ms, at $\delta = 8$ ms and $\tau = 11.2$ ms as obtained by using PFGSE ^{1}H NMR method [60]. It is seen from these plots that the experimental data lie on a straight line. This shows that the diffusion of PDES in the isotropic phase is a single diffusion component. Then, diffusion coefficients D with order of 10^{-11} cm^2/s determined from the slope are summarized in Table 7-5. This shows that such extremely small diffusion coefficients are successfully determined by using the present PFGSE ^{1}H method. From these experimental results, it is seen that the diffusion coefficient of PDES is slowly increased from 6.2 to 7.4×10^{-11} cm^2/s with an increase in temperature from 50 to 80° C. The polymer chains are completely in the isotropic phase. We must consider the cause why the polymer chains are very slowly diffusing. As suggested earlier for any specified viscosity behavior for liquid polymers and high concentrated polymer solution [109], it is thought that extremely slow diffusion of PDES chains in the isotropic phase may come from entanglements between the polymer chains.

Next, we are concerned with diffusion of PDES in the isotropic region of the biphasic phase. In the biphasic phase at 20, 30 and 40° C, as above-mentioned PDES chains are in the isotropic region and LC region, and thus PDES chains in the isotropic region are surrounded by the LC region, and those in the LC region are surrounded by the isotropic region. The ^{1}H T_2 of PDES in the LC region is strongly influenced by strong dipolar interactions and then the ^{1}H T_2 becomes shorter by roughly one thirty-seventh than that in the isotropic region at same temperature. Therefore, the echo signal of the LC component in this PFGSE ^{1}H NMR experiment conditions disappears within the time interval τ of 11.2 ms between the first $\pi/2$ pulse and the second $\pi/2$ pulse, but the signal derived from PDES in the isotropic region of the biphasic phase remains in the finally obtained

(a)   (b)

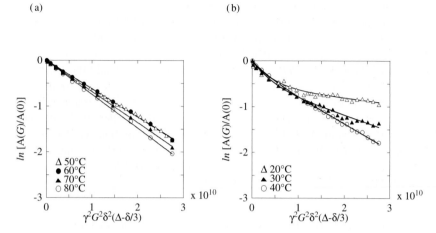

Figure 7-38. The plots of $ln[A(G)/A(0)]$ against $\gamma^2 G^2 \delta^2 (\Delta-\delta/3)$ for determining diffusion coefficients D of PDES by PFGSE ^{1}H NMR method: (a) for the isotropic region in the isotropic phase: at 50 (Δ), 60 (•), 70 (▲) and 80°C (○); (b) for the isotropic region of the biphasic phase: at 20 (Δ), 30 (▲) and 40°C (○)

Table 7-5. Determined diffusion coefficients D of PDES in the liquid crystalline region in the biphasic region by PFGSE ^{13}C NMR method and in the isotropic region of the biphasic phase and the isotropic phase by PFGSE ^{1}H NMR method as a function of temperature

	Diffusion coefficient $D/cm^2/s$			
		Isotropic component[a]		
Temp /°C	$D \times 10^{-9}$ (liquid crystalline component)	$D_1 \times 10^{-10}$ (fast diffusion component)	$D_2 \times 10^{-11}$ (slow diffusion component)	fraction[b]
				f_{D1} f_{D2}
20	9.9[d]	2.6	1.3	0.43 : 0.57
30	75[c] 50[d]	2.8	3.5	0.39 : 0.61
40	–	2.8	5.7	0.21 : 0.79
50	–	–	6.2	0 : 1
60	–	–	6.2	0 : 1
70	–	–	6.8	0 : 1
80	–	–	7.4	0 : 1

[a] for the isotropic region of the biphasic phase at temperatures from 20 to 40°C and for the isotropic phase at temperatures from 50 to 80°C.
[b] D_1 and D_2 for the fast and slow diffusion components, respectively, at temperatures from 20 to 40°C.
[c] for $\Delta = 750$ ms.
[d] for $\Delta = 1000$ ms.

echo. The plots of $ln[A(G)/A(0)]$ against $\gamma^2 G^2 \delta^2 (\Delta - \delta/3)$ consist of two straight lines with different slope as shown in Figure 7-38(b). This shows that the diffusion has two kinds of diffusion components in the isotropic regions such as the slow diffusion component and fast diffusion component. The diffusion coefficients Ds for the slow diffusion component and the fast diffusion component of the isotropic region in the biphasic phase are determined from the slopes as shown in Table 7-5. The diffusion coefficients Ds of PDES for the fast diffusion component in the isotropic region as determined from the large slope of the plots are in order of 10^{-10} cm^2/s and those for the slow diffusion component as determined from the small slope of the plots are in order of 10^{-11} cm^2/s. The magnitude of the diffusion coefficient for the slow diffusion component is close to that in the isotropic phase at 50° C.

Next, we consider whether in the diffusional behavior at 20, 30 and 40°C the partial-restriction effect on diffusion which comes from the obstruction at the interface between the isotropic and LC regions is important or not. The slow diffusion component is more sensitive to the obstruction of the interface. However, the diffusion time Δ of 500 ms used in this experiment is not so long that the diffusion is hindered by the interface. For example, the slow diffusion species with $D = 3.5 \times 10^{-11}$ cm^2/s at 30°C can diffuse only 59 nm as calculated from relation $\sqrt{2D\Delta}$ within the diffusion time of 500 ms. This is much smaller than the domain size of the isotropic region to be over several μm as estimated by an optical micro-scope. Therefore, it is difficult to observe the partial-restriction effect on diffusion. If the observed non-Fickian behavior in the isotropic region of the biphasic phase comes from the partial-restriction effect, the slow diffusion component may be assigned to PDES in the interface region because PDES in the interface region must suffers strongly the partial-restriction effect in diffusion. Thus, the fraction of the slow diffusion component must be small because the fraction of the interface region is very small. Nevertheless, the slow diffusion component is dominantly observed. From these experimental results, it can be said that the slow and fast diffusion components should be assigned to the inside and outside regions of the isotropic region in the biphasic phase. The determined D values and the fractions of the two diffusion components are summarized in Table 7-5. From this table, it is seen that the D values of the slow diffusion components of the isotropic region in the biphasic phase are very largely increased with an increase in temperature, and those in the isotropic phase are very slowly increased. The plots are abruptly changed at 50°C (not shown in figure). The activation energy E for diffusion can be obtained from the Arrhenius plots of lnD against $1/T$ to be 1.7 Kcal/mol and 0.17 Kcal/mol for the slow diffusion component of the isotropic region in the biphasic phase and in the isotropic phase, respectively. The E value for the former is much larger than that for the latter.

The fraction of the fast diffusion component of the isotropic region in the biphasic phase is decreased with an increase in temperature and that of the slow diffusion component of the isotropic region in the biphasic phase is increased. Therefore, it is expected that PDES chains are diffusing from the isotropic region to the LC

region of the biphasic phase through their interface and at the same time from the
LC region from the isotropic region. Its balance may be changed by temperature
change. As above-mentioned, the origin of extremely small diffusion coefficient
in order of $10^{-10} \sim 10^{-11}$ cm^2/s may be due to the entanglements of PDES chains.
In order to recognize the effect of the entanglements on the diffusion process,
the PDES concentration dependence of the diffusion coefficient of the PDES in
PDES/toluene-d_8 solution is measured at 50°C. Figure 7-39 shows the plots of the
D values determined by changing $G = 0 - 1000$ G/cm at $\Delta = 100$ ms, $\delta = 1 - 4$ ms
and $\tau = 11.2$ ms by using PFGSE ^{1}H NMR method against the PDES concentration.
It is seen that as the polymer concentration is increased, the D value of PDES
chains in PDES/toluene-d_8 solution is decreased with an increase in the polymer
concentration, and approaches to the D value of PDES chains in the melt state.
This shows that the entanglements effect exists in the polymer system, which leads
to reduction of the translational diffusion coefficient of the polymer chains.

In order to clarify the diffusional behavior of PDES in the LC region, the PFGSE
^{13}C NMR experiments under ^{1}H CW decoupling are made at 20 and 30° C by
changing $G = 0 - 1150$ G/cm at $\Delta = 750$ and 1000 ms, $\delta = 0.5$ ms and $\tau = 1.7$ ms.
In these temperatures, PDES chains are in the biphasic phase consisting of the LC
region and the isotropic region as characterized by static solid-state ^{29}Si NMR. The
fraction of the LC region is 0.81 at 20°C, and is 0.69 at 30°C as determined by ^{1}H
solid echo method. The typical observed PFGSE ^{13}C NMR spectra of PDES in the
LC region at 30°C and at $\Delta = 1000$ ms are shown in Figure 7-40. An asymmetric
signal appears at about 6.3 ppm. This peak can be straightforwardly assigned to the
methyl carbons in side chains of PDES. In PFGSE experiments, the magnetization
is attenuated by T$_2$ during τ. The ^{13}C T$_2$ value for the methyl carbons in side chains

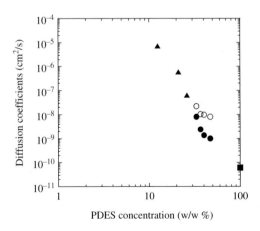

Figure 7-39. The log-log plots of diffusion coefficients D of PDES in PDES/toluene-d_8 solutions against
the PDES concentration at 50°C by PFGSE ^{1}H NMR method: (▲) 12.5, 21.1 and 26.2% w/w; (●)
33.2, 36.9, 40.3 and 47.3% w/w (slow diffusion component); (○) 33.2, 36.9, 40.3 and 47.3% w/w (fast
diffusion component); (■) 100% w/w

of PDES in the LC region is roughly 16 times longer than $^1H\,T_2$ as estimated by using the magnetogyric ratios of $^{13}C(\gamma_{13C})$ and $^1H(\gamma_{1H})$ nuclei and the BPP theory $\left(^{13}C\,T_2/^1H\,T_2 \approx (\gamma_{1H}/\gamma_{13C})^2 \approx 4^2 = 16\right)$ [83]. Therefore, it is better to use ^{13}C nucleus in determining the diffusion coefficient as compared with ^1H nucleus by PFGSE NMR method. The echo signal comes from PDES chains in the liquid crystalline region and does not disappear at the echo time interval during τ. 74% of the initial magnetization is decayed by $^{13}C\,T_2$ during τ and $^{13}C\,T_1$ between the second and third $\pi/2$ pulses at $\Delta = 1000\,ms$ and $\tau = 1.7\,ms$. Then, the remaining magnetization is reduced by diffusion.

The plots of $ln[A(G)/A(0)]$ for PDES as obtained from the PFGSE ^{13}C NMR spectra against $\gamma^2G^2\delta^2(\Delta - \delta/3)$ at 20 and 30°C and at $\Delta = 1000\,ms$ become a straight line as shown in Figure 7-41. Thus, this shows that the diffusion is a single diffusion component with $D \approx 10^{-8}\,cm^2/s$. The reduction of diffusion coefficient of PDES in the LC region with an increase in Δ from 750 to 1000 ms shows that there clearly exists the partial-restriction effect of the diffusion.

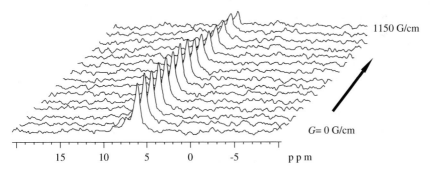

Figure 7-40. Observed PFGSE ^{13}C NMR spectra of PDES in the biphasic phase as a function of field-gradient strength G at 30°C

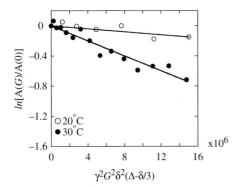

Figure 7-41. The plots of $ln[A(G)/A(0)]$ against $\gamma^2G^2\delta^2(\Delta - \delta/3)$ for determining diffusion coefficients D of PDES at 20 ($\circ$) and 30°C ($\bullet$) by PFGSE ^{13}C NMR method

In the case of PG-18 in chloroform-*d* solution in the lyotropic LC phase and
the isotropic phase [86], the diffusion coefficient of POLG in the LC phase
($\approx 10^{-7}\,\mathrm{cm}^2/\mathrm{s}$) is much smaller D value than POLG in the isotropic phase
($\approx 10^{-6}\,\mathrm{cm}^2/\mathrm{s}$). It is opposite to the results of PDES in the LC region and the
isotropic region. In the case of PG-18, the polymer takes the α-helix form in both
of the LC and isotropic phases, and then diffuses as a rigid-rod. As predicted from
the fact, PG-18 chains in the LC phase may more strongly interact with each other
as compared with those in the isotropic phase. On the other hand, the PDES chain
takes the extended form in the LC region, but takes the random-coil form in the
isotropic region. For this reason, the random-coiled PDES chains have the entanglements in the melt state and thus diffuse more slowly compared with those in the
LC region.

4.   REFERENCES

1. P.J. Collings and M. Hird, *Introduction to Liquid Crystals*, Taylor and Francis, London, 1998.
2. V.P. Shibaeb and L. Lan, *Liquid Crystalline and Mesomorphic Polymers*, Springer-Verlag, Berlin, 1993.
3. M. Ballauff, *Macromol. Chem., Rapid Commun.*, 1986, 7, 407.
4. M. Ballauf, *Angew. Chem., Int. Ed. Engl.*, 1987, 28, 253.
5. M. Ballaff and G.F. Schmidt, *Mol. Crys. Liq. Cryst.*, 1987, 147, 163.
6. R. Stern, M. Ballauff and G. Wegner, *Makromol. Chem., Macromol. Symp.*, 1989, 423, 373.
7. J.M. Rodrigues-Parada, R. Duran and G. Wegner, *Macromolecules*, 1989, 22, 2507.
8. M. Ebert, O. Herrmann-Shenherr, J. Wendorf, H. Ringsdorf and P. Tschirner, *Liq. Cryst.*, 1990, 11, 249.
9. A. Adam and H.W. Spiess, *Makromol. Chem., Rapid Commun.*, 1990, 11, 249.
10. C.H. Frech, A. Adam, U. Falk and H.W. Spiess, *New Polym. Mater.*, 1990, 2, 267.
11. R. Stern, M. Ballaff, G. Lieser and G. Wegner, *Polymer*, 1991, 32, 2079.
12. B.R. Harkness and J. Watanabe, *Macromolecules*, 1991, 24, 6759.
13. J. Watanabe, B.R. Harkness and M. Sone, *Polym. J.*, 1992, 24, 1119.
14. L. Cervinka and M. Ballaff, *Colloid Polym. Sci.*, 1992, 270, 859.
15. M. Sone, B.R. Harkness, J. Watanabe, T. Torii, T. Yamashita and K. Horie, *Polym. J.*, 1993, 25, 997.
16. P. Galda, D. Kistner, A. Martin and M. Ballaff, *Macromolecules*, 1993, 26, 1595.
17. K. Marz, P. Linder, J. Urban, M. Ballaff and E.W. Fischer, *Acta Polym.*, 1993, 44, 139.
18. S.B. Damman and G.J. Vroege, *Polymer*, 1993, 34, 2773.
19. J. Watanabe, B.R. Harkness, H. Ichimura and M. Sone, *Macromolecules*, 1994, 27, 507.
20. J. Watanabe, H. Ono, I. Uematsu and A. Abe, *Macromolecules*, 1985, 18, 2141.
21. T. Yamanobe, M. Tsukahara, T. Komoto, J. Watanabe, I. Ando, I. Uematsu, K. Deguchi, T. Fujito and M. Imanari, *Macromolecules*, 1988, 21, 48.
22. I. Ando, T. Yamanobe, H. Kurosu and G.A. Webb, *Annu. Rep. NMR Spectrosc.*, 1990, 22, 205.
23. I. Ando, T. Yamanobe and T. Asakura, *Prog. NMR Spectrosc.*, 1990, 22, 349.
24. I. Ando and T. Asakura (ed.), *Solid State NMR of Polymers*, Elsevier Science, Amsterdam, 1998.

25. I. Ando, M. Kobayashi, C. Zhao, Y. Yin and S. Kuroki, *Encyclopedia of NMR*, John Wiley & Sons, Ltd., Chichester, UK, 2002, Vol. 3 (Advance in NMR), p. 770.
26. M. Sone, B.R. Harkness, H. Kurosu, I. Ando and J. Watanabe, *Macromolecules*, 1994, 27, 2769.
27. D.L. VanderHart, *J. Magn. Reson.*, 1981, 44, 117.
28. S. Ishikawa, H. Kurosu and I. Ando, *J. Mol. Struct.*, 1990, 248, 361.
29. J. Watanabe, N. Sekine, T. Nematsu and M. Sone, *Macromolecules*, 1996, 29, 4816.
30. K. Fu, N. Sekine, M. Sone, M. Tokita and J. Watanabe, *Polym. J.*, 2002, 34, 291.
31. M. Matsui, Y. Yamane, H. Kimura, S. Kuroki, I. Ando, K. Fu and J. Watanabe, *J. Mol. Struct.*, 2003, 650, 175-185(2003).
32. T. Yamanobe, H. Tsukamoto, Y. Uematsu, I. Ando and I. Uematsu, *J. Mol. Struct.*, 1993, 295, 25.
33. E. Katoh, H. Kurosu and I. Ando, *J. Mol. Struct.*, 1994, 318, 123.
34. M. Matsui, Y. Yamane, S. Kuroki, I. Ando, K. Fu and J. Watanabe, *J. Mol. Struct.*, 2005, 739/1-3, 131.
35. M. Matsui, Y. Yamane, S. Kuroki, I. Ando, K. Fu and J. Watanabe, *Ind. Eng. Chem. Res.*, 2005, 44, 8694.
36. A.B. Brook, *Silicon in Organic, Organometallic, and Polymer Chemistry*, John Wiley & Sons, Inc., New York, 2000.
37. D. Teyssie and S. Boileau, *Silicon-Containing Polymers*, R.G. Jones, W. Ando and J. Chojnowski, Eds. Kluwer Academic Publishers, Dordrecht, 2000, p.593.
38. Y.K. Godovsky and V.S. Papkov, *Adv. Polym. Sci.*, 1989, 88, 129.
39. B.R. Harkness, M. Tatiana, H. Yue and I. Mita, *Chem. Mater.*, 1998, 10, 1700.
40. A. Molenberg and M. Möller, *Macromolecules*, 1997; 30, 8332.
41. G.J.J. Out, A.A. Turetskii and M. Möller, *Macromol. Chem. Rapid Commun.*, 1995, 16, 107.
42. G.J.J. Out, H.A. Klok and M. Möller, *Macromol. Chem. Phys.*, 1995, 196, 196.
43. G.J.J. Out, A.A. Turetskii, M. Möller and D. Oelfin, *Macromolecules*, 1994, 27, 3310.
44. G. Kögler, K. Loufakis and M. Möller, *Polymer*, 1990, 31, 1538.
45. M. Möller, S. Siffrin, G. Kögler and D. Oelfin, *Macromol. Chem. Macromol. Symp.*, 1990, 34, 171.
46. G. Kögler, A. Hasenhindl and M. Möller, *Macromolecules*, 1989, 22, 4190.
47. V.M. Litvinov, A.K. Whittaker, A. Hagemeyer and H.W. Spiess, *Colloid Polym. Sci.*, 1989, 267, 681.
48. J. Friedrich and J.F. Rabolt, *Macromolecules*, 1987, 20, 1975.
49. V.S. Papkov, V.S. Svistunov, Y.K. Godovsky and A.A. Zhdanov, *J. Polym. Sci.* (B), 1987, 25, 1859.
50. Y.K. Godovsky and V.S. Papkov, *Macromol. Chem. Macromol. Symp.*, 1986, 4, 71.
51. D.Y. Tsvankin, V.S. Papkov, V.P. Zhukov, Y.K. Godovsky, V.S. Svistunov and A.A. Zhdanov, *J. Polym. Sci.: Polym. Chem. Ed.*, 1985, 23, 1043.
52. Y.K. Godovsky, N.N. Makarova, V.S. Papkov and N.N. Kuzmin, *Macromol. Chem. Rapid. Commun.*, 1985, 6, 443.
53. V.S. Papkov, Y.K. Godovsky, V.S. Svistunov, V.M. Litvinov and A.A. Zhdanov, *J. Polym. Sci.: Polym. Chem. Ed.*, 1984, 22, 3617.
54. F. Grinberg, R. Kimmich, M. Möller and A. Molenberg, *J. Chem. Phys.*, 1996, 105(21), 9657.
55. R. Kimmich, S. Stapf, M. Möller, R. Out and R.-O. Seitter, *Macromolecules*, 1994, 27, 1505.

56. V.M. Litvinov, V. Macho and H.W. Spiess, *Acta Polym.*, 1997, 48, 471.

57. R.D. Miller, D. Hofer, J.F. Rabolt and G.N. Fickes, *J. Am. Chem. Soc.*, 1985, 107, 2172.

58. E.K. Karikari, A.J. Greso, B.L. Farmer and R.D. Miller, *Macromolecules*, 1993, 26, 3937.

59. R.D. Miller and J. Michl, *Chem. Rev.*, 1989, 89, 1359.

60. V.S. Papkov, M.N. Il'ina, V.P. Zhukov, D.J. Tsvankin and D.R. Tur, *Macromolecules*, 1992, 25, 2033.

61. S. Kanesaka, H. Kimura, S. Kuroki, I. Ando and S. Fujishige, *Macromolecules*, 2004, 37, 453.

62. S. Kuroki, K. Yamauchi, I. Ando, A. Shoji and T. Ozaki, *Cur. Org. Chem.*, 2001, 5, 1001.

63. K. Yamauchi, S. Kuroki and I. Ando, *J. Mol. Struct.*, 2002, 602/603, 171.

64. K. Yamauchi, S. Kuroki, I. Ando, A. Shoji and T. Ozaki, *Chem. Phys. Lett.*, 1999, 302, 331.

65. S. Kuroki, A. Takahashi, I. Ando, A. Shoji and T. Ozaki, *J. Mol. Struct.*, 1994, 323, 197.

66. J.W. Emsley, Liquid Crystal: General Consideration. In *Encyclopedia of NMR*, D.M. Grant and R.K. Harris, Eds., John Wiley & Sons, Ltd., Chichester, UK, 1996; Vol. 3, p.2788.

67. A. Lafuma, F. Fayon, D. Massiot, S.C. Kimmés and C. Sanchez, *Magn. Reson. Chem.*, 2003, 41, 944.

68. T.M. Clark, P.J. Grandinetti, P. Florian and J.F. Stebbins, *J. Phys. Chem. B*, 2001, 105, 12257.

69. M. Mehring, *High Resolution NMR of Solids*, Springer, Berlin, 1983, p.82.

70. C.A. Fyfe, in *Solid State NMR for Chemists*, CFC Press, Guelph, Canada, 1983.

71. H. Kimura, S. Kanesaka, S. Kuroki, I. Ando, A. Asano and H. Kurosu, *Magn. Reson. Chem.*, 2005, 43, 209.

72. T.C. Farrar and T. Becker, *Pulse and Fourier Transform NMR*, Academic Press, New York, 1971.

73. M. Tsukahara, T. Yamanobe, T. Komoto, J. Watanabe, I. Ando and I. Uematsu, *J. Mol. Struct.*, 1987, 159, 345.

74. H. Saito and I. Ando, *Annu. Rep. NMR Spectrosc.*, 1989, 21, 209.

75. A.E. Tonelli and F.C. Schilling, *Acc. Chem. Res.*, 1981, 14, 223.

76. a) T. Sorita, T. Yamanobe, T. Komoto, I. Ando, H. Sato, K. Deguchi and M. Imanari, *Makromol. Chem. Rapid Commun.*, 1984, 5, 657.; b) T. Yamanobe, T. Sorita, T. Komoto, I. Ando and H. Sato, *J. Mol. Struct.*, 1985, 131, 267.

77. J.R. Lyerla and C.S. Yannoni, *IBM J. Res. Dev.*, 1983, 27, 302.

78. C. Zhao, H. Zhang, T. Yamanobe, S. Kuroki and I. Ando, *Macromolecules*, 1999, 32, 3389.

79. K. Takegoshi and K. Hikichi, *J. Chem. Phys.*, 1991, 94, 3200.

80. T. Yamanobe, Y. Naito, Y. Iwakura and T. Komoto, *J. Mol. Struct.*, 2002, 602/603, 449.

81. B. Mohanty, T. Komoto, J. Watanabe, I. Ando and T. Shiibashi, *Macromolecules*, 1989, 22, 4451.

82. B. Mohanty, J. Watanabe, I. Ando and K. Sato, *Macromolecules*, 1990, 23, 4908.

83. N. Bleombergen, E.M. Purcell and R.V. Pound, *Phys. Rev.*, 1948, 73, 679.

84. Y. Yin, C. Zhao, S. Kuroki and I. Ando, *J. Chem. Phys.*, 2000, 113, 7635.

85. Y. Yin, C. Zhao, S. Kuroki and I. Ando, *Macromolecules*, 2002, 35, 2335.

86. Y. Yin, C. Zhao, A. Sasaki, H. Kimura, S. Kuroki and I. Ando, *Macromolecules*, 2002, 35, 5910.
87. S. Sobajima, *J. Phys. Soc. Jpn.*, 1967, 23, 1070.
88. E.T. Samulski and A.V. Tobolsky, *Mol. Cryst. Liq. Cryst.*, 1969, 7, 433.
89. I. Ando, T. Hirai, Y. Fujii, A. Nishioka and A. Shoji, *Makromol. Chem.*, 1983, 184, 2581.
90. E.L. Hahn, *Phys. Rev.*, 1950, 80, 580.
91. E.O. Stejskal and E.J. Tanner, *J. Chem. Phys.*, 1965, 42, 288.
92. J.E. Tanner and E.O. Stejskal, *J. Chem. Phys.*, 1968, 49, 1768.
93. S. Matsukawa, H. Yasunaga, C. Zhao, S. Kuroki and I. Ando, *Prog. Polym. Sci.*, 1999, 24, 995.
94. P.T. Callaghan, *"Principles of Nuclear Magnetic Resonance Microscopy"*, Claredon, Oxford, 1991.
95. R. Kimmich, *"NMR: Tomography, Diffusometry, Relaxometry"*, Springer, Berlin, 1997.
96. J.E. Tanner, *J. Chem. Phys.*, 1978, 69, 1748.
97. E. von Meerwall and R.D. Ferguson, *J. Chem. Phys.*, 1981, 74, 6956.
98. C. Zhao, S. Kuroki and I. Ando, *Macromolecules*, 2000, 33, 4486.
99. S. Matsukawa and I. Ando, *Macromolecules*, 1999, 31, 1865.
100. J.G. Kirkwood, *J. Polym. Sci.*, 1954, 12, 1.
101. M. Doi and S.F. Edwards, *"The Theory of Polymer Dynamics"*, Chapter 8, Clarendon Press, Oxford, 1986.
102. H. Yamakawa, S. Matsukawa, H. Kurosu, S. Kurosu and I. Ando, *J. Chem. Phys.*, 1999, 111, 5129.
103. W. Franklin, *Mol. Cryst. Liq. Cryst.*, 1971, 14, 227.
104. W. Franklin, *Phys. Lett.*, 1974, 48A, 247.
105. W. Franklin, *Phys. Rev.*, 1975, A11, 2156.
106. K.-S. Chu and D.S. Moroi, *J. Phys. Colloq.*, 1975, 36, C1.
107. A.J. Leadbetter, F.P. Temme, A. Heidemann and W.S. Howells, *Chem. Phys. Lett.*, 1975, 34, 363.
108. G.J. Krüger, *Physics Report*, 1982, 82, 229.
109. F. Bueche, *Physical Properties of Polymers*, Interscience Publishers, New York, 1962.
110. G. Kögler, A. Hasenhindl and M. Möller, *Macromolecules*, 1989, 22, 4190.

CHAPTER 8

RECENT EXPERIMENTAL DEVELOPMENTS AT THE NEMATIC TO SMECTIC-A LIQUID CRYSTAL PHASE TRANSITION

ANAND YETHIRAJ

Department of Physics and Physical Oceanography, Memorial University of Newfoundland, St. John's, NL, A1B 3X7, Canada. E-mail: anand@physics.mun.ca

1. INTRODUCTION

Continuous phase transitions are characterized by universality: thermodynamic observables that diverge with power-law exponents whose values are governed by symmetry considerations and are insensitive to other details of the materials [1, 2]. The nematic-smectic-A (NA) transition, where an ordered liquid acquires additional one-dimensional periodicity is one of the outstanding unsolved problems in this field of study [3, 4]. Here the critical behaviour appears non-universal. The order of the transition has been a matter of debate. The complexity of the NA transition arises from an intrinsic coupling between two order parameters. Indeed, even the direct determination of mean-field parameters has been a matter of recent study. Due to this complexity, there are still unresolved issues after more than three decades of research. The subtleties involved have been addressed theoretically via different approximations, leading to a rich addition to the phase transitions literature. Experimentally, these subtleties have inspired precise high-resolution experiments. This article focuses on experimental developments in the last decade that address aspects of the nature of the NA transition.

2. BACKGROUND: THE HIERARCHY OF THEORETICAL APPROXIMATIONS

2.1 Essential Features of the NA Transition

On cooling from the isotropic to the nematic phase, three-dimensional rotational symmetry is spontaneously broken. The average direction of orientation is termed the director. A smectic-A phase has, in addition, one-dimensional positional order

235

A. Ramamoorthy (ed.), Thermotropic Liquid Crystals, 235–248.
© 2007 *Springer.*

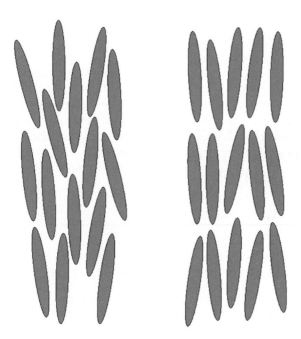

Figure 8-1. A cartoon of the nematic (left) and the smectic-A (right) phase. The nematic phase has only orientational order and no translational periodicity. The smectic-A phase exhibits, in addition, a sinusoidal density modulation ("layering") in the direction of the orientational ordering; this layering suppresses, but does not remove, orientational fluctuations

in the form of layering, with the layer normals being along the director (see cartoon in Figure 8-1). Since the layer normals must lie along the director, the onset of smectic layering has the effect of suppressing fluctuations of the nematic director (see Ref. [3], Chapter 10 for a good introduction).

This intrinsic coupling between nematic and smectic-A order parameters is crucial to the understanding of the NA transition. The nematic order parameter is a symmetric, traceless second-rank tensor, $Q_{ij} = S\left(3\hat{n}_i\hat{n}_j - \delta_{ij}\right)/2$, where S describes the degree of nematic ordering and $\hat{n}$ gives the direction of that ordering.

The smectic phase is envisaged as a one-dimensional density wave of the form

$$(8\text{-}1)\quad \rho(z) = \rho(0)\left[1 + \frac{1}{\sqrt{2}}|\psi(z)|\cos(q_0 z - \phi)\right]$$

where $\frac{2\pi}{q_0}$ is the smectic layer spacing. One can express the smectic order parameter as a complex number $\psi(z) = |\psi(z)|\exp(i\phi)$ whose magnitude is proportional to the amplitude of density modulations in the layered smectic phase and whose phase gives the origin in a given coordinate system.

In the nematic phase, one can induce macroscopic orientation by controlling the surface boundary conditions. However, because nematic ordering is the result of a spontaneously broken symmetry, fluctuations of the director $\hat{n}$ are a soft mode. Indeed a macroscopically oriented nematic phase is much more turbid than a macroscopically oriented smectic-A phase because of light scattering from orientational fluctuation domains. The layered structure of the smectic-A phase suppresses these orientational fluctuations, and it is this coupling that affects the character of the transition (see [5] for a broad survey of such phase transitions). However, smectic phases exhibit one-dimensional orientational order, characterized by the Landau-Peierls fluctuation of the layer spacing [6, 7]. As a result, the essential features needed to capture the NA transition are:

1. Fluctuations of the nematic order parameter S.
2. Fluctuations of the nematic director $\hat{n}$.
3. Phase fluctuations of the complex smectic order parameter ψ.

2.2   Mean-field Theory and Landau–de Gennes Theory

There are several levels of approximation possible in the consideration of the NA transition. First there is the self-consistent mean field formulation due to Kobayashi and McMillan [8–10]. This is an extension to the smectic-A phase of the self-consistent mean-field formulation for nematics ("Maier-Saupe theory" [11]). Kobayashi-McMillan (K-M) theory takes into account the coupling between the nematic order parameter magnitude S with a mean-field smectic order parameter. In Maier-Saupe theory, the key feature of the nematic phase - the spontaneously broken orientational symmetry - is put in by hand by making the pair potential anisotropic. In the same spirit, the K-M formulation puts in by hand a sinusoidal density modulation as well as the nematic-smectic coupling.

In the mean-field picture one expects the phase transition to be generically second order. However, as one adjust materials parameters to reduce the temperature width of the nematic phase, parametrized in K-M theory by the dimensionless ratio T_{NA}/T_{IN}, the coupling drives the transition first order, with the point in material-parameter space where this happens being termed the "Landau Tricritical Point" (LTP).

The Landau-deGennes theory is based on the formal analogy between smectics and superconductors [12, 3]. Within the context of this analogy, there are type-1 and type-2 smectics. As will be seen in the next section, there are well-defined, testable predictions in the extreme type-1 limit. On the other hand, there is also agreement between experiment and the results from a Monte-Carlo simulation by Dasgupta [13] in the extreme type-2 limit. However, the superconducting analogy is imperfect, and in real smectics there appears to be a broad experimental "crossover" region in between these two limits. Varying T_{NA}/T_{IN} (by choosing a variety of liquid crystal preparations with varying nematic range) takes the transition from the small-nematic-range type-1 limit (where the transition is also first order) to the large-nematic-range type-2 limit, where the transition is continuous.

2.3   Thermal Fluctuations

Thermal fluctuations modify critical exponents from the values predicted by mean-field theory [2]. Based on the dimensionality (d = 3) and the order parameter symmetry (n = 2), the NA transition should belong to the isotropic 3DXY (n = 2, d = 3) universality class.

Thermal fluctuations have another effect that is less understood [5]: when two order parameters are simultaneously present and interact with each other, the fluctuations of the additional order parameter may profoundly alter the phase transition of the underlying system. In high-energy physics, for example, such a situation occurs in the Higgs mechanism [2, 14], where the fluctuations of a scalar field can add mass to the soft modes of the underlying transition. In condensed-matter physics, Halperin, Lubensky, and Ma (HLM) [15] predicted 3 decades ago that fluctuations of an additional order parameter could force a system with a second-order phase transition to be first order. The HLM mechanism takes into account the coupling between ψ and the nematic director fluctuations δn. An effective free energy dependent only on ψ is calculated by integrating out the nematic director fluctuations. The theory treats the nematic fluctuations in the Gaussian approximation but is mean field in ψ. They noted two settings where this should occur: the normal-superconducting phase transition in type-1 superconductors and the nematic–smectic-A (NA) transition in liquid crystals [12]. While experiments in superconductors are well-described by mean-field theory, the importance of fluctuations in soft condensed matter makes the effect observable at the NA transition when material parameters have been chosen so as to be in the "type-1" limit.

Order parameters that exhibit large-distance correlations can be influenced strongly by external fields. A recent theoretical extension of the HLM study considered a liquid crystal in the presence of an external field [16] and found that the term in the free energy that is responsible for the HLM first-order character can be suppressed at relatively moderate fields; i.e. 10 T for a magnetic field and 1 V/μm for an electric field. The testing of these predictions is an important recent experimental development, discussed in the next section. It has also been shown [17] that allowing for ψ fluctuations in a type-I superconductor or a liquid crystal always increases the size of the first-order transition. So the HLM prediction that the NA transition is always first order is on firm theoretical ground in the type-1 limit.

With one-dimensional periodicity, the smectic phase cannot exhibit true long-range order due to the Landau-Peierls instability [6, 7]. An anisotropic scaling analysis [18] (see [3], page 521 for a summary) predicts the divergence of the layer compression modulus $B \propto \xi_\parallel / \xi_\perp^2$, thus a divergence with exponent $\phi = \nu_\parallel - 2\nu_\perp$. In addition anisotropic scaling allows for three possibilities for the fixed point of the nematic splay elastic constant: $K_1^* = 0$, implying $A_\nu \equiv \nu_\parallel / \nu_\perp = 1$, K_1^* finite, implying $A_\nu = 2$, and K_1^* infinite, implying $A_\nu > 2$. The Nelson-Toner anisotropic analysis [19] predicts $A_\nu = 2$. All options imply either a strong anisotropy ($A_\nu \equiv \nu_\parallel / \nu_\perp \geq 2$ or no anisotropy at all $A_\nu \equiv \nu_\parallel / \nu_\perp = 1$. Experimentally, a *weak anisotropy* $1.08 < A_\nu < 1.6$ is observed, in contradiction with all the above possisibilities!

The main difficulty with the analogy to superconductors is that the the the additional length associated with the nematic splay elastic constant K_1 breaks gauge symmetry. Thus the splay elastic constant does not diverge at the transition. Results based on the superconducting(SC) analogy need to be transformed back into the liquid crystal (LC) gauge in order to be valid. Even isotropic exponents in the SC gauge could become anisotropic after the transformation [20] to the LC gauge. Patton and Andereck [21, 22] analyzed a model based on the Landau-deGennes free energy functional without invoking the analogy to superconductors. They found that the correlation length exponents parallel and perpendicular to the nematic director have different renormalizations, resulting in different values for the exponents $\nu_\parallel$ and $\nu_\perp$. They found too a curious temperature-dependent crossover from weak to strong anisotropy as one came closer the the transition. While weakly anisotropic critical exponents are clearly the norm in experiments at the NA transition, and this is the only theory to predict them, no quantitative test of Patton-Andereck theory has been achieved. The theory also makes no statements about phase transition order.

Another field-driven fluctuation effect recently predicted is that the fluctuation-induced Casimir-like force between two surfaces bounding a smectic film - one a solid surface and the other a nematic–smectic interface - can be driven from repulsive to attractive in the presence of an external field [23]. Experimentally this should manifest itself by a transition from complete to incomplete wetting at the nematic–smectic interface on increasing an electric field from zero to $10\,\text{kV/mm}$. The presence of this effect has not yet been probed experimentally.

3. EXPERIMENTS PROBING MEAN FIELD PARAMETERS

Experimental developments in the field of liquid crystals are well discussed in a recent collection of reviews [24]. In what follows various recent experimental developments at the NA transition are presented in a form that follows the hierarchy of theoretical approximations within which they were analyzed, beginning with mean-field behaviour.

Realistically, the mean-field K-M formulation is only complete when coupled with experimentally measured values for the model parameters. NMR spectroscopy probes local molecular order. Thus data from NMR spectra are well suited for quantitative fitting to molecular models [25, 26].

3.1 Dissolved Solutes

Molecular order in liquid crystals has been studied extensively by NMR spectroscopy [25, 26] using dissolved solutes as probes of the local environment. Investigations into this problem have used rigid and flexible solutes as probes of the anisotropic environment [26–28] as well as liquid crystal molecules themselves [25]. The dipolar interaction, which averages to zero when the solutes are free to tumble in an isotropic solvent, is non-zero in a liquid-crystalline (anisotropic) solvent giving rise to complex spectra even for simple molecules [29]. These dipolar coupling induced splittings present a wealth of information from which one may calculate the order matrix [26, 30].

The concentration of dissolved solute in NMR studies is typically 1 mole % or larger. Such concentrations can have an effect on more subtle effects such as phase transition order in weakly first-order phase transitions. However, it is an excellent probe of mean-field effects, especially when the effects are probed for different solutes and the results found solute independent. Orientational order in some smectic-A phases changes less rapidly as a function of temperature than in the nematic phase with no break at the phase transition [31–34].

Molecular order in the smectic-A phase was probed recently via a proton NMR study of three aromatic solutes in the liquid crystal 8CB [35]. The results were analyzed in the context of a simple modification of K-M theory for dissolved non-uniaxial solutes. The smectic solute Hamiltonian was written in the form:

$$(8\text{-}2) \quad H_A = -\tau' \cos(2\pi Z/d) + H_N(1 + \kappa' \cos(2\pi Z/d))$$

where κ' and τ' are the nematic-smectic-A coupling and the smectic order parameter magnitude - fit parameters in the experiment. This tested the applicability of a mean-field Hamiltonian that accounted for both nematic and smectic-A ordering and fully determine the Hamiltonian prefactors κ' and τ'. While τ' was temperature-dependent and solute-dependent, the nematic–smectic-A coupling was found to be a solute-independent liquid-crystal quantity: $\kappa' \approx 0.9$. While measurements of the smectic order parameter have been made since McMillan [10] this is the first measurement of the coupling term found by the author.

Recent optical experiments have probed the effect of flexible polymeric solutes in both lyotropic [36] and thermotropic liquid-crystal [37, 38] environments. Long-chain solutes feel an average effective anisotropy, and are found to exhibit much stronger ordering than smaller solutes. The synthesis of the bent-core mesogens ("banana" molecules) has resulted in a novel biaxial smectic-A (both pure phase [39] and solute-induced [40]) as well as nematic phases [41, 42].

An intriguing experiment with a long flexible polymer dissolved in the liquid crystal 8CB [43] has found that while the polymer predominantly aligns along the primary orientation direction (perpendicular to the smectic layers or along the layer normal), a sizeable fraction ($\approx 10\%$) lies in the direction perpendicular to the layer normal. While this would not be surprising if one pictured the smectic structure as a "bookshelf" geometry, the surprise arises because it has been well established that the smectic density is a square-wave but a sinusoidal modulation [44]. The field of solutes in thermotropics is thus seeing a resurgence.

3.2   Carbon-13 and Deuterium NMR Studies

The cleanest way to eliminate the effects of a solute is to have no solute. Recently a method which uses carbon-13 NMR [45] and one that uses deuterium NMR [46] to study the reorientation of a smectic liquid crystal has been reported. Both studies concentrate on macrodomain reorientation in a single phase. The carbon-13 NMR method, which turns off and on near-magic-angle spinning in order to achieve reorientation, can be used in both nematic and smectic-A phases, and can thus be used as a way to probe the NA transition.

The NA transition in a series of homologous liquid-crystalline compounds, nO.m (4-n-alkoxybenzilidene-4'-n-alkylanilines), has also been probed by carbon-13 NMR [47]. The order parameters were determined by a two dimensional technique called separated local field spectroscopy combined with off-magic angle spinning. NMR probes local order parameters of different molecular segments, and so the behaviour of different molecular segments can be followed simultaneously. Changes in the order parameters of the different molecular segments were quantitatively related to the McMillan ratio, and the Landau Tricritical Point in this series of mixtures was located by this technique.

4.   EXPERIMENTS PROBING CRITICAL BEHAVIOUR

A comprehensive review of the study of critical phenomena at the NA transition is not attempted here. The reader is pointed to excellent reviews of this literature (see e.g. Ref. [48] and Ref. [49] and references therein). High-resolution calorimetry results produced a consensus that the NA transition in most liquid crystals is second order. The main issue that remained was to rationalize the large spread in critical exponents observed on different liquid crystals with varying nematic range (characterized by the McMillan ratio T_{NA}/T_{IN}, which is a material-dependent number), with the critical exponents approaching 3DXY values for large nematic range. At smaller nematic range, it appeared (see [48]) that there was a broad crossover in critical exponents from 3DXY to tricritical. The critical exponent anisotropy $A_\nu \equiv \nu_\parallel/\nu_\perp \approx 1.08 - 1.6$ depending on T_{NA}/T_{IN} (Figure 8-2).

This was qualitatively consistent with predictions by Patton and Andereck. But the main prediction of Patton-Andereck theory was that the crossover would be observable on approaching the phase transition in a *single liquid crystal* [21, 22]. Recent high-resolution xray scattering studies have corrected for the effect of mosaicity [50] and were qualitatively consistent with previous results with minor quantitative changes. The removal of mosaicity with a 5 T magnetic field allowed an increase in the range of the study of critical behaviour. Contrary to the Patton-Andereck prediction, the temperature dependence *in any one experiment* was still fit well by a single critical exponent.

A possible resolution of the critical exponent anisotropy was proposed by Bouwman and de Jeu [51]. By not holding the splay correlation length to $\xi_s^4 = c\xi_\perp^4$ in their fits to high-resolution x-ray structure factors, they find larger (and perhaps more realistic) estimates of the uncertainties in the correlation length exponents. Within these larger uncertainties, the spread in experimental values of critical exponents and discrepancies with the 3DXY model are less significant. Errors quoted in the recent experiments by Primak et al are significantly less [50] – yet

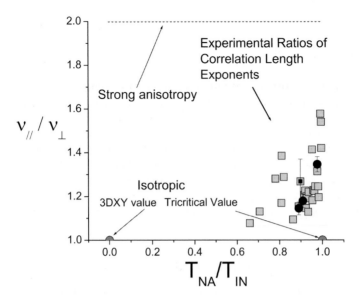

Figure 8-2. Correlation exponent anisotropy ratio $A_\nu \equiv \nu_\parallel / \nu_\perp$ as a function of the ratio T_{NA}/T_{IN}. The grey-filled squares are values from numerous experiments (tabulated in [48]). The square with the large error bar (x-ray scattering experimental result of Bouwman and de Jeu [51]) with errors modified due to a reconsideration of the fitting procedure. The black circles with errors are from recent scattering studies with mosaicity corrections taken into account[50]. Both the 3DXY and the tricritical models predict isotropic exponents ($A_\nu = 1$). Strong anisotropy would predict $A_\nu = 2$

there is appreciable difference ($A_\nu = 1.14 \pm 0.03 - 1.35 \pm 0.05$) in the anisotropy measured at different T_{NA}/T_{IN}.

Yet another technique, measuring ultrasonic velocity anisotropies in the vicinity of the NA transition, however, also find anisotropies consistent with crossover behaviour. Sonntag et al [52] studied the divergence of three elastic constants, the bulk compression constant A, the layer compression constant B and the bulk-layer coupling constant C. B and C have critical exponents that are unequal and are in between that of the 3DXY values and anisotropic scaling values.

Given the above situation with static critical exponents, it is understandable that studies of dynamical critical exponents have been fewer. Dynamical exponents were recently measured in liquid crystals whose nematic range puts them in the "crossover" region. Marinelli et al [53] found that the thermal transport critical exponents did not show any orientational dependence, i.e. they were isotropic. In addition, they found [54] the values obtained to be consistent with a dynamical model also with n = 2 and d = 3.

Much recent work has addressed the vexing question of phase transition order, reviewed in the following section. While clearly the measurements of critical exponents is predicated on the non-existence of a discontinuity, the discontinuities being argued over are small enough so as not to suppress the pre-transitional effects characteristic of a second-order phase transition. However, one cannot rule out subtle effects of one on the other.

5.   EXPERIMENTS PROBING PHASE TRANSITION ORDER

Discussed in this section are the effects of fluctuations and of external fields on the phase transition. Experimentally, the NA transition is usually indistinguishable from second order [48]; however, for materials with a small nematic range, calorimetric measurements have detected a small latent heat associated with the phase change, which is interpreted as a mean-field, second-order phase transition that is driven first order by the coupling to a second, strongly fluctuating order parameter. The second order parameter is associated with a nearby transition, between the isotropic and nematic phases (IN).

One suggested way [55, 56] to compare different experimental probes of phase transition order is to express the phase transition discontinuity in terms of the dimensionless quantity $t_0 = (T_{NA} - T^*)/T^*$, where T_{NA} is the equilibrium NA transition temperature. T^* is the spinodal temperature, where the nematic phase would become unstable. One can measure T^* via almost any physical quantity by extrapolating to the temperature at which critical effects diverge. Because t_0 is positive for a first-order transition and zero in a second-order transition, it is a useful dimensionless measure of the strength of a first-order transition. One may also then use t_0 to estimate latent heats, allowing direct comparison with calorimetry experiments.

5.1   The Landau Tricritical Point and the HLM Effect

Experimental studies in the small-nematic-range limit were carried out by various groups [57–62]. Systematic measurements of the latent heat as a function of T_{NA}/T_{NI} were possible because one can tune the nematic range by mixing two almost-similar liquid crystals with slightly different aliphatic chain lengths. The alkyl-cyanobiphenyl liquid crystals "8CB" and "10CB", and alkyoxy-cyanobiphenyls "8OCB" and "10OCB" have been used extensively (8 and 10 here refer to the length of the aliphatic chain, and the series of liquid crystal with varying n is referred to as a homologous series). In the de Gennes-McMillan theory (i.e., taking only the δS-ψ coupling into account), the latent heat should vary linearly with $\delta x = x - x^*$, for small δx, where x is the mixture concentration and x^* is the concentration where the latent heat vanishes and is thus, in the context of this Landau theory, a tricritical point. By convention, x^* is known as the Landau tricritical point (LTP). Brisbin et al [57] and Thoen et al [58] showed that this is true well above x^*. Reanalyzing data [60] from mixtures of 8CB and 10CB, Anisimov et al [61] showed that the latent heat did not go to zero at the LTP but crossed over nonlinearly to a measurable, non-zero value. This indicated that something other than $\delta S - \psi$ coupling was also important, consistent with the HLM predictions. Tamblyn et al [63, 64] demonstrated a similar

Table 8-1. Latent Heats measured in pure 8CB by different techniques. A comparison of latent heats (error bars in parentheses) from Intensity fluctuation microscopy (IFM; latent heats are estimated from the measured ($t_0 = 1.2 \pm 0.1) \times 10^{-5}$ [66] and Landau parameters in [65]), Adiabatic Scanning Calorimetry (ASC; errors from [69]) and Modulated differential scanning calorimetry (MDSC; errors estimated from smaller error bar). Also listed is the molar fraction X of 10CB in 8CB-10CB (alkyl-cyanobiphenyl) mixtures at which the latent heat appears to vanish

Technique	Latent heat	Tricritical point
IFM [66]	5.25 (0.50) J/kg	NONE
ASC [60, 69]	0.0 (1.8) J/kg	X = 0.314
MDSC [67]	0.0 (17.0) J/kg	X = 0.22

dependence via capillary length (ratio of the surface tension to the latent heat) measurements on 8CB-10CB mixtures.

In addition, Anisimov et al and Cladis et al [61, 62] introduced a different technique for studying weakly first order transitions. They measured the propagation speed of a NA front after a step jump in temperature. The velocity was a linear function of the temperature jump ΔT. For second-order transitions, the front propagation speed should be proportional to $\Delta T^{1/2}$. This result implied that the NA transition was first order for all 8CB-10CB mixtures, including pure 8CB.

The above result was confirmed by Yethiraj et al [56, 66] using an optical technique that quantified intensity fluctuations in the liquid crystal. Since the intensity fluctuations in the nematic phase are caused by nematic director fluctuations, which in turn are suppressed in the smectic-A phase, this is a very sensitive probe of the phase transition. In addition, the temperature dependence of the phase transition was obtained optically by placing the sample in a well-calibrated temperature gradient. Indeed the existence of a sharp static interface is itself a qualitative indicator of first-order character. Not only was the transition in the 8CB-10CB system always first order, the strength of the discontinuity was *larger* not smaller than the HLM prediction. This result is consistent with a subsequent calculation by Herbut et al which takes into account ψ fluctuations.

Lafouresse et al [67] and Sied et al [68] have used Modulated Differential Scanning Calorimetry (MDSC) to measure latent heats as a function of concentration in two liquid crystal mixtures (8CB-10CB and 8OCB -10OCB). In both cases, they obtain results quantitatively differing from previous calorimetry results.

Experimental measurements of zero (or non-zero) latent heat should always be coupled with an estimate for experimental resolution (see Table 8-1). A recent adiabatic scanning calorimetry ("ASC") study [69] reiterates the results of an older study [58] and puts an upper bound on the NA transition in 8CB at < 2 J/kg. In the presence of relatively large impurity concentrations (~5 mole % of cyclohexane in 8CB) ASC does measure a discontinuity of 17 J/kg [70] (here the estimated errors are 5 J/kg). The stated errors in the MDSC data are large ($\approx 17\,J/kg$ as estimated from the smallest error bar on the graph) and thus consistent with both the intensity fluctuation microscopy (IFM) results of Yethiraj et al and the ASC result within errors. The ASC results are, however, clearly at odds with the IFM results of Yethiraj et al. The source of the discrepancy is unclear. It seems safe to say that the last word on phase transition order has not been spoken.

5.2   External Fields

External fields (surfaces, electric, or magnetic fields) can have an important effect on liquid crystal alignment. However, strong enough fields can also have an effect on liquid crystal phase behaviour. The work of Lelidis et al [71, 72] has demonstrated that nematic phases can be induced in a system that exhibits an isotropic and a smectic-A phase (and close in the phase diagram to the emergence of a nematic phase). Such a nematic phase has been termed "non-spontaneous" (NSN). It turns out that the NSN-A transition can also be either first or second order, and that the discontinuity can be suppressed by increasing the electric field even more! Comparable results have been found by Basappa et al [73] who were able to suppress the range of the smectic phase via an external electric field. In both cases, care had to be taken to use pulsed alternating electric fields to simultaneously prevent charge migration and prevent heating effects. The heating effects put a limit on the experimental resolution of the measurement.

Lelidis also reports [74] confirmation of the electric-field-assisted suppression of the HLM effect (predicted by Mukhopadhay et al [16]) with an important point of departure. The theory predicted a critical field of 10 T (magnetic) or 1 V/μm (electric). Since this is the point when the discontinuity is completely suppressed, one would expect to see an effect for much smaller fields. However, Yethiraj et al found no trace of a field-induced suppression in magnetic fields upto 1.5 T, suggesting a minimum critical field of 33 T. This would correspond to a critical electric field of atleast 3.3 V/μm. The critical field reported in Ref. [74] was 20 V/μm, i.e. an order of magnitude higher. This result is indeed *experimentally* consistent with the earlier findings of Yethiraj et al. However, given the strength of the field required, there exists a question if the effect observed is truly the subtle HLM effect. The modified HLM theory [16] also predicts an increase in the critical field as the zero-field phase transition discontinuity (characterized, for example, by the dimensionless temperature t_0) increases. This can be achieved experimentally in a homologous mixture such as 8CB-10CB where increasing 10CB concentration increases the zero-field discontinuity. Measuring the critical electric field as a function of the concentration in mixtures might be a good test of the HLM mechanism in this case.

6.   THE LANDAU-PEIERLS INSTABILITY IN SMECTICS

A smectic-A liquid crystal is expected to exhibit algebraic decay of the layer correlations rather than true long-range order [6, 7]. In x-ray scattering, the smectic Bragg peaks would be expected to be power-law singularities of the form $q_\parallel^{-2+\eta}$ and $q_\perp^{-4+\eta}$. Distinguishing between power-law singularities and "delta function" peaks broadened by diffuse scattering from acoustic modes requires an instrument with a resolution function whose wings drop off much faster than the power law/s. Als-Nielsen et al [44] found in a high-resolution x-ray diffraction study of the density wave of the smectic-A phase that the smectic-A Bragg peaks in the liquid-crystal 8OCB were indeed consistent with the predicted power-law singularity form. While a non-vanishing (i.e. a non-divergence to zero) of the layer compression modulus has been reported in second-sound measurements [75], the validity of the interpretation of the measurements has been contested via other measurements using the same technique [76]. In addition, Martinoty et al [77] found, using dynamic compression measurements, that the layer compression modulus exhibits a single power law divergence (to zero) on approaching the phase transition over 4 decades in reduced temperature.

7.   SUMMARY

The nematic-smectic-A (NA) transition is one that has been studied theoretically with fluctuations being accounted for within different levels of approximation. It has also been studied experimentally by a diverse array of high-resolution techniques in laboratories around the world. While much has been understood about the transition, almost every probe of the NA transition, whether mean-field behaviour of solutes, nature of divergences and values of critical exponents or phase transition order, has met with conflicting experimental results. It appears that another generation of resolution and precision enhancement is required before the complete story is told. As remarked several years ago by deGennes and Prost [3]: "It seems that we almost understand, but not quite".

8.   ACKNOWLEDGEMENTS

The author thanks John Bechhoefer, Elliott Burnell, Ronald Dong and Ranjan Mukhopadhyay for several discussions over the years on issues related to this review. This work was supported by NSERC.

9.   REFERENCES

1. S.K. Ma, *Modern Theory of Critical Phenomena*, Frontiers in Physics; 46 (Addison-Wesley,1976).
2. P. Pfeuty and G. Toulouse, *Introduction to the Renormalization Group and to Critical Phenomena*, 1st ed. (Wiley-Interscience, Chichester, 1978).
3. P.G. de Gennes and J. Prost, *The Physics of Liquid Crystals*, 2nd ed. (Clarendon Press, Oxford, 1993).
4. S. Chandrasekhar, Liquid Crystals (Cambridge University Press, 1992).
5. D. Belitz, T.R. Kirkpatrick, T. Vojta, Rev. Mod. Phys. **77** , 579 (2005).
6. R.E. Peierls, Helv. Phys. Acta Suppl. **7**, 81 (1934).
7. L.D. Landau, in Collected Papers of L.D. Landau, edited by D. ter Haar (Gordon and Breach, New York, 1965), p. 209.

8. K.K. Kobayashi, Mol. Cryst. Liq. Cryst. 13 (1971) 137.

9. W.L. McMillan, Phys. Rev. A 4 (1971)1238 .

10. W.L. McMillan, Phys. Rev. A 6 (1972) 936.

11. W. Maier, A. Saupe, Z. Naturforsch. **A13**, 564 (1958); **A14** 882 (1959); **A15** 287 (1960).

12. P.G. de Gennes, Solid State Comm. **10**, 753 (1972).

13. C. Dasgupta, Phys. Rev. Lett. **55**, 1771 (1985); J. Phys. (Paris) **48**, 957 (1987).

14. C. Itzykson and J.B. Zuber, *Quantum Field Theory*, McGraw Hill, N.Y., 1980. pp. 612–614.

15. B. I. Halperin, T.C. Lubensky, and S.K. Ma, Phys. Rev. Lett. **32**, 292 (1974).

16. R. Mukhopadhyay, A. Yethiraj, and J. Bechhoefer, Phys. Rev. Lett. **83**, 4796 (1999).

17. I.F. Herbut, A. Yethiraj, J. Bechhoefer, Europhys. Lett., **55**, 317 (2001).

18. T.C. Lubensky, J.H. Chen, Phys. Rev. B **17**, 366 (1978).

19. D.R. Nelson, J. Toner, Phys. Rev. B **24**, 363 (1981).

20. T.C. Lubensky, J. de Chimie Physique **80**, 32 (1983).

21. B.R. Patton and B.S. Andereck, Phys. Rev. Lett. **69**, 1556 (1992).

22. B.S. Andereck and B.R. Patton, Phys. Rev. E. **49**, 1393 (1994).

23. I.N. de Oliveira and M.L. Lyra, Phys. Rev. E /bf 70, 050702 (2004).

24. S. Kumar, Liquid Crystals: Experimental Study of Physical Properties and Phase Transitions (Cambridge, 2000).

25. R.Y. Dong, Nuclear Magnetic Resonance of Liquid Crystals (Springer Verlag, New York, 1994).

26. NMR of Ordered Liquids, ed. E.E. Burnell, C.A. de Lange (Kluwer Academic, Dordrecht, The Netherlands, 2003).

27. G. Celebre, G. De Luca, M. Longeri, Mol. Phys. **98**, 559 (2000).

28. E.E. Burnell, C.A. de Lange, Chem. Rev. **98**, 2359 (1998).

29. A. Saupe, G. Englert, Phys. Rev. Lett. **11**, 462 (1963).

30. P. Diehl, C.L. Khetrapal, NMR, Basic Principles and Progress. Vol. 1 (Springer, Berlin, 1969).

31. Z. Luz, S. Meiboom, J. Chem. Phys. **59**, 275 (1973).

32. N.A.P. Vaz, J.W. Doane, J. Chem. Phys. **79**, 2470 (1983).

33. D. Catalano, C. Forte, C.A. Veracini, J.W. Emsley, G.N. Shilstone, Liquid Cryst. **2**, 357 (1987).

34. F. Barbarin, J.P. Chausse, C. Fabre, J.P. Germain, B. Deloche, J. Charvolin, J. Phys-Paris **44**, 45 (1983).

35. A. Yethiraj, Z. Sun, R. Y. Dong, E. E. Burnell, Chem. Phys. Lett. **398**, 517 (2004).

36. Z. Dogic et al, Phys. Rev. Lett., **92**, 125503 (2004).

37. R.K. Lammi, K.P. Fritz, G.D. Scholes, P.F. Barbara, J.Phys.Chem. B **108**, 4593 (2004).

38. S. Link, D. Hu, W.-S. Chang, G.D. Scholes, P.F. Barbara, Nano Lett., **5**, 1758 (2005).

39. C.V. Yelamaggad et al Angew. Chem. Int. Ed. **43**, 3429 (2004).

40. R. Pratibha, N.V. Madhusudana, B.K. Sadashiva, Science **288**, 2184 (2000).

41. L.A. Madsen, T.J. Dingemans, M. Nakata, E.T. Samulski, Phys. Rev. Lett. **92**, 145505 (2004).

42. B.R. Acharya, A. Primak, S. Kumar, Phys. Rev. Lett. **92**, 145506 (2004).

43. S. Link, W.–S. Chang, A. Yethiraj, P.F. Barbara, Phys. Rev. Lett. **96**, 17801, (2006).

44. J. Als-Nielsen, J.D. Litster, R.J. Birgeneau, M. Kaplan, C.R. Safinya, A. Lindegaard-Andersen, S. Mathiesen, Phys. Rev. B, **22**, 312 (1980).

45. M.L. Magnuson, B.M. Fung, J. Chem. Phys. **100**, 1470 (1994).

46. J.W. Emsley, J.E. Long, G.R. Luckhurst, P. Pedrielli, Phys. Rev. E **60**, 1831 (1999).

47. B.M. Fung, M.L. Magnuson, T.H. Tong, M.S. Ho, Liquid Crystals, **14**, 1495 (1993).

48. C.W. Garland and G. Nounesis, Phys. Rev. E **49**, 2964 (1994).

49. W.G. Bouwman and W.H. de Jeu, in *Modern Topics in Liquid Crystals*, edited by Agnes Buka (World Scientific Pub. Co. Pte. Ltd., Singapore, 1993), pp. 161–186.

50. A. Primak, M. Fisch, S. Kumar, Phys. Rev. E **66**, 051707 (2002).

51. W.G. Bouwman, W.H. de Jeu, Phys. Rev. Lett. **68**, 800 (1992).

52. P. Sonntag, D. Collin, P. Martinoty, Phys. Rev. Lett. **85** 4313 (2000).

53. M. Marinelli, F. Mercuri, S. Foglietta, U. Zammit, F. Scudieri, Phys. Rev. E **54**, 1 (1996).

54. M. Marinelli, F. Mercuri, U. Zammit, F. Scudieri, Phys. Rev. E **53**, 701 (1996).

55. A. Yethiraj, J.L. Bechhoefer, Mol. Crystals Liquid Crystals **304**, 301 (1997).

56. A. Yethiraj and J. Bechhoefer, Phys. Rev. Lett. **84**, 3642 (2000).

57. D. Brisbin, R. DeHoff, T.E. Lockhart, and D.L. Johnson, Phys. Rev. Lett. **43**, 1171 (1979).

58. J. Thoen, H. Marynissen, and W. Van Dael, Phys. Rev. A **26**, 2886 (1982).

59. J. Thoen, H. Marynissen, and W. Van Dael, Phys. Rev. Lett. **52**, 204 (1984).

60. H. Marynissen, J. Thoen, and W. Van Dael, Mol. Cryst. Liq. Cryst. **124**, 195 (1985).

61. M.A. Anisimov, V.P. Voronov, E.E. Gorodetskii, and V.E. Podnek, JETP Lett. **45**, 425 (1987).

62. P.E. Cladis, W. van Saarloos, D.A. Huse, J. S. Patel, J. W. Goodby, and P. L. Finn, Phys. Rev. Lett. **62**, 1764 (1989).

63. N. Tamblyn, P. Oswald, A. Miele, and J. Bechhoefer, Phys. Rev. E **51**, 2223 (1995).

64. N. Tamblyn, M.Sc. thesis, Simon Fraser University, 1994.

65. M.A. Anisimov, P.E. Cladis, E.E. Gorodetskii, D.A. Huse, V.E. Podneks, V.G. Taratuta, W. van Saarloos, and V.P. Voronov, Phys. Rev. A **41**, 6749 (1990).

66. A. Yethiraj, R. Mukhopadhyay, J. Bechhoefer, Phys. Rev. E **65**, 021702 (2002).

67. M.G. Lafouresse et al, Chem. Phys. Lett. **376** 188 (2003).

68. M.B. Sied et al, J. Phys. Chem. B , 109, 16284 (2005).

69. P. Jamée, G. Pitsi, J. Thoen. Phys. Rev. E **67**, 031703 (2003).

70. K. Denolf, B. Van Roie, C. Glorieux, J. Thoen, Phys. Rev. Lett **97**, 107801 (2006).

71. I. Lelidis, Phys. Rev. Lett. **73**, 672 (1994).

72. I. Lelidis, J. Phys. II **6**, 1359 (1996).

73. G. Basappa, A.S. Govind, N.V. Madhusudana, J. Phys. II, 7, 1693 (1997).

74. I. Lelidis, Phys. Rev. Lett. **86**, 1267 (2001).

75. M. Benzekri, T. Claverie, J.P. Marcerou. J.C. Rouillon, Phys. Rev. Lett. **68**, 2480 (1992).

76. P. Martinoty, P. Sonntag, L. Benguigui, D. Collin, Phys. Rev. Lett., **73**, 2079 (1994).

77. P. Martinoty, J.L. Gallani, D. Collin, Phys. Rev. Lett. **81**, 144 (1998).

CHAPTER 9

LIQUID CRYSTALLINE CONJUGATED POLYMERS – SYNTHESIS AND PROPERTIES

KAZUO AKAGI

Department of Polymer Chemistry, Graduate School of Engineering,
Kyoto University, Katsura, Nishikyo-ku, Kyoto 615-8510, Japan

1.   INTRODUCTION

Among a variety of electrically conducting polymers reported so far, polyacetylene is the most typical conjugated polymer [1] and it shows the highest electrical conductivity with an order of $10^4 \sim$ to 10^5 S/cm upon an iodine doping [2]. However, it is insoluble and infusible, which makes it difficult to evaluate molecular weight and chemical structure and also to process the polymer in solution such as casting. Introduction of alkyl or aromatic substituents into the polyacetylene chain makes the polymer soluble in organic solvents when the alkyl chain length is sufficiently long, i.e., hexyl, pentyl and octyl groups [3]. Electrical conductivity of substituted polyacetylene, however, is significantly lower than that of non-substituted polyacetylene. This is due to a less coplanarity of the main chain arising from steric repulsions between substituents, as well as a higher ionization potential and a lower electron affinity. In addition, even in the substituted polyacetylene the main chain is still randomly oriented, which depresses the observed electrical conductivity of the polymer.

   If the substituent is a liquid crystalline (LC) group, the polymer would be not only soluble in organic solvents, but also easily aligned by spontaneous orientation of the LC group. Besides, it could be macroscopically aligned by an external perturbation such as shear stress, electric or magnetic force field. This situation means that a mono domain structure of the LC phase is constructed on a macroscopic level. Under such a circumstance, the polymer is expected to show a higher electrical conductivity, compared with the case of random orientation. At the same time, one can control the molecular orientation and hence the electrical conductivity of the polymers by the external force. In Figure 9-1 are schematically described both

A. Ramamoorthy (ed.), Thermotropic Liquid Crystals, 249–275.
© 2007 *Springer.*

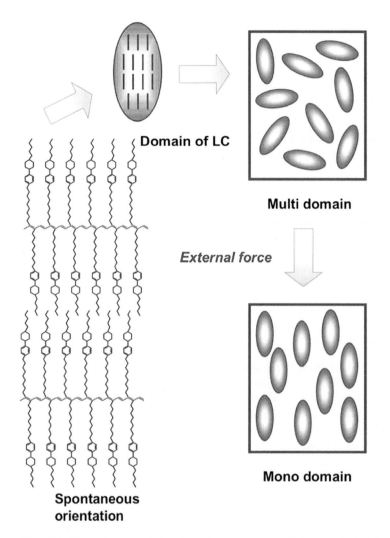

Domain of LC

Multi domain

External force

Mono domain

Spontaneous
orientation

Figure 9-1. Schematic representation of spontaneous and magnetically forced orientations of side chain liquid crystalline conjugated polymer

spontaneous orientation and externally forced macroscopic alignment of side chain type LC conjugated polymers, where the former and the latter generate multi and mono domains of LC phases, respectively.

Currently, a series of LC polyacetylene derivatives have been synthesized and characterized from aspects of thermal and electrical properties [4]. For these novel LC conducting polymers, the macroscopic alignment has been first achieved under magnetic force field [4e]. Since a concept of LC conducting polymers is versatile, it is straightforward to apply the concept to other kinds of conjugated polymers. In fact, following LC polyacetylene derivatives, which are to be reviewed in more details,

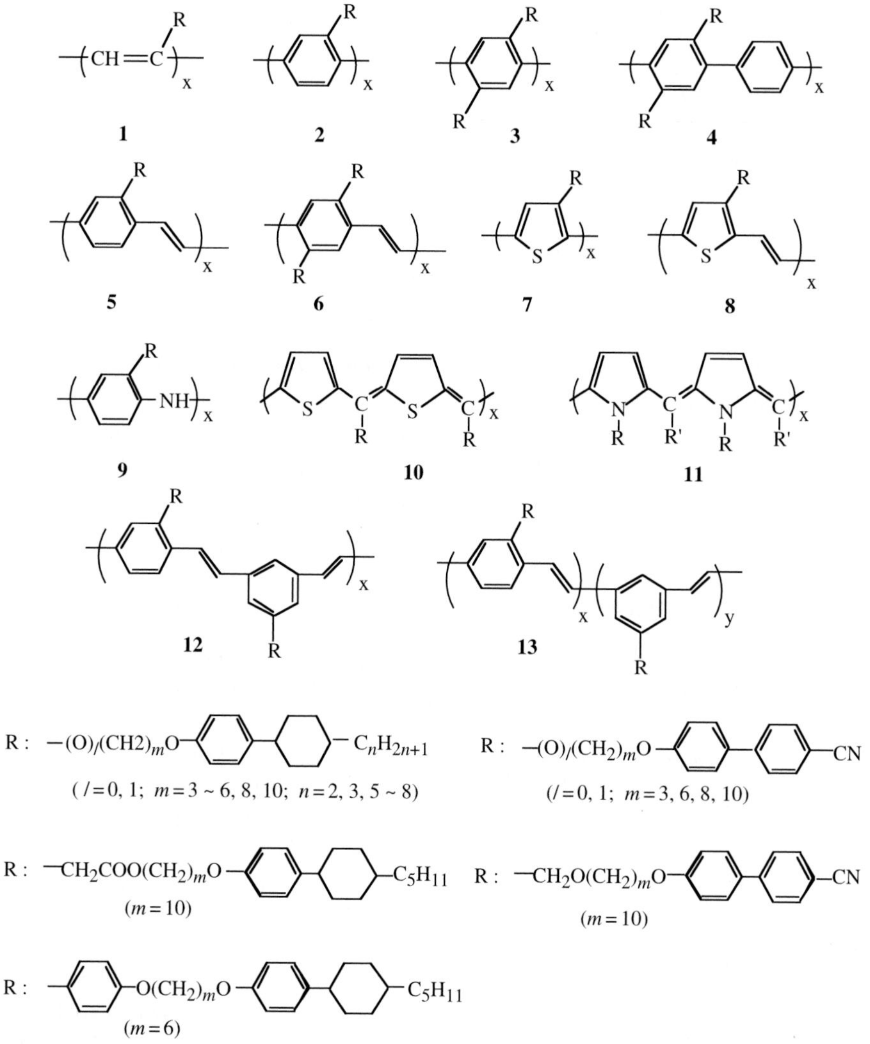

Scheme 1

various types of LC polymers have been synthesized in these decades. Scheme 1 shows representative LC conducting polymers, e.g., LC polyene derivatives [5] with a six-membered cyclic backbone structures such as poly(1,6-heptadiyne) and poly(dipropargylamine), polythiophene derivatives [6] with mesogenic groups at 3-position of thiophene rings, LC polypyrrole derivatives [7] with mesogenic groups at 3-position or N-position of pyrrole rings, LC poly(*para*-phenylene) [8] derivatives, LC poly(*para*-phenylenevinylene) derivatives [8] and LC polythienylenvinylene derivatives [9] with mesogenic groups at phenylene or thiophene rings, LC polyaniline derivatives [10] with mesogenic groups at ortho-position of aniline ring.

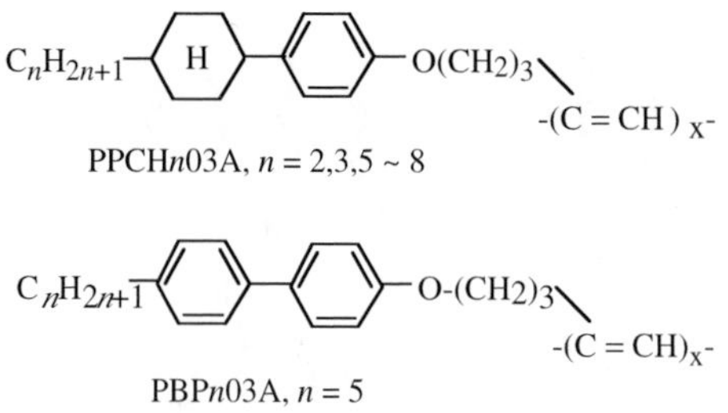

C_nH_{2n+1} —(H)—( )— $O(CH_2)_3$ ⟍
                                        $-(C = CH)_x-$

PPCH*n*03A, $n = 2,3,5 \sim 8$

C_nH_{2n+1} —( )—( )— $O-(CH_2)_3$ ⟍
                                         $-(C = CH)_x-$

PBP*n*03A, $n = 5$

Scheme 2

Subsequently, small-band gap LC conjugated polymers [11] based on benzonoid and quinoid resonance structures were synthesized. Subsequently, ferroelectric LC (FLC) conjugated polymers [12] consisting of acetylene, thiophene, phenylene, or phenylenevinylene unit cell were synthesized in order to achieve highly quick response to an electric force field during macroscopic alignment. These FLC conjugated polymers, however, will not be further mentioned here and reviewed elsewhere.

1.1   LC Polyacetylene Derivatives [4–18]

The liquid crystalline substituent of the monomer is composed of the phenylcyclohexyl (PCH) moiety as a mesogenic core, methylene chain linked with an ether-type oxygen atom, $-(CH_2)_3O$, as a spacer, and an alkyl chain ($-C_nH_{2n+1}$, $n = 2, 3, 5 \sim 8$) as a terminal group. This monomer is abbreviated as PCH*n*03A, where A stands for acetylene segment. Another kind of substituent consists of a biphenyl (BP) mesogenic, a methylene chain linked with oxygen as a spacer, and an n pentyl group as a terminal moiety. This type of monomer is abbreviated as BP*n*03A ($n = 5$). The monomers were polymerized with Ziegler-Natta, metathesis, and rhodium-based catalysts. The corresponding polymers are abbreviated as PPCH*n*03A and PBP*n*03A, as shown in Scheme 2.

The present monomers and polymers were characterized by means of infrared (IR), UV-Vis, ^{1}H- and ^{13}C-nuclear magnetic resonance (NMR), elemental analyses, DSC, and polarizing optical microscopy. Molecular weights of the polymers were evaluated by means of gel-permeation chromatography (GPC) using polystyrene standards, and electrical conductivities upon iodine doping for the cast films of the polymers were measured with a four-probe method. Morphologies were examined through measurements of scanning electron microscopy (SEM).

2.   PREPARATION

2.1   Syntheses of Monomers

(i) The monomer of *para*-(trans-4-alkylcyclohexyl)phenoxy-1-pentyne, PCH*n*03A, was synthesized through reaction of *para*-(trans-4-alkylcyclohexyl)phenol

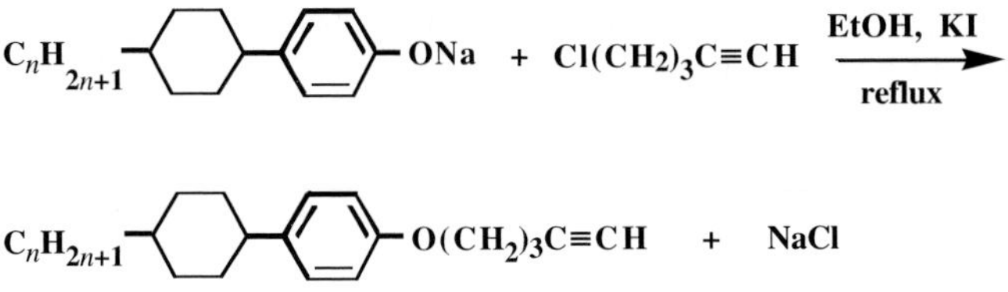

C_nH_{2n+1} ⬡—⬢—ONa + Cl(CH$_2$)$_3$C≡CH $\xrightarrow[\text{reflux}]{\text{EtOH, KI}}$

C_nH_{2n+1} ⬡—⬢—O(CH$_2$)$_3$C≡CH + NaCl

PCHn03A (n = 2, 3, 5 ~ 8)

Scheme 3.1

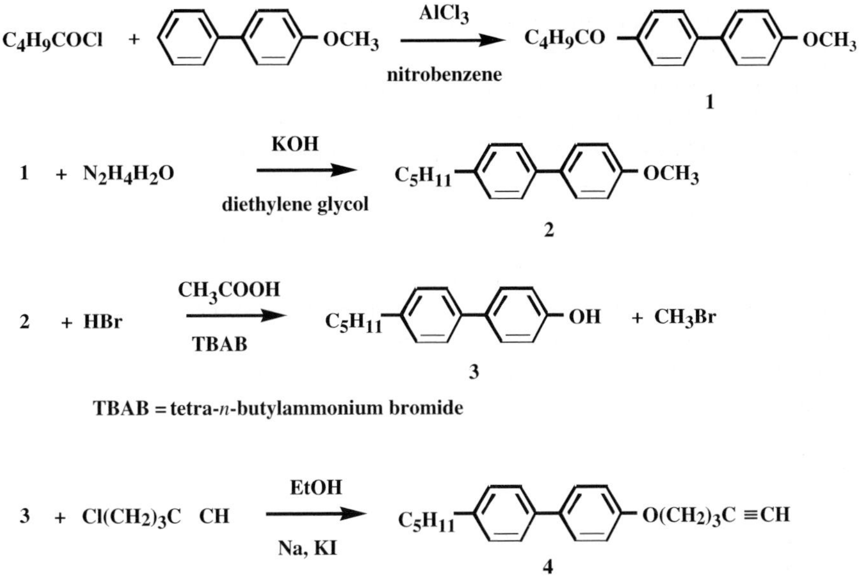

C$_4$H$_9$COCl + ⬢—⬢—OCH$_3$ $\xrightarrow[\text{nitrobenzene}]{\text{AlCl}_3}$ C$_4$H$_9$CO—⬢—⬢—OCH$_3$

1

1 + N$_2$H$_4$H$_2$O $\xrightarrow[\text{diethylene glycol}]{\text{KOH}}$ C$_5$H$_{11}$—⬢—⬢—OCH$_3$

2

2 + HBr $\xrightarrow[\text{TBAB}]{\text{CH}_3\text{COOH}}$ C$_5$H$_{11}$—⬢—⬢—OH + CH$_3$Br

3

TBAB = tetra-n-butylammonium bromide

3 + Cl(CH$_2$)$_3$C CH $\xrightarrow[\text{Na, KI}]{\text{EtOH}}$ C$_5$H$_{11}$—⬢—⬢—O(CH$_2$)$_3$C ≡CH

4

BPn03A (n = 5)

Scheme 3.2

and 5-chloro-1-pentyne in sodium ethoxide by using potassium iodide as a catalyst (Scheme 3).

(ii) Four steps of synthesis for BP503A were necessary. First, 4-methoxy-4'-pentanoylbiphenyl (**1**) was synthesized through Friedel-Crafts reaction between 4-methoxybiphenyl and pentanoyl chloride in nitrobenzene using aluminium chloride as a catalyst. The carbonyl moiety of **1** was reduced into a methylene group by hydrazine and potassium hydroxide in diethylene glycol to yield 4-methoxy-4'-pentylbiphenyl, (**2**). The methoxy group of **2** was changed into a hydroxy one using hydrogen bromide in acetic acid, where tetra-n-butylammonium bromide (TBAB) was employed. The etherification of the product, 4'-pentyl-4-biphenlyol (**3**), was carried out through coupling reaction with 5-chloro-1-pentyne in sodium ethoxide using potassium iodide as a catalyst, yielding 5-(4-pentyl-4'-biphenylol)-1-pentyne (**4**), that is BP503A (Scheme 3).

2.2 Polymerizations of Monomers

The polymerizations of PCHn03A and BP503A monomers were carried out by using Ziegler – Natta catalyst [19], Fe(acac)$_3$-AlEt$_3$, and metathesis catalyst [20], MoCl$_5$ – Ph$_4$Sn in toluene for 21 h at room temperature. For some monomers, chlorine-bridged rhodium complex [21], [Rh(NBD)Cl]$_2$ was also used with NEt$_3$ as a co-catalyst, in which the polymerization time was within 6 minutes at room temperature [4(f)–(g)]. The polymerizations were terminated by pouring the reaction mixture into a large amount of methanol. The polymers were filtered off and washed with methanol and then dried under vacuum. The polymerization yield and molecular weight evaluated from GPC measurement are summarized in Table 9-1. The MoCl$_5$-Ph$_4$Sn catalyst gave higher yields of polymers than the Fe(acac)$_3$-AlEt$_3$ catalyst. The high number-average molecular weight of $M_n = 10^5$ of the Fe-based PPCH803A polymer is worth to be emphasized. Note that although the Fe-based PBP503A polymer was insoluble in tetrahydrofuran (THF) and benzene in the range of room temperature to 90° C, it became soluble into cumene by heating up to 120° C for 30 min. The polymer dissolved has a low viscosity and relatively small molecular weight of $3000 < M_n < 4000$. This may be due to degradation during the heating process.

Table 9-1. Polymerization yields and molecular weights of PPCHn03A and PBP503A prepared with Fe(acac)$_3$ – AlEt$_3$ and MoCl$_5$ – Ph$_4$Sn catalysts (acac = acetylacetonate)

Polymer[a]	Fe(acac)$_3$ - AlEt$_3$[b]				MoCl$_5$ - Ph$_4$Sn[c]			
	Yield	M_n	M_w	M_w/M_n (%)	Yield	M_n	M_w	M_w/M_n (%)
PPCH303A	71	5.5×10^5	2.6×10^6	4.7	61	1.5×10^4	3.4×10^4	2.3
PPCH503A	70	4.5×10^5	1.8×10^6	4.0	67	1.4×10^4	3.3×10^4	2.4
PPCH803A	67	5.2×10^5	2.6×10^6	5.0	73	1.2×10^4	2.7×10^4	2.3
PBP503A	65	3.5×10^3	1.0×10^4	2.9	85	7.5×10^3	6.8×10^4	9.1

[a]: Polymerized in toluene for 21 h at room temperature. [Monomer] = 3 mmol, [Fe(acac)$_3$] = 0.01 mol/l, [AlEt$_3$]/[Fe(acac)$_3$] = 6, [MoCl$_5$] = 0.01 mol/l, [Ph$_5$Sn]/[MoCl$_5$] = 0.5.
[b]: Polymers are soluble in organic solvents such as CHCl$_3$, THF, and benzene at $50 \sim 60°$ C. For the measurements of GPC spectra, the polymers were first dissolved in THF with heating and then allowed to cool to room temperature.
[c]: Polymers are soluble in the same solvents as mentioned above at room temperature.

3. PROPERTIES

3.1 Liquid Crystalline Phases of Monomers

Liquid crystallinity of the PCH*n*03A monomer was examined with a miscibility test. Figure 9-2 shows a phase diagram of the mixture of PCH803A monomer and 1-hexyloxy-4-(*trans*-4-propylcyclohexyl)benzene, PCH306. The latter is known to have a nematic liquid crystalline phase and therefore is used as a reference.

In the miscibility phase diagram the mixture is homogeneous, irrespective of the mole ratio between the two compounds. This indicates that the liquid crystalline phase of the PCH803A monomer is nematic. The result is supported by the experimental facts that the PCH803A monomer shows a mesophase in DSC and it exhibits a Schlieren texture characteristic of the nematic phase in a polarizing optical microscope. Similarly, the PCH*n*03A monomers with long terminal alkyl chain lengths ($n = 5 \sim 7$) were also confirmed to have nematic liquid crystalline phases. It should be noted that the above mentioned mesophases of the monomers were observed only in the cooling process but not in the heating one, indicating that the monomers are monotropic liquid crystals.

3.2 Liquid Crystalline Phases of Polymers

All PCH and BP polymers with liquid crystalline groups in the side chain exhibited fan-shaped textures and uniaxial conoscopic patterns in a polarizing optical microscope that are characteristic of the smectic A phase. Typical fan-shaped textures of the polymers prepared with Fe-based and Mo based catalysts are shown in Figure 9-3.

It was found that the domain size of the liquid crystalline phase increased with lengthening of terminal alkyl chain length in the side-chain of the PCH polymer. Meanwhile, the domain size of the Mo-based PBP503A polymer was larger that of

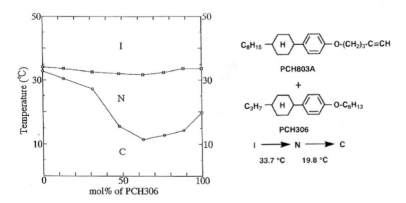

Figure 9-2. Phase diagram of mixture of PCH803A and PCH306 (I = isotropic; N = nematic; C = crystalline)

(a)

(b)

(c)

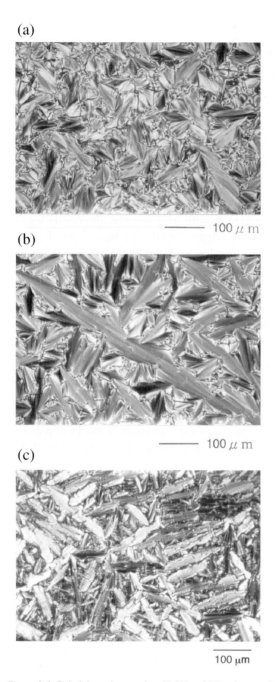

Figure 9-3. Polarizing micrographs of PCH and BP polymers showing fan-shaped textures. (a) Fe-based PPCH803A polymer at 132°C; (b) Mo-based PPCH803A polymer at 130°C; (c) Mo-based PBP503A polymer at 120°C

the Fe-based polymer. This is consistent with the experimental fact that the latter is thermally more stable than the former.

DSC measurements of the polymers were carried out, as shown in Figure 9-4. It has been recently elucidated that the PCH and BP polymers prepared with Ziegler-Natta catalyst such as $Fe(acac)_3$-$AlEt_3$ assume a cis form, while those obtained with a metathesis catalyst such as $MoCl_5$-Ph_4Sn exhibit a trans form [13]. The Fe-based PPCH803A exhibited a large exothermic peak due to a cis-trans isomerization at $150 \sim 190°$ C in the first heating process. An endothermic peak around $170°$ C was hidden by the isomerization peak. In the first cooling, a peak of isotropic to liquid crystal phase transition was observed at $146°$ C, and that of liquid crystal to solid phase at $98°$ C. In the second heating, the liquid crystal phase was observed in a temperature range of $100 \sim 128°$ C. Phase transition temperatures and enthalpy changes of some representative polymers are shown in Table 9-2.

In the PCH polymer, the enthalpy change due to the smectic to isotropic phase increases with an increase of the terminal alkyl chain length. This means that the liquid crystal state is more stabilized by lengthening the alkyl chain as a terminal moiety of the liquid crystalline group.

Meanwhile, the stabilization of the liquid crystal state in the BP polymer is relatively small compared with the PCH polymer. Mesophases in the PPCH*n*03A and PBP503A polymers were observed in both heating and cooling processes. This indicates that the polymers, irrespective of the catalyst used in the polymerization, are enantiotropic liquid crystals. Here it is of interest to emphasize that the PCH*n*03A monomers ($n = 5 \sim 8$) show nematic phases, while all the polymers exhibit smectic phases where an order is higher than in the nematic phase. This can be rationalized with the so-called polymerization effect, i.e., a higher order in molecular arrangement is generated by the polymerization of the liquid crystalline monomer.

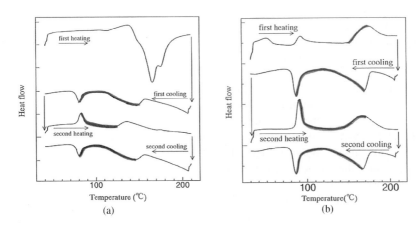

Figure 9-4. DSC curves of PPCH803A polymers prepared with (a) $Fe(acac)_3$ − $AlEt_3$ and (b) $MoCl_5$ − Ph_4Sn catalysts. The rate of temperature change in the heating or cooling process is 5°C/ min

Table 9-2. Phase transition temperature (°C) and enthalpy changes (J/g) in PPCH*n*03A and PBP503A polymers

Cat.[a] Polymer	First cooling			Second heating		
	k	s_A	i	k	s_A	i
Fe	PPCH303A	60 (0.7)	107 (4.5)	70 (0.7)		102 (3.3)
	PPCH503A	98 (1.1)	146 (6.3)	100 (1.3)		128 (6.2)
	PPCH803A	81 (3.2)	151 (8.2)	83 (3.3)		142 (5.8)
	PBP503A	126 (3.2)	180 (3.1)	129 (4.9)		171 (1.0)
Mo	PPCH303A[b]		133 (4.2)			134 (3.7)
	PPCH503A[b]		160 (6.2)			160 (4.5)
	PPCH803A	86 (3.4)	171 (6.7)	91 (3.7)		166 (6.4)
	PBP503A	132 (4.4)	189 (4.5)	135(6.7)		180 (-)

k = crystalline s_A = smectic A   i = isotropic
[a]: Fe and Mo denote catalysts of $Fe(acac)_3$ - $AlEt_3$ and $MoCl_5$ - Ph_4Sn, respectively.
[b]: No distinct DSC peak corresponding to the phase transition between crystalline and smectic A phases was observed.

3.3 Higher Order Structure and Stereospecific Configuration

Figure 9-5 shows an XRD pattern of Fe-based PPCH503A. Two diffraction peaks, sharp and broad ones, are observed in small- and wide-angle regions, respectively.

Such a profile of XRD is usually encountered in a smectic LC phase. Similar XRD patterns were also observed in other polymers subjected here. The results of XRD analyses are summarized in Table 9-3. Diffraction angles of sharp and broad peaks, 2.35 and 18.6 degrees in 2θ, were evaluated to be 37.5 and 4.76 Å, respectively. These values well correspond to an inter-layer distance of layered structure of smectic state and a distance between side chains, as shown in Figure 9-6(a).

These assignments were supported by molecular mechanics calculations based on finite systems of LC polyacetylene derivatives; The inter-layer and side-chain distances were calculated to be 41 and 5 Å, respectively. The present analyses enabled us to draw a polymer structure of Fe-based PPCH503A where the LC side-chains are alternatively located at both sides of polyene chain to form a stereoregular configuration of head-head-tail-tail linkage, as described also in Figure 9-6(b). At the same time, the polymer structure implies that the LC side-chains have no tilt to the main chain, leading to a typical picture of smectic A phase. As found in Table 9-3, all of the polymers have the side-chain distances of 4.6 ~ 4.8 Å. This implies that the side-chain distances are irrespective of differences of LC substituents. Meanwhile, the inter-layer distance increases as the terminal alkyl chain or the methylene chain is lengthened; It increases by 6.4 Å from Mo-based PPCH503A to PPCH803A, or by 5.6 Å from Rh-based PPCH503A to PPCH506A. This also supports the present assignment that the sharp peak at small angle region is due to a diffraction between inter-layers of smectic A form.

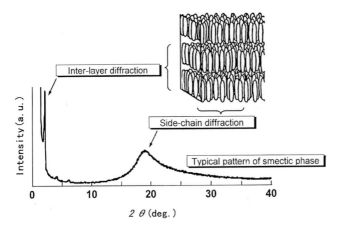

Figure 9-5. X-ray diffraction pattern of Fe-based PPCH503A

Here it is worthwhile to figure out how the above-mentioned stereoregular config-uration is formed. The previous studies using ¹H-NMR, DSC, and ESR [13(b)] confirmed that as-prepared LC polyacetylene derivatives with Fe- or Rh-based catalyst and Mo-based one have cis and trans forms, respectively. The cis form is thermally unstable and is easily isomerized into the trans form during the first heating process in DSC or polarizing optical microscope measurements. Taking account of it, we propose a possible polymerization mechanism followed by a cis-trans isomerization, as described in Figure 9-7. It depicts that the LC-substituted acetylene monomers approaching to a catalytically active site are polymerized with

Table 9-3. XRD results of LC polyacetylene derivatives

Polymer	Tail [a]	Spacer [a]	Catalyst [b]	Isomerization [c]	LC phase	Inter-layer distance (Å)	Side-chain distance (Å)
PPCH503A	5	3	Fe	cis → trans	S_A	37.5	4.76
			Rh	cis → trans	S_A	38.0	4.73
			Mo	trans	S_A	40.2	4.55
			(cal.) [d]	–	–	$(\sim 41)^d$	$(5.0)^d$
PPCH506A	5	6	Rh	cis → trans	S_A	43.6	4.71
PPCH803A	8	3	Mo	trans	S_A	46.6	4.75

[a] Number of the carbon atoms in the terminal alkyl moiety or methylene spacer moiety of LC side chain.
[b] Fe, Mo, and Rh denote catalysts of $Fe(acac)_3 - AlEt_3$, $MoCl_5 - Ph_4Sn$, $[Rh(NBD)Cl]_2 - NEt_3$, respectively.
[c] Fe- and Rh-based catalysts and Mo-based one give cis-rich and trans-rich polyacetylene derivatives, respectively.
[d] Values in the parenthesis are based on molecular mechanics calculations for finite systems.

(a)

(b)

(c)

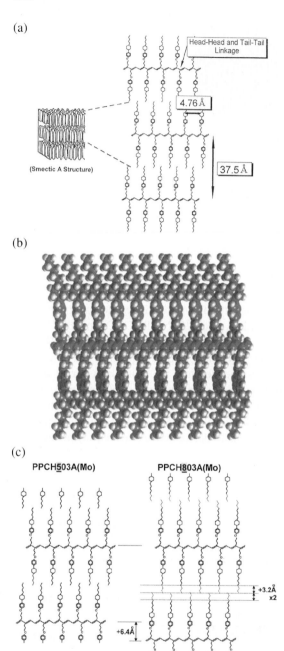

Figure 9-6. (a) Higher order structure of Fe-based PPCH503A and smectic A structure. (b) Structure of PPCH503A optimized by molecular mechanics (MM) calculation on finite system. (The bowls of red, gray, and purple indicate carbon, hydrogen, and oxygen atoms, respectively.) (c) Comparison of interlayer distance between PPCH503A and PPCH803A

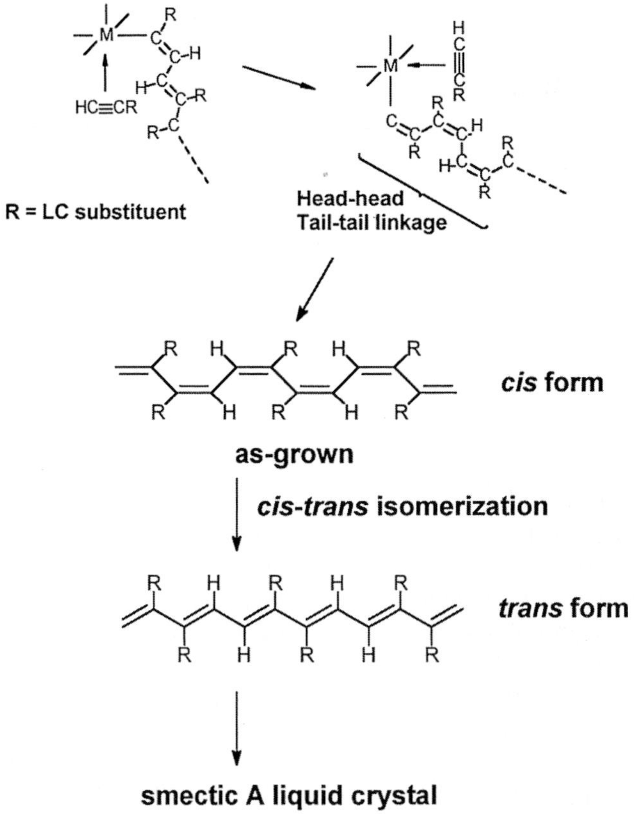

Figure 9-7. Possible polymerization mechanism of Ziegler-Natta catalyst for LC-substituted acetylene and cis-trans isomerization

a sequence of head-head and tail-tail linkages, yielding a stereoregular cis form. In cases of Fe- or Rh-based catalysts, the cis form is a kinetic product and hence it is converted into the trans form that is a thermodynamic product. In case of Mo-based catalyst, the cis to trans isomerization should occur simultaneously during the polymerization.

3.4   Absorption Spectra and Chemical Doping

Next, UV-Vis absorption spectra of PCH503A monomer and PPCH503A polymer prepared with Fe-based catalyst were examined. An absorption band at 270 ~ 280 nm, observed in both the monomer and the polymer, was assigned to the $\pi \rightarrow \pi^*$ transition of the benzene moiety of the liquid crystalline group. The absorption band at 320 ~ 330 nm in the polymer was assigned to the $\pi \rightarrow \pi^*$ transition in

Table 9-4. Absorption band (λ_{max}) and intensity (ε_{max}) and proton chemical shift (δ) of PPCHn03A and PBP503A polymers

Catalyst	Polymer	λ_{max} (nm)[a]	ε_{max} ($\times 10^3$)[a]	δ (ppm)[b]
	PPCH303A	324	2.0	5.98
Fe(acac)$_3$ –	PPCH503A	325	1.9	5.97
AlEt$_3$	PPCH803A	324	2.5	5.97
	PBP503A	insoluble		
	PPCH303A	301	1.4	
MoCl$_5$ –	PPCH503A	301	1.4	no value
Pn$_4$Sn	PPCH803A	301	1.2	
	PBP503A	305	2.6	5.94 ∼ 6.14

[a]: Absorption band due to $\pi \rightarrow \pi^*$ transition in conjugated polyene chain.
[b]: Chemical shift of olefinic proton in cis form.

the conjugated polyene chain. Broadening of the band implies that the polymer is composed of finite polyenes with various conjugation lengths. The results from the absorption spectra for the representative polymers are summarized in Table 9-4.

The Mo-based polymer gave an absorption band located at shorter wave length (around 300 nm) than the Fe-based polymer, indicating that the former has a shorter conjugation length in the polyene chain than the latter. Here it should be noted that the microscopic UV-Vis measurement of the inside of the liquid crystalline domain in the Fe-based PPCH803A polymer gave an absorption band at 403 nm, which is shifted by 70 nm towards a longer wavelength, compared with the band observed in solution of the polymer [14]. The result demonstrates that the polyene chains are aligned owing to the spontaneous orientation of the liquid crystalline side chains, and that the effective conjugation lengths of polyene chains inside of each domain are longer than those in a randomly oriented case such as in solution, probably due to an increase of co-planarity of the polyene chains. Therefore, from a view point of enhancement of electrical conductivity, it is of particular importance to perform a macroscopic alignment of the polymer that is a so-called mono domain structure of the liquid crystalline state by using an external perturbation, as will be mentioned in the next section (see also Figure 9-1).

Absorption spectra of the Fe-based PPCH803A polymer were measured in situ with iodine doping, for which the polymer had been cast onto an inner wall of a quartz cell. The results are shown in Figure 9-8. Since the ionization potential of substituted polyacetylene is generally larger than that of non-substituted polyacetylene, it might take more time to complete the iodine doping of the polymers. Actually, there was no notable change in the absorption spectrum even after 3 hr from an exposure of iodine to the PPCH803A polymer.

After about 5 hr, the absorption band at 320 ∼ 330 nm, assigned to a $\pi \rightarrow \pi^*$ transition in conjugated polyene, slightly decreased in intensity. At the same time, new bands appeared at 420 and 700 nm. Interestingly, these bands were also observed in gas-phase doping with sulfuric acid. It is therefore evident that the

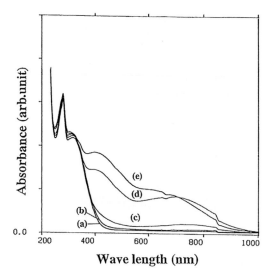

Figure 9-8. Changes of UV-Vis absorption spectra upon gas-phase iodine doping for the cast film of Fe-based PPCH503A polymer. (a) intact ; (b) 30 min; (c)171 min; (d) 291 min; (e) 1131 min

two bands originate from a chemically doped conjugated polyene chain and can be assigned to two kinds of electronically allowed transitions in positively charged polyene segments. Thus the feasibility of chemical doping to the present polymers was confirmed.

3.5 Molecular Orientation and Electrical Conductivity

3.5.1 Orientation by shear stress

Figure 9-9 shows SEM photographs of the Fe-based PPCH503A polymer. Figure 9-9(a) shows as grown randomly oriented fibrils. Figures 9-9(b) ~ (d) show aligned fibrils that are generated by a shear stress (along the horizontal direction of paper) in the liquid crystalline state of the polymer.

Figure 9-10(a) shows polarizing micrographs of the Mo-based PPCH503A polymer. It is evident that an optical texture of batonet, characteristic of a smectic liquid crystal, is oriented along the direction of the shear stress. From these results, we can confirm that the side chain liquid crystalline conjugated polymers are macroscopically aligned by the shear stress as an external force.

3.5.2 Orientation by magnetic field

Electrical conductivity of the polymer was measured with the four-probe method. The polymer was first melted by heating on the substrate where gold had been vaporized with a form of four leads, and it was gradually cooled down to the liquid

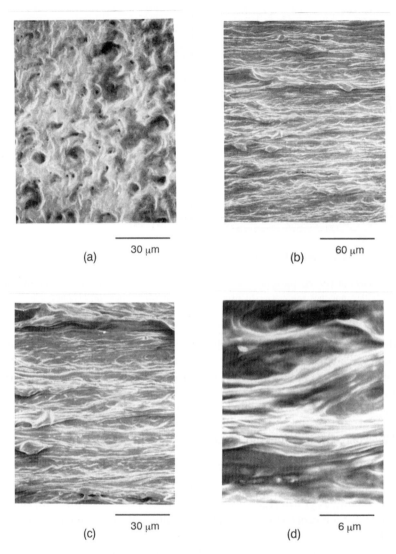

(a) 30 μm

(b) 60 μm

(c) 30 μm

(d) 6 μm

Figure 9-9. SEM photographs of Fe-based PPCH503A polymer. (a) no shear stress ; (b) ∼ (d) aligned with shear stress

crystalline temperature by applying an external magnetic field of 0.7 ∼ 1.0 Tesla. After the alignment of the polymer was completed in the liquid crystalline state, the polymer was further cooled down to room temperature to give a solid sample available for conductivity measurement. Schematic representation of magnetically forced alignment of the polymer and current direction in the four-probe measurement are shown in Figure 9-11, where $\sigma_{//}$ and $\sigma_{\perp}$ stand for the conductivities parallel

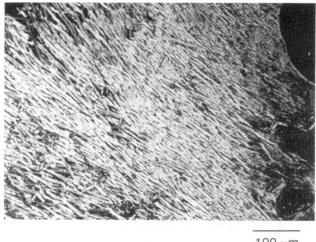

100 μm

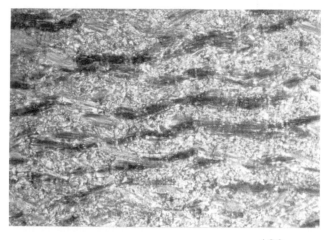

——— 100 μ m

Figure 9-10. Polarizing micrographs of (a) Mo-based PPCH503A polymer aligned by shear stress, and (b) Fe-based PPCH803A polymer aligned by magnetic force field of 0.7 Tesla, where magnetic field is parallel to the horizontal direction

and perpendicular to the direction of the polymer main chain. The results are summarized in Table 9-5. Conductivities of undoped and iodine doped samples were $10^{-11} \sim 10^{-9}$ and $10^{-8} \sim 10^{-7}$ S/cm, respectively. Through the magnetically forced alignment, the conductivity of the doped sample was further enhanced by 1–2 orders of magnitude, giving a value of 10^{-6} S/cm. Especially the aligned

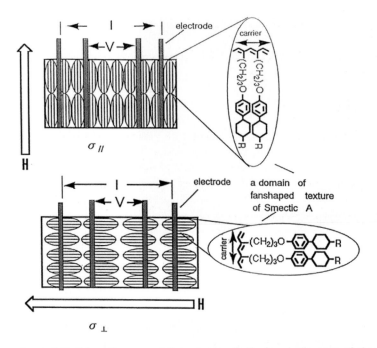

Figure 9-11. Schematic representation of magnetically forced alignment of the side chain liquid crystalline mono-substituted polyacetylene and sample cell for the four-probe method of electrical conductivity measurement (**H** is magnetic field)

sample of the Fe-based PPCH503A polymer showed an increase of 10^2 times in parallel conductivity, i.e., from 10^{-8} to 10^{-6} S/cm. As to the PBP503A polymer, the increase due to the alignment was only about 2 times.

Meanwhile, the perpendicular conductivity of the Mo-based PPCH503A and PBP503A polymers decreased by $10^2 \sim 10^4$ as expected. However, those of the Fe-based PPCH503A polymer showed a reverse trend; it increased by 70 times. Such a sample dependency of conductivity resulted in a diverse of the electrical anisotropy ($\sigma_{//}/\sigma_{\perp}$) ranging from $2 \sim 10^5$. It may be worthy noting that the increase

Table 9-5. Electrical conductivities (σ) of PPCH503A and PBP503A polymers before and after iodine doping; unit of σ : S/cm

Catalyst [a]	Polymer	Undoped	Iodine doped		
		σ	σ_{random}	$\sigma_{//}$	$\sigma_{\perp}$
Fe	PPCH503A	$10^{-11} \sim 10^{-10}$	1.0×10^{-8}	1.3×10^{-6}	7.2×10^{-7}
Mo	PPCH503A	$10^{-11} \sim 10^{-9}$	1.3×10^{-7}	1.5×10^{-6}	1.4×10^{-11}
Fe	PBP503	$10^{-11} \sim 10^{-9}$	1.0×10^{-8}	2.3×10^{-8}	1.2×10^{-10}

a) Fe : Fe(acac)$_3$ - AlEt$_3$, Mo : MoCl$_5$ - Ph$_4$Sn

of perpendicular conductivity after macroscopic alignment, as mentioned in the case of the Fe-based PPCH503A polymer, is partly due to an enhancement of coplanarity of the conjugated polyene chain accompanied by the alignment. This suggests that in analyzing the change of conductivity upon the molecular orientation, an in-plane alignment of the polyene chain must be considered at the same time, as well as a so-called parallel alignment of the polyene chain [15].

Figure 9-10(a) shows polarizing optical micrographs of the Fe-based PPCH803A polymer after the magnetically forced alignment. The magnetic field is applied to the horizontal direction of the photograph. One can see that the texture is aligned parallel to the magnetic field, although the alignment is not homogeneous. The incomplete alignment is attributed to an insufficient strength of the magnetic field such as $0.7 \sim 1.0$ Tesla, which may cause a variation of the electrical conductivity and sample dependency, especially in the case of perpendicular conductivity. In order to achieve a complete orientation of the polymer in the highly viscous smectic liquid crystalline state, the magnetic strength of more than 2 Tesla should be required at least.

3.6 Spin State and Chemical Doping

Figure 9-12 shows ESR spectra of the PCH and BP polymers in non-doped and iodine-doped states [16b]. The results are also summarized in Table 9-6. In the non-doped state, the Fe-based PCH and BP polymers showed no signal, meanwhile the Mo-based PCH polymer showed two kinds of signals with g-values of 2.014 and 2.005. This is consistent with the above mentioned result that the Fe-based Ziegler-Natta and the Mo-based metathesis catalysts produce cis and trans forms of

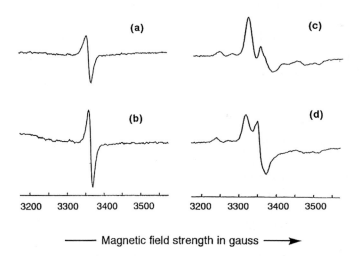

Figure 9-12. ESR spectra measured at room temperature. (a) I_2-doped PPCH503A (Fe); (b) I_2-doped PBP503A (Fe); (c) non-doped PPCH503A (Mo); (d) I_2-doped PPCH503A (Mo)

Table 9-6. ESR results of the iodine-doped LC polyacetylene derivatives

Catalyst	Polymer	g-Value	ΔH_{pp}[a] (gauss)	$\Delta H_{1/2}/\Delta H_{pp}$	Spin density (spins / g)	y-Value [b]
Fe	PPCH503A	2.006	15.1	2.05	3×10^{15}	0.065
Fe	PBP503A	2.003	11.0	2.36	1×10^{15}	0.097
Mo	PPCH503A	2.014, 2.005	—[c]	—[c]	2×10^{17}	0.028

[a] ΔH_{pp} means a full width at maximum slope, corresponding to a peak to peak width in differential curve of signal. $\Delta H_{1/2}$ means a full width at half maximum.
[b] Dopant concentration defined as $[(CHCR)^{+3y}(I_3{}^-)_y]_x$, where R is a liquid crystal group.
[c] Two peaks are overlapped.

mono-substituted polyacetylenes, respectively. Namely, the cis form is a kinetically favored product and it would be easily isomerized into a thermally more stable trans form during a thermal heating, chemical doping, or even polymerization using the metathesis catalyst. Such a cis-trans isomerization should cause defects in polyene chains, giving rise to unpaired electrons responsible for paramagnetic behavior. In practice, upon the iodine doping, the Fe-based PCH and BP polymers showed sharp ESR signals with g-values of 2.006 and 2.003, respectively. On the other hand, in the Mo-based polymer doped with iodine, the lower field signal ($g = 2.014$) slightly decreased in intensity and the higher field signal ($g = 2.005$) increased, resulting in overlapped signals as seen in Figure 9-12.

The peak width, ΔH_{pp}, is larger by 3 ~ 5 times and spin density is smaller by 2 ~ 4 orders of magnitude than those of iodine-doped polyacetylene [22]. Such a large peak width suggests that the unpaired electrons in the substituted polyacetylene tend to be more localized, probably due to an increase of non-planarity of the polyene chain. From the view point of electrical transport phenomenon, it can be argued that the present polymers have lower mobility and lower concentration of carriers. This may account for the lower electrical conductivity of the substituted polyacetylenes including the present PCH and BP polymers.

3.7   Orientation Behavior and Anisotropy under Magnetic Field

We measured fused state ^{13}C-NMR spectra of side-chain liquid crystalline poly acetylene derivatives as well as a liquid crystalline acetylene monomer in isotropic and liquid crystalline states to elucidate their orientation behaviors and anisotropies in chemical shifts [16]. Measurements of the fused state ^{13}C-NMR spectra were carried out with Bruker AMX-40 equipped with a high magnitude of proton decoupler. The strength of magnetic field was 9.4 Tesla. Procedure of the measurement was as follows; The sample was first melted by heating and measured in the isotropic state. Then by applying the magnetic field, the sample was gradually cooled down with a temperature controller and measured in the liquid crystalline state.

Figure 9-13 shows ^{13}C-NMR spectra of PCH803A monomer in the isotropic and nematic phases. Upon the phase transition from the isotropic to liquid crystalline phase, chemical shifts of aromatic carbons (C6 ~ C9) of phenyl ring shifted to a lower magnetic field, which will be called a down-field shift. Especially, those of C6 and C9 carbons showed relatively larger down-field shifts. The results are described in Figure 9-14. On the other hand, chemical shifts of alkynyl carbons of acetylene segments (C1 and C2) shifted to a higher magnetic field, which is called a up-field shift. The up-field shifts are also found in alkyl carbons of terminal group and methylene chain, although the degrees of the shifts are quite small. These tendencies

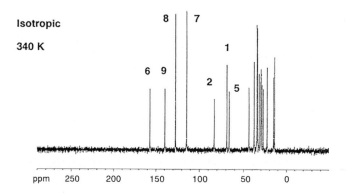

PCH803A Monomer

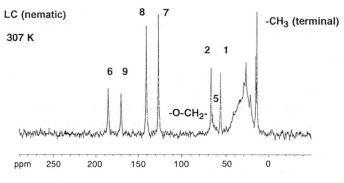

Figure 9-13. ^{13}C-NMR spectra of PCH803A monomer in the isotropic (upper) and nematic (lower) states

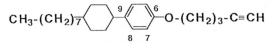

PCH803A Monomer

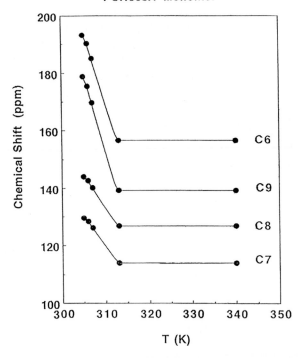

Figure 9-14. Carbon chemical shifts of phenylene ring of PCH803 monomer in the isotropic and nematic states

in chemical shift become more remarkable as temperature decreases from 307 K to 305 K, indicating an increase in degree of molecular orientation. Interestingly, the methylene carbon (C5) neighboring with the phenoxy moiety shows almost no change in chemical shift even after the phase transition. This suggests that the C5 carbon atom might be located in an axis which forms an angle of about 53.5 Å (magic angle) to the molecular axis.

Figure 9-15 shows ^{13}C-NMR spectra of PPCH803A polymer in the isotropic and smectic phases. The polymer in the smectic phase has a high viscosity and low fluidity, which causes a broadening of signals. Upon the phase transition, the phenyl carbons (C6 ~ C9) also show the down-field shifts. It is intriguing that the down-field shifts in the polymer are larger than in the case of the monomer, as summarized in Table 9-7.

That is, the down-field shifts of C6, C9 and C7, C8 carbons of the polymer are 48 ~ 50 ppm and 19 ~ 20 ppm, respectively. Whereas, those of the monomer are 37 ~ 40 ppm and 15 ~ 17 ppm, respectively. The results imply that the liquid

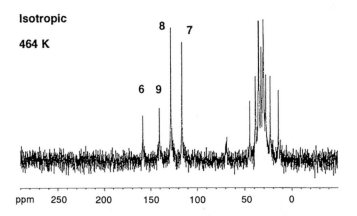

PPCH803A Polymer

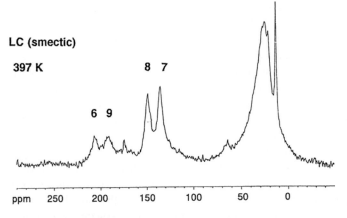

Figure 9-15. ^{13}C-NMR spectra of PPCH803A polymer in the isotropic (upper) and smectic (lower) states

crystalline state of the polymer gives a larger anisotropic environment than that of the monomer. This can be rationalized with a difference of liquid crystalline phase between the polymer and the monomer where the former and the latter are respectively smectic and nematic phases. That is, higher order and hence larger anisotropy are generated in the smectic phase, compared with the nematic phase.

Table 9-7. Changes of chemical shifts upon transition from isotropic to LC phase for PCH803A monomer and Fe-based PPCH803A polymer (units of ppm)

	C6	C7	C8	C9	Terminal (Alkyl)
σ_{aniso}	206.6	135.0	148.4	190.5	12.4
σ_{iso}	158.3	115.8	127.7	140.2	14.1
$\Delta\sigma$	48.3	19.2	20.7	50.3	− 1.7
$\Delta\sigma$	36.6	15.5	17.0	39.5	− 0.5
(Monomer)					

$\Delta\sigma = \sigma_{aniso} - \sigma_{iso}$

$$CH_3\text{-}(CH_2)_7 \hspace{-0.2em}\underset{8 \quad 7}{\overset{9 \quad 6}{\bigcirc\!\!-\!\!\bigcirc}}\hspace{-0.2em}\text{-O-}(CH_2)_3\diagdown$$

$$-(C\!\!=\!\!CH)_n\text{-}$$

PPCH803A Polymer

We evaluated chemical shifts semi-quantitatively in the isotropic and anisotropic states by focusing on shielding tensors of di-substituted benzene, 1-alkoxy-4-alkylbenzene, used as a model compound. The calculated isotropic (σ_{iso}) and anisotropic (σ_{aniso}) chemical shifts were 157.4 ppm and 191.3 ppm, respectively. The difference between them, defined as $\Delta\sigma = \sigma_{aniso} - \sigma_{iso}$, was 39.6 ppm. The tendency of the down-field shift and its magnitude well account of the experimental results for the PCH803A monomer, where δ of the C6 carbon for instance was 39.5 ppm. Taking account of electron density distributions of sp^3, sp^2 and sp hybrid carbon atoms in a cartesian coordinate and chemical shift tensors reported so far, one can find that the largest shielding for aromatic carbons lies in the perpendicular direction to the aromatic ring. For aliphatic carbons, however, the smallest shielding lies in this direction. Such an anisotropy of the shielding produces changes of chemical shifts observed upon the phase transition from the isotropic to liquid crystalline states. The changes of chemical shifts indicate that the monomer and the side chain of the polymers subjected here are oriented parallel to the direction of magnetic field employed in NMR measurement, which means that liquid crystalline domains are macroscopically aligned to give a mono domain structure as shown in Figure 9-1.

We also evaluated orientational orders of liquid crystalline state of the monomer and polymer. By correlating the order parameter with the change in chemical shift associated with isotropic – anisotropic phase transition, we obtained the order parameter of the monomer, $S_{mono} = 0.38(307\,K) \sim 0.49(305\,K)$, and then derived a conventional formulae as follows,

$$S_{polym} = 1.26 \cdot S_{mono}.$$

Thus, the order parameter of the polymer, S_{polym}, was calculated to be 0.62 when the $S_{\text{mono}} = 0.49$. The value of 1.26 in the above equation means that the orientational order increases upon the polymerization of liquid crystalline monomer, and therefore this value can be regarded as an index of polymerization effect in the orientational order.

4. CONCLUSION

We have synthesized two kinds of liquid crystalline polyacetylene derivatives by using Fe-based Ziegler-Natta, Mo-based metathesis and Rh-based catalysts. The polymers showed enantiotropic smectic A phases by virtue of a spontaneous orientation of liquid crystalline side chain composed of phenylcyclohexyl or biphenyl mesogenic moieties. It was found from XRD measurements that the LC side chains are alternatively located at both sides of the polyene chain, giving rise to a stereoregular sequence such as head-head-tail-tail linkage. Electrical conductivities measured with the four-probe method were $10^{-8} \sim 10^{-7}$ S/cm upon iodine doping to the cast films of the polymers. The alignment of the main chain accompanied with the side chain orientation using the external magnetic force of $0.7 \sim 1.0$ Tesla enhanced the electrical conductivity up to 10^{-6} S/cm and gave rise to a notable electrical anisotropy. It was shown from the fused state ^{13}C NMR measurements using a superconducting magnet of 9.4 Tesla that the liquid crystalline side chains of the polymer were completely oriented along the magnetic field, giving a mono domain structure.

5. ACKNOWLEDGEMENTS

The author is grateful to Professor Emeritus H. Shirakawa and Dr. H. Goto (University of Tsukuba) and Dr. K. Araya (Hitachi Ltd.) for their collaborations. Acknowledgements are also given to Prof. J. Isoya (University of Tsukuba) for ESR measurements, Prof. A. Kawaguchi (Ritsumeikan University) for XRD analyses, and Dr. T. Nishizawa and Dr. K. Masuda (Kobe Steel Ltd.) for the fused state ^{13}C NMR measurements. This work was supported by a Grant-in-Aid for Science Research in a Priority Area, "Super-Hierarchical Structures" (No. 446) from the Ministry of Education, Culture, Sports, Science and Technology, Japan.

6. REFERENCES

1. (a) T. Ito, H. Shirakawa, and S. Ikeda, *J. Polym. Sci., Polym. Chem. Ed.*, **12**, 11 (1974); (b) J.C.W. Chien, "Polyacetylene – Chemistry, Physics and Material Science", Academic Press, Orlando, FL (1984), Chap. 2.
2. (a) H. Naarrmann and N. Theophilou, *Synth. Met.*, **22**, 1 (1987); (b) K. Akagi, M. Suezaki, H. Shirakawa, H. Kyotani, M. Shimomura, and Y. Tanabe, *Synth. Met.*, **28**, D1 (1989); (c) J. Tsukamoto, A. Takahashi, and K. Kawasaki, *Jpn. J. Appl. Phys.*, **29**, 125 (1990).

3. (a) W. J. Trepka and R. J. Sonnenfeld, *J. Polym. Sci. A-1*, **8**, 2721 (1970); (b) R.J. Kern, *J. Polym. Sci. A-1*, **7**, 621 (1969); (c) C. I. Simionescu, and V. Percec, *J. Polym. Sci.: Polym. Symp.*, **67**, 43 (1980).

4. (a) S.-Y. Oh, R. Ezaki, K. Akagi, and H. Shirakawa, *J. Polym. Sci., Part A: Polym. Chem.*, **31**, 2977 (1993); (b) S.-Y. Oh, K. Akagi, H. Shirakawa, and K. Araya, *Macromolecules*, **26**, 620 (1993); (c) K. Akagi, S.-Y. Oh, H. Goto, Y. Kadokura, and H. Shirakawa, *Trans. Mat. Res. Soc. Jpn.*, **15A**, 259 (1994); (d) H. Shirakawa, Y. Kadokura, H. Goto, S.-Y. Oh, K. Akagi, and K. Araya, *Mol. Cryst. Liq. Cryst.*, **255**, 213 (1994); (e) K. Akagi, H. Goto, Y. Kadokura, H. Shirakawa, S.-Y. Oh, and K. Araya, *Synth. Met.*, **69**, 13 (1995); (f) H. Goto, K. Akagi, H. Shirakawa, S.-Y. Oh, and K. Araya, *Synth. Met.*, **71**, 1899 (1995); (g) K. Akagi, H. Goto, S. Fujita, and H. Shirakawa, *J. Photopolym. Sci. Tech.*, **10**, 233 (1997); (h) K. Akagi, H. Goto, J. Murakami, S. Silong, and H. Shirakawa, *J. Photopolym. Sci. Tech.*, **12**, 269 (1999); (i) H. Goto, S. Nimori, and K. Akagi, *Synth. Met.*, **155**, 576 (2005).

5. (a) S.-H. Jin, S.-J. Choi, W. Ahn, H.-N. Cho, and S.-K. Choi, *Macromolecules*, **26**, 1487 (1993); (b) S.-J. Choi, S.-H. Jin, J.-W. Park, H.-N. Cho, and S.-K. Choi, *Macromolecules*, **27**, 309 (1994); (c) S.-J., Choi, S.-H., Kim, W., Ahn, H.-N. Cho, and S.-K. Choi, *Macromolecules*, **27**, 4871 (1994).

6. (a) K. Akagi, H. Tanaka, R. Toyoshima, and H. Shirakawa, *Trans. Mat. Res. Soc. Jpn.*, **15A**, 513 (1994); (b) R. Toyoshima, M. Narita, K. Akagi, and H. Shirakawa, *Synth. Met.*, **69**, 289 (1995); (c) N. Koide, and H. Iida, *Mol. Cryst. Liq. Cryst.*, **261**, 427 (1995); (d) R. Toyoshima, K. Akagi, and H. Shirakawa, *Synth. Met.*, **84**, 431 (1997); (e) K. Akagi, M. Narita, R. Toyoshima, and H. Shirakawa, *Mol. Cryst. Liq. Cryst.*, **318**, 157 (1998).

7. (a) F. Vicentini, J. Barrouillet, R. Laversanne, M. Mauzac, F. Bibonne, and J. P. Parneix, *Liq. Cryst.*, 19, 235 (1995): (b) P. Ibison, P. J. S. Foot, and W. Brown, *Synth. Met.*, **76**, 297 (1996).

8. (a) K. Akagi, J. Oguma, and H. Shirakawa, *J. Photopolym. Sci. Tech.*, **11**, 249 (1998); (b) K. Akagi, J. Oguma, S. Shibata, R. Toyoshima, I. Osaka, and H. Shirakawa, *Synth. Met.*, **102**, 1287 (1999); (c) J. Oguma, R. Kawamoto, H. Goto, K. Itoh, and K. Akagi, *Synth. Met.*, **119**, 537 (2001); (d) J. Oguma, K. Akagi, and H. Shirakawa, *Synth. Met.*, **101**, 86 (1999); (e) J. Oguma, X.-M. Dai, and K. Akagi, *Mol. Cryst. Liq. Cryst.*, **365**, 331 (2001).

9. (a) R. Toyoshima, K. Akagi, and H. Shirakawa, *Synth. Met.*, **84**, 431 (1997); (b) I. Osaka, S. Shibata, R. Toyoshima, K. Akagi, and H. Shirakawa, *Synth. Met.*, **102**, 1437 (1999); (c) I. Osaka, H. Goto, K. Itoh, and K. Akagi, *Synth. Met.*, **119**, 541 (2001); (d) I. Osaka, H. Goto, K. Itoh, and K. Akagi, *Mol. Cryst. Liq. Cryst.*, **365**, 339 (2001).

10. (a) H. Goto, K. Itoh and K. Akagi, *Synth Met*, **119**, 351 (2001); (b) H. Goto and K. Akagi, *Macromolecules*, **35**, 2545 (2002).

11. (a) K. Akagi, H. Goto, M. Okuda, and H. Shirakawa, *Mol. Cryst. Liq. Cryst.*, **316**, 201 (1998); (b) H. Goto and K. Akagi, *Synth. Met.*, **102**, 1292 (1999); (c) R. Kiebooms, H. Goto and K. Akagi, *Synth. Met.*, **119**, 117 (2001); (c) H. Goto and K. Akagi, *Mol. Cryst. Liq. Cryst.*, **365**, 491 (2001); (c) H. Goto and K. Akagi, *J. Polym. Soc, Part A : Polym. Chem.*, **43**, 616 (2005); (d) R. Kiebooms, H. Goto, and K. Akagi, *Macromolecules*, **34**, 7989 (2001).

12. (a) K. Akagi, H. Goto, and H. Shirakawa, *Synth. Met.*, **84**, 313 (1997); (b) X.-M. Dai, H. Goto, K. Akagi and H. Shirakawa, *Synth. Met.*, **102**, 1289 (1999); (c) X.-M. Dai, H. Goto, K. Akagi, and H. Shirakawa, *Synth. Met.*, **102**, 1291 (1999); (d) X.M. Dai,

H. Narihiro, H. Goto, K. Akagi, and H. Yokoyama, *Synth. Met.*, **119**, 397 (2001); (e) X.-M. Dai, H. Goto, and K. Akagi, *Mol. Cryst. Liq. Cryst.*, **365**, 347 (2001); (f) H. Narihiro, X.-M. Dai, H. Goto, and K. Akagi, *Mol. Cryst. Liq. Cryst.*, **365**, 363 (2001); (g) X.-M. Dai, H. Narihiro, H. Goto, K. Akagi, and H. Yokoyama, *Mol. Cryst. Liq. Cryst.*, **365**, 355 (2001); (h) H. Goto, X. Dai, H. Narihiro, and K. Akagi, *Macromolecules*, **37**, 2353 (2004); (i) H. Goto, X. Dai, T. Ueoka, and K. Akagi, *Macromolecules*, **37**, 4783 (2004).

13. (a) S.-Y. Oh, F. Oguri, K. Akagi, and H. Shirakawa, *J. Polym. Sci., Part A :Polym. Chem.*, **31**, 781 (1993); (b) K. Akagi, H. Goto, K. Iino, H. Shirakawa, and J. Isoya, *Mol. Cryst. Liq. Cryst.*, **267**, 277 (1995); (c) K. Iino, H. Goto, K. Akagi, H. Shirakawa, and A. Kawaguchi, *Synth. Met.*, **84**, 967 (1997).

14. (a) K. Yoshino, K. Kobayashi, T. Kawai, M. Ozaki, K. Akagi and H. Shirakawa, *Synth. Met.*, **69**, 49 (1995); (b) K. Yoshino, K. Kobayashi, K. Myojin, T. Kawai, H. Moritake, M. Ozaki, K. Akagi, H. Goto, and H. Shirakawa, *Mol. Cryst. Liq. Cryst.*, **261**, 637 (1995); (c) K. Yoshino, K. Kobayashi, K. Myojin, M. Ozaki, K. Akagi, H. Goto, and H. Shirakawa, *Jpn. J. Appl. Phys.*, **35**, Part 1, 3964 (1996).

15. K. Akagi and H. Shirakawa, *Synth. Met.*, **60**, 85 (1993).

16. (a) K. Akagi, H. Goto, H. Shirakawa, T. Nishizawa, and K. Masuda, *Synth. Met.*, **69**, 33 (1995); (b) K. Masuda, T. Nishizawa, K. Akagi, and H. Shirakawa, *J. Mol. Struct.*, **441**, 173 (1998).

17. (a) H. Kuroda, H. Goto, K. Akagi, and A. Kawaguchi, *Macromolecules*, **35**, 1307 (2002); (b) Y. Kurauchi, H. Miki, K. Akagi, and A. Kawaguchi, *Polymer*, **45**, 303 (2004).

18. (a) K. Akagi and H. Shirakawa, *Macromol. Symp.*, **104**, 137 (1996); (b) K. Akagi and H. Shirakawa, "The Polymeric Materials Encyclopedia. Synthesis, Properties and Applications", CRC Press, (1996), 5, pp. 3669-3675; (c) K. Akagi and H. Shirakawa, "Electrical and Optical Polymer Systems: Fundamentals, Methods, and Applications", ed. by D. L. Wise, et al., Marcel Dekker, (1998), Chap. 28, pp. 983–1010.

19. C. I. Simionescu, and S. Dumitrescu, *J. Polym. Sci. Polym. Chem.*, **15**, 2479 (1977).

20. P. S. Woon, and M. F. Farona, *J. Polym. Sci. Polym. Chem. Ed.*, **12**, 1749 (1974).

21. Y. Yang, M. Tabata, S. Kobayashi, K. Yokota, and A. Shimizu, *Polym. J.*, **23**, 1135 (1991).

22. D. Davidov, S. Roth, W. Neuman, and H. Sixl, *Solid State Comm.*, **52**, 375 (1984).

CHAPTER 10

FAST SWITCHING OF NEMATIC LIQUID CRYSTALS BY AN ELECTRIC FIELD: EFFECTS OF DIELECTRIC RELAXATION ON THE DIRECTOR AND THERMAL DYNAMICS

YE YIN, SERGIJ V. SHIYANOVSKII AND OLEG D. LAVRENTOVICH
Liquid Crystal Institute and Chemical Physics Interdisciplinary Program,
Kent State University, Kent, OH 44242

1.   INTRODUCTION

Liquid crystals (LCs) represent a special state of soft matter in which the anisotropic molecules or their aggregates demonstrate a long-range orientational order but little or no positional long-range order [1]. In the simplest case of a uniaxial nematic LC (NLC), the long-range positional order is absent altogether; the molecules form a uniaxial centrosymmetric medium, as they are predominantly aligned along a single direction called a director and described by a unit vector $\hat{\mathbf{n}}$ (such that $\hat{\mathbf{n}} \equiv -\hat{\mathbf{n}}$). The director can be realigned by electric and magnetic fields that thus change the optical properties of sample. The orientational dynamics of director in the applied electric field is a fundamental physical phenomenon that is at the heart of modern display technologies. Discovered in 1930s by Frederiks (initially for the magnetic case) [2], this phenomenon, rather surprisingly, still poses a fundamental unanswered question regarding the behavior of NLCs with the finite rate of dielectric relaxation. The problem can be understood as follows.

The dielectric torque that reorients the director has the density $\mathbf{M}(t) = \mathbf{D}(t) \times \mathbf{E}(t)$, where $\mathbf{E}(t)$ is the electric field and $\mathbf{D}(t)$ is the electric displacement at the moment of time t. In the widely accepted standard approach [1, 3, 4], the director response is assumed to be *instantaneous*, i.e., the displacement $\mathbf{D}(t)$ is determined by the electric field at *the very same moment* $\mathbf{D}(t) = \varepsilon_0 \boldsymbol{\varepsilon} \mathbf{E}(t)$, where ε_0 is the free space

277

A. Ramamoorthy (ed.), Thermotropic Liquid Crystals, 277–295.
© 2007 *Springer.*

permittivity, and ε is the relative permittivity tensor. Such an approximation is certainly valid when the director reorientation time in NLC is much longer than the time of dielectric relaxation. However, it might be violated when the driving field is switched quickly. The dielectric relaxation is determined by a number of processes, some of which (electronic polarizability, intra-molecular vibrations) are very fast, with characteristic less than a nanosecond time, while others, such as reorientation of the permanent molecular dipoles are slow, with the typical relaxation time τ of the order of 0.01-1000 μs [5, 6]. If the external field changes over the time comparable to τ, then it is intuitively clear that the dielectric response should depend not only on the present value of the electric field, but also on its past values. This effect is well known for isotropic fluids and solid crystals, as seen, for example, in the classic book by Fröhlich [7], but not for the NLCs. The effect of dielectric relaxation in NLCs is intrinsically more complex than in isotropic fluids and solid crystals, because the dielectric tensor changes with the applied field (as the result of director reorientation).

Is there any experimental evidence that the phenomenon of non-instantaneous relationship between the displacement **D** and the electric field **E** can indeed be relevant in NLCs? Some time ago, N. Clark's group demonstrated that the typical nematic material pentylcyanobiphenyl (5CB) can be switched very quickly, within a few tens of nanosecond, if the applied voltage is large (hundreds of volts) [8]. According to the dielectric spectroscopy data [9, 10], this switching time is in fact of the same order as τ for 5CB, which might suggest that the NLC response is influenced by the dipole relaxation dynamics. A direct evidence comes from the recent experiments with dual frequency nematics (DFNs) [11, 12], in which τ is much larger, 10-100 μs, and thus easier to study. We demonstrated that when the electric field changes over time close to the time of dielectric relaxation, the instantaneous and "past" contributions to the dielectric torque can indeed be comparable; moreover, they can cause opposite directions of the director reorientation [11, 12]. There is thus no doubt that the effect causes a profound influence on the way a liquid crystal responds to the applied electric field.

The objective of this chapter is to overview the physics of time-dependent dielectric response of NLCs. We discuss how the electric displacement depends on the present and the past values of the electric field and director. We also discuss a closely related problem of dielectric heating that is most pronounced in the region of dielectric relaxation.

The review is organized as follows. Section 2 introduces the basics of dielectric relaxation and dielectric heating. A general model of the time-dependent dielectric response of NLCs during the reorientation dynamics is described in section 3. The model is applicable for all types of dielectric relaxations related to the reorientation of the permanent molecular dipoles. We present an experiment that is well described by the proposed theory but which cannot be described by the standard "instantaneous" model of NLC switching. In section 4, we explore the dielectric heating effect taking into account the finite thermal conductivity of the bounding plates and heat transfer to surrounding media of different nature.

2.   DIELECTRIC RESPONSE IN NEMATIC LIQUID CRYSTALS

Under a changing electric field, a dielectric adjusts its properties over a certain period of time. Because of a finite speed of this dielectric relaxation process, the electric displacement $\mathbf{D}(t)$ depends not only on the current electric field $\mathbf{E}(t)$ but also the past electric field $\mathbf{E}(t')$, where $-\infty \le t' < t$. If the electric field is not very strong, then $\mathbf{D}(t)$ can be represented using the superposition rule [13],

$$(10\text{-}1) \quad \mathbf{D}(t) = \varepsilon_0 \mathbf{E}(t) + \mathbf{P}(t) = \varepsilon_0 \mathbf{E}(t) + \varepsilon_0 \int_{-\infty}^{t} \alpha(t, t') \mathbf{E}(t') dt',$$

where $\mathbf{P}(t)$ is the polarization, the tensor $\alpha(t, t')$ is the step response function describing the influence of the past field $\mathbf{E}(t')$ on the current electric displacement $\mathbf{D}(t)$. For solid crystals, or liquid crystals with the fixed director $\hat{\mathbf{n}}(t') = const$, $\alpha(t, t')$ depends only on the time difference $\bar{t} = t - t'$,

$$(10\text{-}2) \quad \alpha(t, t') = \alpha(t - t').$$

If, in addition, the electric field is harmonic, $\mathbf{E}(t) = \mathbf{E}_\omega e^{-i\omega t}$, where ω is the angular frequency, then the electric displacement becomes harmonic as well,

$$(10\text{-}3) \quad \mathbf{D}(t) = \varepsilon_0 \left(\mathbf{I} + \int_0^\infty \alpha(\bar{t}) e^{i\omega \bar{t}} d\bar{t} \right) \mathbf{E}_\omega e^{-i\omega t},$$

and can be represented as $\mathbf{D}(t) = \mathbf{D}_\omega e^{-i\omega t}$, where

$$(10\text{-}4) \quad \mathbf{D}_\omega = \varepsilon_0 \varepsilon^*(\omega) \mathbf{E}_\omega,$$

and the frequency dependent dielectric permittivity tensor $\varepsilon^*(\omega)$ is

$$(10\text{-}5) \quad \varepsilon^*(\omega) = \mathbf{I} + \int_0^\infty \alpha(\bar{t}) e^{i\omega \bar{t}} d\bar{t}.$$

Equation (10-4), the well-known dielectric response equation in the frequency domain, is equivalent to Eq. (10-1) if the assumption (2) holds. However, the assumption (2) is generally not valid for a reorienting NLC, when $\hat{\mathbf{n}}(t') \ne const$.

There are three different mechanisms contributing to the electric polarization $\mathbf{P}(t)$ in Eq. (10-1) and thus to $\mathbf{D}(t)$: electronic polarization, molecular polarization, and orientational polarization . These three have very different characteristic relaxation times [5,14–16]. Electronic polarization is the fastest, with the characteristic time $\sim 10^{-15}$ s. Molecular polarization, caused by the relative displacement of the atomic groups within the molecule, is slower, with the characteristic time $\sim 10^{-12}$ s. Finally, the orientational polarization, associated with the reorientation of permanent molecular electric dipoles, is the slowest, with the time scales ranging from 10^{-12} s (for small-molecule substances in a gaseous state) to 10 s (for large molecules or aggregates) [17]. With more than one relaxation mechanism involved, the step response tensor function can be represented as

$$(10\text{-}6) \quad \boldsymbol{\alpha}(t, t') = \sum_k \boldsymbol{\alpha}_k(t, t'),$$

where $\boldsymbol{\alpha}_k(t, t')$ is the step response tensor function of a certain relaxation mode; $\boldsymbol{\alpha}_k(t, t')$ vanishes when $t' \to -\infty$.

If a certain relaxation process is sufficiently fast, so that both the electric field and the director can be considered constant during the relaxation period, then

$$(10\text{-}7) \quad \int_{-\infty}^{t} \boldsymbol{\alpha}_k(t, t') \mathbf{E}(t') dt' = \mathbf{E}(t) \int_{-\infty}^{t} \boldsymbol{\alpha}_k(t, t') dt'.$$

We now split $\boldsymbol{\alpha}(t, t')$ into a "fast" component $\boldsymbol{\alpha}_f(t, t')$ and a "slow" component $\boldsymbol{\alpha}_s(t, t')$ (with respect to the rate with which the electric field and director change),

$$(10\text{-}8) \quad \boldsymbol{\alpha}(t, t') = \boldsymbol{\alpha}_f(t, t') + \boldsymbol{\alpha}_s(t, t').$$

Any $\boldsymbol{\alpha}_k(t, t')$ satisfying Eq. (10-7) is regarded as a part of the fast $\boldsymbol{\alpha}_f(t, t')$ contribution. Therefore, the electric displacement $\mathbf{D}(t)$ of the NLC can be written as

$$(10\text{-}9) \quad \mathbf{D}(t) = \varepsilon_0 \boldsymbol{\varepsilon}_f(t) \mathbf{E}(t) + \varepsilon_0 \int_{-\infty}^{t} \boldsymbol{\alpha}_s(t, t') \mathbf{E}(t') dt',$$

where

$$(10\text{-}10) \quad \boldsymbol{\varepsilon}_f(t) = \mathbf{I} + \int_{-\infty}^{t} \boldsymbol{\alpha}_f(t, t') dt'.$$

The electronic polarization ($\sim 10^{-17}$ s) and molecular polarization ($\sim 10^{-12}$ s) contribute to the fast $\boldsymbol{\alpha}_f(t, t')$. The slow part is related to the orientational relaxation, as discussed below.

Consider first an anisometric molecule with the longitudinal $\mathbf{p}_l$ and transversal $\mathbf{p}_t$ permanent dipole moments in an isotropic phase. There are two relaxation modes: mode 1, rotations of $\mathbf{p}_t$ around the long axis, and mode 2, reorientation of $\mathbf{p}_l$, Figure 10-1. The mode 1 has a smaller relaxation time, $\tau_1 < \tau_2$, because of the smaller moments of inertia involved. When this isotropic fluid is cooled down into the NLC phase, the dynamics is affected by the appearance of the "nematic" potential associated with the orientational order along the director $\hat{\mathbf{n}}$. The mode 1 remains almost the same as in the isotropic phase, and contributes to both the parallel and perpendicular components of dielectric polarization (determined with respect to $\hat{\mathbf{n}}$). Mode 2 is associated with small changes of the angle between $\mathbf{p}_l$ and $\hat{\mathbf{n}}$; it contributes to the parallel component of dielectric polarization. Mode 3 is associated with conical rotations of $\mathbf{p}_l$ around the director (as the axis of the cone); it is effective when the applied electric field is perpendicular to $\hat{\mathbf{n}}$ and contributes

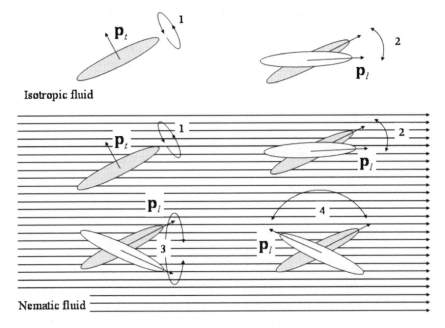

Figure 10-1. Orientational relaxation modes for a molecule in an isotropic fluid (top) and in a nematic medium (bottom)

to the perpendicular component of the dielectric polarization. Finally, mode 4 is a qualitatively different mode related with the flip-flops of $\mathbf{p}_l$. The molecular reorientations in mode 4 are hindered by the nematic surrounding of the molecule as in the middle point of the flip-flop it finds itself perpendicular to $\hat{n}$. The relaxation time for the mode 4 is thus significantly larger than for any other mode. Typically, $\tau_1 < \tau_2, \tau_3 << \tau_4$; τ_4 might be as large as few milliseconds in some NLCs [18].

We now turn to the discussion of the so-called Debye model that allows one to account for the relaxation processes with a finite τ in the expressions for the electric displacement and the dielectric permittivity. To obtain the explicit dependencies, one should know the step response function which is generally unknown. Debye assumed that, once a constant external electric field is applied, the equilibrium state of an isotropic dielectric is achieved through an exponential relaxation, i.e.,

$$(10\text{-}11) \quad \alpha\,(t-t') = \bar{\alpha}\,exp\left(-\frac{t-t'}{\tau}\right),$$

where $\bar{\alpha}$ is the coefficient to be determined later. The hypothesis (10-11), combined with Eq. (10-5), predicts how the dielectric permittivity should depend on the frequency of the applied harmonic field:

$$(10\text{-}12) \quad \varepsilon^*\,(\omega) = 1 + \frac{\bar{\alpha}\tau}{1 - i\omega\tau}.$$

The last equation tells us that the real part of the dielectric permittivity decreases from the "static" value $\varepsilon_s = 1 + \bar{\alpha}\tau$ at low frequencies to $\varepsilon_h = 1$ at $\omega \to \infty$. In reality, ε_h is different from 1 because of the electronic and intramolecular processes at high frequencies. The steepest decrease of the real part of $\varepsilon^*(\omega)$, according to Eq. (10-12), should be observed near the frequency $\omega = 1/\tau$. The imaginary part of $\varepsilon^*(\omega)$ should reach a maximum at the same frequency $\omega = 1/\tau$. As we shall see later, the last feature implies a maximum dielectric heating in the region near $\omega = 1/\tau$.

Let us now apply this Debye model to the orientational relaxation modes of a nematic with fixed director, $\hat{\mathbf{n}}(t') = const$, assuming that each $\boldsymbol{\alpha}_k$ corresponding to the orientational modes in Eq. (10-6) follows the Debye behavior (11). Using Eq. (10-5) and (10-6), we find

$$(10\text{-}13) \quad \varepsilon^*(\omega) = \varepsilon_h + \sum_{k=1}^{4} \frac{\Delta\varepsilon_k}{1 - \omega\tau_k},$$

where $\Delta\varepsilon_k = \bar{\alpha}_k\tau_k$ is the maximum contribution by the kth mode to the real part of the dielectric tensor, the summation is taken over the orientational modes 1-4 only and ε_h is generally different from the unit tensor, as discussed above. By choosing an appropriate mutual orientation of the probing electric field and the director, one can measure the dielectric relaxation spectra for the parallel geometry, $\varepsilon_\parallel^*(\omega)$, $\mathbf{E}\|\hat{\mathbf{n}}$, or for the perpendicular geometry, $\varepsilon_\perp^*(\omega)$, $\mathbf{E}\perp\hat{\mathbf{n}}$. The relaxation modes 1, 2 and 4 contribute to the component $\varepsilon_\parallel$ of the dielectric permittivity tensor that is parallel to the director, and modes 1 and 3 contribute to the perpendicular component. Later, $\varepsilon_\parallel(\omega)$ and $\varepsilon_\perp(\omega)$ are used to represent the real part of the parallel and perpendicular dielectric permittivity components, respectively. $\varepsilon_\parallel^i(\omega)$ and $\varepsilon_\perp^i(\omega)$ are used to represent the imaginary part of the parallel and perpendicular dielectric permittivity components, respectively. Figure 10-2 illustrates a typical frequency dependence of the dielectric components for a nematic material.

As seen in Figure 10-2, the relaxation frequencies correspond to the regions of the significant increase of the imaginary part of dielectric permittivities. If the electric field is applied at a frequency close to the relaxation one, then one would expect a significant dielectric heating effect. The effect is caused by an efficient absorption of electromagnetic energy through reorientation of the molecular dipoles and the associated molecular friction [13]. Although the effect is common for all types of dielectrics, it acquires especially interesting facets for the case of a liquid crystal dielectric, most notably dual frequency materials with large relaxation times. The effect can greatly influence the performance of NLC devices as the NLC properties are sensitive to the temperature. Note that the components $\varepsilon_\parallel^*$ and $\varepsilon_\perp^*$ have different frequency dependencies and thus the dielectric heating in the NLC should strongly depend on the director orientation. The orientation dependent heating effect is absent in isotropic dielectrics, and is also of no interests in solid crystals due to the fixed orientation. However, since the director orientation of NLC can be realigned by

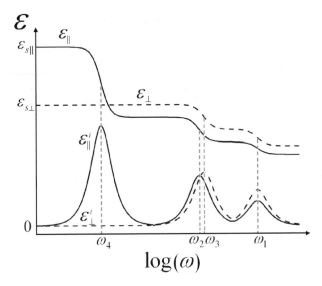

Figure 10-2. A typical frequency dependence of real and imaginary parts of the components of the dielectric permittivity tensor for a nematic material

the electric field, one might find a broader variety of interest to explore dielectric heating phenomena in the NLC medium. The dielectric heating effect in NLC will be further discussed in section 4.

3. SWITCHING MEMORY EFFECT

3.1 Theory

Understanding the dynamics of dielectric response of nematic liquid crystals (NLCs) at time scales typical of dielectric relaxation is a challenging problem not tackled so far. As discussed in section 2, the widely used equation to calculate the displacement, $\mathbf{D}(t) = \varepsilon_0 \varepsilon \mathbf{E}(t)$, is only a simplification of Eq. (10-9) that ignores the contribution of the slow step response function $\alpha_s(t, t')$. Such an omission is justified when the NLC switching dynamics takes much longer time than the dielectric relaxation process. For example, the current nematic displays have the characteristic switching time of the order of 10 ms, which is longer than the relaxation time of all the relaxation modes discussed above, including the slowest relaxation mode 4 caused by the flip-flops of the longitudinal molecular dipoles. The industrial need, however, is to reduce the switching time by an order or two. The much needed short response time ~ 0.1 ms of the nematic cells is indeed possible [12], for both directions of the director reorientation (towards the field and away from the field) when the dual-frequency materials are used. In the latter case, the relaxation times and the switching times become comparable; therefore, further understanding of the fast switching should be based on the complete form of Eq.(10-9) that would predict how the reorienting torque depends on the present and past values of the field and the director.

According to the orientational symmetry, one can divide NLCs into two types: uniaxial and biaxial NLCs. Most nematics are uniaxial: they have a single axis (the director $\hat{\mathbf{n}}$), around which the system is rotationally symmetric. The biaxial NLC has three orthogonal directors ($\hat{\mathbf{n}}_1$, $\hat{\mathbf{n}}_2$, and $\hat{\mathbf{n}}_3$) [19, 20], and one might expect that the biaxial NLCs might offer a broader variety of geometries of interest to explore and utilize the dielectric electro-optical response. We start with the discussion of a uniaxial case.

As mentioned above, we have to determine the slow part $\boldsymbol{\alpha}_s(t, t')$ caused by reorientation of the permanent molecular dipoles in the NLC. Let $\mathbf{P}(t, t')$ be a contribution to the electric polarization $\mathbf{P}(t)$ determined by the "past" electric field $\mathbf{E}(t')$, $t' \leq t$, namely,

(10-14) $\mathbf{P}(t, t') = \varepsilon_0 \boldsymbol{\alpha}_s(t, t') \mathbf{E}(t')$.

The main contribution to $\mathbf{P}(t, t')$ is provided by the mode 4 associated with the molecular flip-flops of the long axes. These flip-flops occur through unfavorable molecular orientations perpendicular to $\hat{\mathbf{n}}$; they thus need to overcome high potential barriers of intermolecular interactions. Although each individual flip-flop is fast (the typical rate is $\nu_{ff} \sim 10^7 s^{-1}$[21]), the overall dielectric relaxation is slow because the probability of flip-flops (or the relative number of molecules experiencing it) is low. In a uniaxial NLC, the characteristic relaxation time $\tau_\parallel = \tau_4$ for polarization component parallel to $\hat{\mathbf{n}}$ is in microsecond range, $\tau_\parallel \sim 1 - 100\,\mu s$, whereas the barrier-free relaxation of the perpendicular component $\tau_\perp$ is much shorter, $1 - 100\,ns$, and does not contribute to $\mathbf{P}(t, t')$. The potential barriers should keep the polarization parallel to $\hat{\mathbf{n}}$ and thus when $\hat{\mathbf{n}}$ reorients, it drags the polarization parallel to itself. This director-mediated rotation should not affect the reorientational relaxation of the individual molecules. Even when the director rotation velocity is high, $\Omega \sim 10^4 s^{-1}$, and approaches the relaxation rate of polarization $\tau_\parallel^{-1}$, the director-imposed slow rotation of all molecules with the velocity Ω should not affect substantially the fast individual flip-flops with $\nu_{ff} \sim 10^7 s^{-1}$. The latter assumption implies that $\mathbf{P}(t, t')$ can be expressed through the step response tensor component $\alpha_\parallel(t - t')$ along the director, when the director is fixed:

(10-15) $\mathbf{P}(t, t') = \varepsilon_0 \alpha_\parallel(t - t')\hat{\mathbf{n}}(t)\,(\hat{\mathbf{n}}(t')\mathbf{E}(t'))$.

Therefore, $\boldsymbol{\alpha}_s(t, t')$ in a uniaxial NLC has contribution from the relaxation mode 4 only,

(10-16) $\boldsymbol{\alpha}_s(t, t') = \alpha_\parallel(t - t')\hat{\mathbf{n}}(t) \otimes \hat{\mathbf{n}}(t')$,

where $\otimes$ stands for the external product of two vectors: $[\hat{\mathbf{n}}(t) \otimes \hat{\mathbf{n}}(t')]_{ij} = n_i(t)n_j(t')$.

In the case of biaxial NLC, we also assume that the potential barriers around all three directors $\hat{\mathbf{n}}_i$ keep the corresponding polarization components parallel to these directors: When $\hat{\mathbf{n}}_i$ reorients, the director reorientation drags the polarization component $\mathbf{P}_i(t, t')$ keeping it parallel to $\hat{\mathbf{n}}_i(t)$ all the time. Thus,

$$(10\text{-}17) \quad \mathbf{P}(t, t') = \sum_{i=1}^{3} \mathbf{P}_i(t, t') = \varepsilon_0 \sum_{i=1}^{3} \alpha_i(t - t')\hat{\mathbf{n}}_i(t) \left(\hat{\mathbf{n}}_i(t')\mathbf{E}(t')\right),$$

where $\alpha_i(t - t')$ is the diagonal element of the step response tensor along the fixed director $\hat{\mathbf{n}}_i(t)$. The step response tensor function $\boldsymbol{\alpha}_s$ for biaxial NLC is

$$(10\text{-}18) \quad \boldsymbol{\alpha}_s(t, t') = \sum_{i=1}^{3} \alpha_i(t - t')\hat{\mathbf{n}}_i(t) \otimes \hat{\mathbf{n}}_i(t').$$

To analyze the switching process in a uniaxial NLC, we have to use Eq. (10-9), Eq. (10-16), and explicit expressions for $\boldsymbol{\varepsilon}_f(t)$ and $\alpha_\|(t - t')$. $\boldsymbol{\varepsilon}_f(t)$ contains contribution of all modes except of the orientational mode 4, thus in an uniaxial NLC:

$$(10\text{-}19) \quad \boldsymbol{\varepsilon}_f(t) = \varepsilon_\perp \mathbf{I} + \left(\varepsilon_{h\|} - \varepsilon_\perp\right)\hat{\mathbf{n}}(t) \otimes \hat{\mathbf{n}}(t)$$

where $\varepsilon_{h\|} = \varepsilon_\|(\omega_h)$ and $\varepsilon_\perp = \varepsilon_\perp(\omega_h)$ are equal to the components of the dielectric tensor at the frequency ω_h, $\tau_\|^{-1} \ll \omega_h \ll \tau_2^{-1}, \tau_3^{-1}$.

We assume that the contribution of mode 4 satisfies the Debye's model with the exponential decay of $\alpha_\|(t - t')$ and Lorenzian behavior for $\alpha_\|(\omega)$ [7, 22]:

$$(10\text{-}20) \quad \alpha_\|(t - t') = \frac{\varepsilon_{l\|} - \varepsilon_{h\|}}{\tau_\|} \exp\left(-\frac{t - t'}{\tau_\|}\right),$$

$$(10\text{-}21) \quad \alpha_\|(\omega) = \frac{\varepsilon_{l\|} - \varepsilon_{h\|}}{1 - i\omega\tau_\|},$$

where $\varepsilon_{l\|} = \varepsilon_\|(\omega_l)$ is the dielectric permittivity components at low frequency $\omega_l \ll \tau_\|^{-1}$. In this case the resulting dielectric torque density for the uniaxial NLC is:

$$(10\text{-}22) \quad \begin{aligned} \mathbf{M}(t) = {} &\varepsilon_0 \hat{\mathbf{n}}(t) \times \mathbf{E}(t) \cdot \\ &\left[\left(\varepsilon_{h\|} - \varepsilon_\perp\right)\hat{\mathbf{n}}(t) \cdot \mathbf{E}(t) + \frac{\varepsilon_{l\|} - \varepsilon_{h\|}}{\tau_\|} \int_{-\infty}^{t} \exp\left(-\frac{t - t'}{\tau_\|}\right) \hat{\mathbf{n}}(t') \cdot \mathbf{E}(t')dt' \right] \cdot \end{aligned}$$

The memory effect is described by the integral term of Eq. (10-22), which is absent in the standard approach. Equation (10-22) holds for a liquid crystal with a Debye type of relaxation but it can be easily modified or generalized to other relaxation models (such as Cole-Davidson, Havriliak-Negami and other models with different functional forms of $\boldsymbol{\alpha}(t - t')$ [5, 17]).

3.2   Experiment and Analysis

To verify experimentally the switching memory effect, the best approach is to use a dual frequency nematic (DFN). In the DFN, the dielectric anisotropy

$\Delta\varepsilon(f) = \varepsilon_\parallel(f) - \varepsilon_\perp(f)$ changes its sign from positive at $f < f_c$ to negative at $f > f_c$, where the critical frequency $f_c \approx \left(2\pi\tau_\parallel\right)^{-1}$ corresponds to the region of dielectric relaxation of $\varepsilon_\parallel(f)$ with a characteristic time $\tau_\parallel$, here $f = \omega/2\pi$ is the field frequency. The dielectric tensor of DFN obeys the relation $\varepsilon_{l\parallel} < \varepsilon_\perp < \varepsilon_{h\parallel}$, which allows one to produce opposite signs of the instantaneous and memory contributions to the torque (10-22). It is important to bear in mind that many conventional NLCs are actually DFNs at sufficiently high frequencies. For example, in the most popular nematic material pentylcyanobiphenyl (5CB), $f_c \approx 15\,\text{MHz}$ [9], while in heptylcyanobiphenyl (7CB), $f_c \approx 9\,\text{MHz}$ [23] (both measured at room temperature). Usually in commercially available DFN mixtures, f_c is adjusted to be relatively low, from 1 kHz to 1 MHz [18, 24, 25].

The DFN material chosen to verify Eq. (10-22) was the mixture MLC2048 (EM Industries, NY), as its dielectric properties are well described by the Debye model, see Refs. [11, 12] and Figure 10-3. The relaxation time has been determined to be $\tau_\parallel = 13.4\,\mu\text{s}$ at room temperature [11].

To test the basic feature of Eq. (10-22), namely, the competition between the instantaneous and "memory" contributions to the total torque, one can trace the director dynamics by measuring the optical phase retardation of the cell filled with MLC2048. The dielectric torque is maximized by choosing a high value of the angle between the director and the normal to the bounding plates, $\theta_0 \sim 45°$ [11, 12]. The cell thickness was $10\,\mu\text{m}$.

The electric field reorients the director which is manifested by the change of the optical phase retardation $\Delta\varphi$ between the extraordinary and ordinary waves.

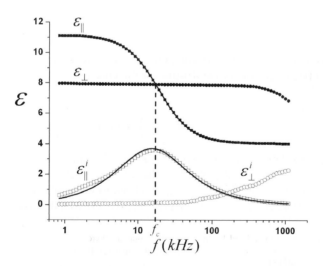

Figure 10-3. Real and imaginary parts of the dielectric permittivities $\varepsilon_\parallel^*$ and $\varepsilon_\perp^*$ of MLC2048 as a function of the field frequency at 23 °C. The experimental measured data of the frequency dependent parallel permittivities $\varepsilon_\parallel^*$ are well fitted by the Debye model (solid line)

The latter, in its turn, changes the measured intensity $I \propto \sin^2(\Delta\varphi/2)$ of laser light transmitted through the cell and the pair of crossed polarizers. Figure 10-4 shows the transmitted intensity (top trace) versus the applied voltage (bottom traces) at two frequencies (100 kHz and 1 kHz) when the amplitude of the voltage varies slowly with the rate 2.4 V/s. For such a slow rate, the dielectric behavior of DFN can be regarded as a quasi-static dielectric response, where the standard description with an instantaneous relation between the displacement and the field is valid.

The memory effect described by the last term in Eq. (10-22) becomes evident when the voltage changes abruptly, for a step-like pulse of a low frequency 1 kHz, Figure 10-5 (the behavior for 100 kHz for slowly and abruptly changing voltage is practically the same [11]). The behavior is exactly opposite to what is expected from the quasi-static model and experiment with a slowly increased voltage in Figure 10-4. Namely, Figure 10-4 suggests that the light intensity should increase when the voltage is increased at 1 kHz, while Figure 10-5 demonstrates that the light intensity actually decreases (towards point Y in Figure 10-5) at the beginning of director reorientation. This anomalous decrease is not related to the possible parasitic effects such as light scattering losses, as verified by placing an additional π phase retarder is between the cell and the polarizer, which changes the optical signal amplitude, see the insert in Figure 10-5. Therefore, the experiment suggests that the reason for the different response of the director to the quasi-static, Figure 10-4, and abrupt, Figure 10-5, voltage increase at 1 kHz is the dielectric memory effect.

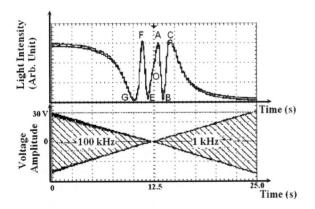

Figure 10-4. Transmitted light intensity modulated by the changes of optical retardation (top curve) vs. slowly changing voltage (bottom trace) applied at two different frequencies 100 kHz (left part) and 1 kHz (right part) to the MLC2048 cell of thickness $d = 10\,\mu$m. Point "O" corresponds to light transmittance at zero voltage. Reprinted with permission from Y. Yin, S. V. Shiyanovskii, A. B. Golovin and O. D. Lavrentovich, Phys. Rev. Lett, 95, 087801. Copyright (2006) by the American Physical Society

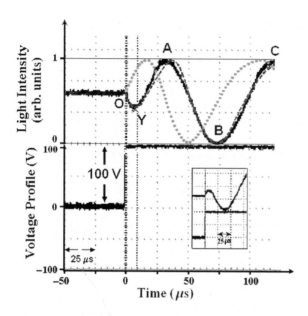

Figure 10-5. Transmitted light intensity modulated by the changes of optical retardation for the same DFN cell as in Figure 10-4 but now driven by an abruptly applied voltage pulse at 1 kHz; at the time scale shown, the pulse looks like a DC pulse, as the time scale is 25 μs/sqr. In the top part, the solid line is the oscilloscope's trace for the experimentally determined light transmittance, the dashed line represents the transmitted intensity as calculated from the theoretical model, and dotted line represents the standard approach with $D(t) = \varepsilon_0 \varepsilon E(t)$. The insert is the optical transmission when a π phase retarder is inserted between the polarizer and the cell. Reprinted with permission from Y. Yin, S. V. Shiyanovskii, A. B. Golovin and O. D. Lavrentovich, Phys. Rev. Lett, 95, 087801. Copyright (2006) by the American Physical Society

Numerical simulations of the director response demonstrate that indeed the model with Eq. (10-22) describes the experimental data in Figure 10-5 well. The polar angle $\theta(z, t)$ and its changes caused by the field are described by the Erickson-Leslie equation [26, 27]:

$$(10\text{-}23) \quad -\gamma \frac{\partial \theta(z, t)}{\partial t} = M(t) - \left[K_1 \sin^2 \theta(z, t) + K_3 \cos^2 \theta(z, t) \right] \frac{\partial^2 \theta(z, t)}{\partial z^2},$$

where, γ is the rotational viscosity, $M(t)$ is the magnitude of the dielectric torque density represented by Eq. (10-22) K_1 and K_3 are the nematic elastic constants of splay and bend, respectively. The backflow effects are neglected, as one is interested in the very beginning of field-induced reorientation; the initial condition is $\theta(z, t = 0) = \theta_0 = 45^0$ at $M(t = 0) = 0$. The optical phase shift is calculated as

$$(10\text{-}24) \quad \Delta\varphi = \frac{2\pi n_o}{\lambda} \int_0^d \left(\frac{n_e}{\sqrt{n_o^2 \sin^2 \theta(z, t) + n_e^2 \cos^2 \theta(z, t)}} - 1 \right) dz,$$

where n_o and n_e are the ordinary and extraordinary refractive indices of NLC, respectively.

The numerical model (with the material parameters of MLC2048 determined independently) reproduces the beginning of the director reorientation process very well, Figure 10-5, while the standard model (dotted lines) with an instantaneous relationship $\mathbf{D}(t) = \varepsilon_0 \varepsilon \mathbf{E}(t)$ shows a drastic disagreement. When the voltage amplitude increases slowly, as in Figure 10-4, the difference in the new and standard approaches vanishes. In addition, the behavior of the DFN driven at a high frequency 100 kHz, is practically the same for the slow and abruptly increasing voltage [11].

The main conclusion of this part is that the dielectric response of a nematic material is strongly influenced by the dielectric relaxation processes. One of the consequences is that the sharp front of an applied voltage pulse can be perceived by the NLC as a high-frequency field for which the anisotropy of dielectric permittivity is different than that for the (low) frequency of the driving field, or even be of the opposite sign. Of course, the effect should manifest itself not only for DFN, but for all nematic materials; the difference would be only in the time/frequency domain where the effect is most pronounced. For example, if would be of interest to verify whether the delay effects observed by Clark's group in Ref. [8] at the scale of tens of nanoseconds are caused by the dielectric dispersion effect described above.

4. DIELECTRIC HEATING EFFECT

Dielectric heating is caused by the absorption of electromagnetic energy in the dielectric medium through reorientation of the molecular dipoles and the associated molecular friction. The absorption is especially strong when the frequency of the applied field is close to the frequency region of dielectric relaxation, as described in section 2. The heating mechanism in isotropic dielectrics is well known, and the major application of the dielectric heating, namely, the microwave heating, is used for food processing. The food exposed to the microwave at the frequency 2.5 GHz is heated effectively since $f = 2.5$ GHz is the relaxation frequency of water. Dielectric heating effects in the nematic liquid crystals (NLCs) are less useful since the induced temperature changes cause the change of the liquid crystal properties and thus affect the NLC switching behavior.

The first model of dielectric heating in NLC has been developed by M. Schadt [25]. In his model, the temperature decrease across the bounding (glass) plates of the NLC cell was considered negligibly small. The corresponding experiment on dielectric heating was performed for NLC cells of thickness 15 μm by placing a 80 μm thermocouple at one of the silver electrodes (apparently outside the cell) [25]. These studies were expanded in our previous work by using a much smaller thermocouple inserted directly into the NLC slab [28]. Recently, Wu et al. proposed a new, noncontact, method [29] based on the measurements of phase retardation changes caused by the temperature-induced changes of the birefringence Δn of the

NLC. This technique is applicable when the electric field causes no reorientation of the director (otherwise the phase retardation would change because of the director reorientation), i.e. when the following conditions are satisfied: (1) the material has a negative dielectric anisotropy, $\Delta\varepsilon(f) < 0$, (2) the cell yields a strict planar alignment (zero pretilt angle) and (3) the electric field causes no hydrodynamics. Because of these limitations, the technique cannot address the issue of dielectric heating in the region of dielectric relaxation for $\varepsilon_{\parallel}(f)$, mode 4 in Figure 10-2.

Below, we explore the dielectric heating effects caused by the dielectric relaxation of $\varepsilon_{\parallel}(f)$ by expanding the theoretical model to include the finite thermal conductivity of the bounding plates. Evidently, the temperature changes should depend not only on the electric field, properties of the NLC and bounding plates, but also on the thermal properties of the medium in which the NLC cell is placed; however, to the best of our knowledge, this issue has not been explored in the prior work. To directly measure the temperature of the nematic slab in the broad range of field frequencies, one can use the direct method with a tiny thermocouple smaller than the cell thickness, as described in Ref. [28].

When a nematic cell is driven by a harmonic electric field $\mathbf{E}(t) = \mathbf{E}_0 \cos \omega t$, the dielectric heating power density P can be written as [13]:

$$(10\text{-}25) \quad P = \frac{\omega\varepsilon_0 \mathbf{E}_0 \cdot \boldsymbol{\varepsilon}^i(\omega) \cdot \mathbf{E}_0}{2},$$

where $\boldsymbol{\varepsilon}^i$ is the imaginary part of the dielectric tensor $\boldsymbol{\varepsilon}^*$. The dielectric tensor $\boldsymbol{\varepsilon}^*$ and thus the heating power density P depend on the orientation of the NLC director $\hat{\mathbf{n}}$. In what follows, for the sake of simplicity, we assume that the electric field is homogeneous inside the cell. This assumption is valid when the director field is uniform or only weakly distorted across the cell (including the case of a very strong voltage when most of the NLC is reoriented) or when the dielectric anisotropy of the LC material is weak. Therefore, we represent the electric field through the applied voltage U acting across a NLC cell of a thickness d, $\mathbf{E}_0 = U\hat{\mathbf{z}}/d$, so that

$$(10\text{-}26) \quad P = \frac{\pi f \varepsilon_0 \varepsilon^i_{zz} U^2}{d^2},$$

where $\varepsilon^i_{zz} = \varepsilon^i_{\perp} \sin^2 \theta + \varepsilon^i_{\parallel} \cos^2 \theta$ is the imaginary part of the effective dielectric permittivity, and is the angle between $\hat{\mathbf{n}}$ and the normal $\hat{\mathbf{z}}$ to the bounding plates, as described in Eq. (10-23).

Let us discuss now the scheme of the dielectrically-induced heat production and its transfer in the multilayered system comprised of an NLC layer, two bounding plates of finite thickness (but of infinite size in the two other directions), placed into some surrounding medium of infinite extension, Figure 10-6.

The heat conduction equation for each of the layers is of a generic form [30]:

$$(10\text{-}27) \quad c_p\rho\frac{\partial T(z, t)}{\partial t} = G\frac{\partial^2 T(z, t)}{\partial z^2} + P_0,$$

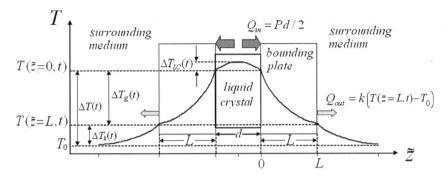

Figure 10-6. The scheme of the dielectric heating of an NLC cell. Reprinted with permission from J. Appl. Phys. 100, 024906. Copyright 2006, American Institute of Physics

where c_p is the heat capacity of the layer, ρ is the material density, G is the layer's thermal conductivity, P_0 is the specific heat production inside the layer. Equation (10-27) should be accompanied with the boundary conditions that guarantee the heat flow balance.

For the case of the NLC layer, $P_0 = P$. The temperature drop across the NLC slab is maximum for a stationary regime. For a stationary case, Eq. (10-27) (with the left hand side being zero) allows us to estimate the temperature drop as $\Delta T_{LC}(t) \leq Pd^2/8G = \pi f \varepsilon_0 \varepsilon^i_{zz} U^2/8G$. Consider a typical situation, $G \sim 0.2\,\mathrm{W/mK}$ [31], $\varepsilon^i_{zz} \sim 1$, $U \sim 50\,\mathrm{V}$, $f \sim 10^6\,\mathrm{Hz}$, the temperature drop is less than the order of 0.05°C. Therefore, we can assume that the temperature drop across the (relatively thin) NLC layer is small and concentrate on the temperature changes in the bounding plates. This small temperature difference in the NLC layer mitigates the possible measurement error caused by the finite size of the thermocouple which is comparable to the cell thickness.

In what follows, we consider the system being symmetric with respect to the middle plane of the NLC cell and apply Eq. (10-27) to describe the heat conduction through one of the two bounding plates, for example, the one located within $z \in [d/2, d/2+L]$ or $\tilde{z} \in [0, L]$, where $\tilde{z} = z - d/2$ is introduced to simplify the notations. Equation (10-27) needs to be supplemented by two boundary conditions for the heat transfer, at the NLC-bounding plate interface and at the bounding plate-surrounding medium interface. Because the NLC layer is thin as compared to the bounding plates, one can neglect the heat stored in the NLC layer itself, so that the heat flux from the NLC layer to the bounding plate is $Q_{in} = Pd/2$. We assume that the heat flux Q_{out} at the bounding plate-surrounding medium interface obeys the Newton's cooling law, i.e., $Q_{out} = k\left(T\left(\tilde{z} = L, t\right) - T_0\right)$, where $T\left(\tilde{z} = L, t\right)$ is the temperature of the plates outer boundary, T_0 is the temperature of the surrounding medium at infinity, and k is the heat transfer coefficient of the surrounding medium.

The boundary conditions are then written as:

$$(10\text{-}28) \quad \frac{\partial T(\tilde{z}=0,t)}{\partial z} = -\frac{Pd}{2G}, \quad \frac{\partial T(\tilde{z}=L,t)}{\partial z} = -\frac{k}{G}\left(T(\tilde{z}=L,t)-T_0\right),$$

where $T(\tilde{z}=0,t)$ is the temperature of the NLC-bounding plate interface. The solution of Eq. (10-27), supplemented with the boundary conditions in Eq. (10-28) and the initial condition $T(\tilde{z}=0,t)=T_0$, is

$$(10\text{-}29) \quad T(\tilde{z},t) = T_0 + \Delta \bar{T}_k \left[1+\xi\left(-\frac{\tilde{z}}{L}+1\right)-\sum_{n=1}^{\infty} a_n \cos q_n \frac{\tilde{z}}{L} \exp\left(-\frac{t}{\tau_n}\right)\right]$$

where

$$(10\text{-}30) \quad \Delta \bar{T}_k = \frac{\pi f \varepsilon_0 \varepsilon_{zz}^i U^2}{2kd}$$

is some characteristic temperature drop; as we shall see later, $\Delta \bar{T}_k$ is the stationary value of the temperature difference between the bounding plate's external boundary $\tilde{z} = L$ and the point $\tilde{z} \to \infty$ in the surrounding medium. The constant $\xi = kL/G$ is called the Biot number [32], $a_n = 2\tan q_n/(\sin q_n \cos q_n + q_n) > 0$ are the dimensionless coefficients with the eigenvalues q_n satisfying the equation $q_n \tan q_n = \xi$, and $\tau_n = c_p \rho L^2/Gq_n^2$ are the characteristic time constants determining the transition to the stationary regime. For the convenience of further analysis, we will consider the temperature change $\Delta T(t)$ of NLC measured as the difference between the temperature of the NLC bulk and the temperature T_0 at infinity, $\tilde{z} \to \infty$ (which is also the initial temperature of the whole system),

$$(10\text{-}31) \quad \Delta T(t) = \Delta \bar{T}_k \left[1+\xi-\sum_{n=1}^{\infty} a_n \exp\left(-\frac{t}{\tau_n}\right)\right],$$

as the sum of the following two contributions,

$$(10\text{-}32) \quad \Delta T(t) = \Delta T_k(t) + \Delta T_g(t),$$

where

$$(10\text{-}33) \quad \Delta T_k(t) = T(\tilde{z}=L,t) - T_0 = \Delta \bar{T}_k \left[1-\sum_{n=1}^{\infty} a_n \cos q_n \exp\left(-\frac{t}{\tau_n}\right)\right].$$

is the difference between the temperature $T(\tilde{z}=L,t)$ of the bounding plate's external boundary, and T_0, and $\Delta T_g = T(\tilde{z}=0,t) - T_0(\tilde{z}=L,t)$ is the temperature difference across the bounding (say, glass) plate, Figure 10-6 .

As one can see from Eqs (10-31)—(10-33), the stationary value $\Delta T_k(t \to \infty)$ is indeed equal to $\Delta \bar{T}_k$; besides, $\Delta T_g(t \to \infty) = \Delta \bar{T}_g = \xi \Delta \bar{T}_k$. Therefore, the temperature increase of the NLC slab, $\Delta \bar{T} = \Delta T(t \to \infty)$ can also be represented by the sum of two terms:

$$(10\text{-}34) \quad \Delta \bar{T} = \Delta \bar{T}_k(1 + \xi) = \frac{\pi f \varepsilon_0 \varepsilon_{zz}^i U^2}{2kd} \left(1 + \frac{kL}{G}\right).$$

The model above, Eqs (10-31)—(10-34), suggests that one can distinguish two regimes of dielectric heating. When $\xi \gg 1$, the temperature increase of the NLC is controlled mainly by the temperature gradient in the bounding (glass) plates. For $\xi \ll 1$, the dominating control parameter is the heat transfer coefficient k of the surrounding medium, $\Delta T(t) \approx \Delta T_k(t)$. Note that in the second case, the model above and the Schadt' model [25] predict different dynamics of the temperature increase.

Equation (10-34) indicates that the NLC temperature change is influenced by many factors, including the liquid crystal's imaginary dielectric permittivity component ε_{zz}^i, the electric field (voltage U and frequency f), the thermal properties of the surrounding medium (heat transfer coefficient k), and the properties of the bounding plates (thermal conductivity G, heat capacity c_p, thickness L). Some of these predictions have been recently verified experimentally [33]. It would be of interest to expand the experiments [33] to the study of how the temperature changes affect the state of director and vice-versa; such a work is in progress.

5. SUMMARY

Dielectric properties of nematic liquid crystals (NLCs) depend on the frequency of the applied field through the dielectric relaxation. There are two important general consequences of the dielectric relaxation phenomena: (1) the switching memory effect, i.e., the non-local time relationship between the electric displacement and the electric field; (2) dielectric heating effect.

The switching memory effect is a reflection of the fact that the electric displacement, being the function of both the applied field and the material's properties, needs some finite time to adjust to the value of the electric field. The widely accepted model of the instantaneous relationship between the electric displacement and the electric field in the NLC is invalid when the characteristic times of the director dynamics are close to the relaxation times for molecular permanent dipoles. This time scale is typically in the submillisecond range which is of great interest for modern fast-switching devices. The electric displacement (as well as the dielectric torque density) becomes a function of the static dielectric properties of the NLC, the present and past electric field, and the present and past director. We discussed the recently proposed theory and experimental verification of the phenomenon [11]. The model in Ref [11] should be applicable to dynamic reorientation of other LC phases in the appropriate range of times/frequencies. In the case of ferroelectric LCs, the theory should be supplemented by the consideration of spontaneous electric polarization. A similar approach should be also applied to other systems, including those of biological significance, with tensor order parameters and nonstationary dielectric properties.

In the section 4, we presented an expanded model of dielectric heating effects of NLC in an external AC electric field. This part, especially the possible interplay between the temperature changes and director reorientation deserve further experimental studies.

The work was partially supported by NSF grants DMS 0456286 and DMR 0504516 and by DOE grant DE-FG02-06ER 46331.

6.   REFERENCE

1. P. G. de Gennes and J. Prost, *The Physics of Liquid Crystals* (Oxford University Press, New York, 1997).
2. V. Fréedericksz and V. Zolina, Trans. Faraday Soc. **29**, 919 (1933).
3. L. M. Blinov and V. G. Chigrinov, *Electrooptical Effects in Liquid Crystal Materials* (Springer, New York, 1994).
4. M. Kleman and O. D. Lavrentovich, *Soft Matter Physics: An Introduction* (Springer-Verlag, New York, 2003).
5. C. J. F. Böttcher and P. Bordewijk, *Theory of Electric Polarization* (Elsevier, New York, 1978).
6. W. H. de Jeu, *Physical properties of liquid crystalline materials* (Gordon and Breach, New York, 1980).
7. H. Fröhlich, *Theory of Dielectrics* (Oxford University Press, London, 1958).
8. H. Takanashi, J. E. Maclennan, and N. A. Clark, Jpn. J. Appl. Phys **37**, 2587 (1998).
9. B. A. Belyaev, N. A. Drokin, V. F. Shabanov, and V. A. Baranova, Physics of the Solid State **46**, 574 (2004).
10. H. G. Kreul, S. Urban, and A. Wurflinger, Phys. Rev. A. **45**, 8624 (1992).
11. Y. Yin, S. V. Shiyanovskii, A. B. Golovin, and O. D. Lavrentovich, Phys. Rev. Lett. **95**, 087801 (2005).
12. A. B. Golovin, S. V. Shiyanovskii, and O. D. Lavrentovich, Appl. Phys. Lett. **83**, 3864 (2003).
13. J. D. Jackson, *Classical Electrodynamics* (John Willey & Sons, Inc, New York, 1975).
14. N. Hill, W. E. Vaughan, A. H. Price, and M. Davies, *Dielectric Properties and Molecular Behavior* (Van Nostrand, London, 1969).
15. A. K. Jonscher, *Dielectric Relaxation in Solids* (Chelsea Dielectric Press, London, 1983).
16. B. K. P. Scaife, *Principle of Dielectrics* (Clarendon Press, Oxford, 1998).
17. W. Haase and S. Wrobel, *Relaxation Phenomena* (Springer, New York, 2003).
18. M. Schadt, Mol. Cryst. Liq. Cryst. **89**, 77 (1982).
19. L. A. Madsen, T. J. Dingemans, M. Nakata, and E. T. Samulski, Phys. Rev. Lett. **92**, 145505 (2004).
20. B. R. Acharya, A. Primak, and S. Kumar, Phys. Rev. Lett. **92**, 145506 (2004).
21. The estimate ν_{ff} follows from the dielectric relaxation time τ_{iso} ~13 ns measured in the isotropic phase of our material.
22. P. Debye, *Polar Molecules* (Dover, New York, 1929).
23. T. K. Bose, B. Campbell, S. Yagihara, and J. Thoen, Phys. Rev. A. **36**, 5767 (1987).
24. X. Ming and D. Yang, Appl. Phys. Lett **70**, 720 (1997).

25. M. Schadt, Mol. Cryst. Liq. Cryst. **66**, 319 (1981).
26. F. M. Leslie, Arch. Ration. Mech. Anal. **28**, 265 (1968).
27. J. L. Erickson, Trans. Soc. Rheol. **5**, 23 (1961).
28. Y. Yin, M. Gu, A. B. Golovin, S. V. Shiyanovskii, and O. D. Lavrentovich, Mol. Cryst. Liq. Cryst. **421**, 133 (2004).
29. C. H. Wen and S. T. Wu, Appl. Phys. Lett **86**, 231104 (2005).
30. J. Fourier, *The analytical theory of heat* (Dover, New York, 1955).
31. Hiroshi-Ono and Kazuaki-Shibata, J. Phys. D: Appl. Phys. **33**, L137 (2000).
32. A. V. Luikov, *Analytical Heat Diffusion Theory* (Academic Press, New York, 1968).
33. Y. Yin, S. V. Shiyanovskii and O. D. Lavrentovich, J. Appl. Phys. **100**, 024906 (2006).

CHAPTER 11

PHOTOCONDUCTING DISCOTIC LIQUID CRYSTALS

QUAN LI* AND LANFANG LI

Liquid Crystal Institute, Kent State University, Kent, OH 44242, USA
E-mail: quan@lci.kent.edu

1.   INTRODUCTION

Photoconducting discotic liquid crystals (DLCs) as a relatively new class of photo-conductors are attracting tremendous attention since they can form unique columnar nanostructures which exhibit remarkable mechanical, optical and electric properties. The history of photoconducting DLCs dates back to 1993 when Haarer and his coworkers reported that the discotic liquid crystalline hexapentyloxy triphenylene (HAT$_5$) showed high charge carrier mobilities although the first DLCs were discovered in 1977 [1, 2]. Over the past decade, great progress has been made on these fascinating photoconducting DLCs.

For liquid crystal display application, the conductivity of liquid crystals is detri-mental. Interestingly, liquid crystals are intrinsically conductive. This conductivity arises from two origins: ionic contamination which is almost inevitable, and conduc-tivity originating from the molecular structure and the phase of liquid crystal. The liquid crystals widely used in display application are composed of rod-like molecules in the nematic phase, resulting in low conductivity, i.e. at the order of ionic conduction. DLCs, on the other hand, typically have a much higher intrinsic conductivity than their display application counterpart. Instead of reducing DLC conductivity for use in aforementioned application, they are a good candidate for use in applications which require high conductivity. The disc- or plank-like structure of DLC molecule, in contrast with rod-like one, can facilitate face-to-face molecular stacking and lead to self-assembled superstructure by strong intermolecular inter-actions. Meanwhile, similar to liquid crystals comprised of conventional rod-like molecules, DLCs are also "soft" matter and are susceptible to external stimuli.

---

* Corresponding Author

A. Ramamoorthy (ed.), Thermotropic Liquid Crystals, 297–322.
© 2007 *Springer.*

2.   PHOTOCONDUCTIVITY

Photoconductivity is a property of some materials in which the material becomes more conductive upon electro-magnetic irrdiation, and the conductivity increases orders of magnitude. The irridiation could be visible light, ultraviolet light, X-ray or gamma radiation. Photoconductive materials are called photoconductors which are semiconductors in nature. When light is absorbed by the semiconductor, the configuration of electrons and holes changes and raises the electrical conductivity of the semiconductor, or the photogenerated charge carriers appear and render the material to be more conductive. Quantum mechanics requires the incident photon energy to match the intrinsic energy band gap of the photoconductor. A bias voltage is usually applied to characterize the photocurrent. The photosensitive conductivity, in another word, photo-tunable current at constant bias voltage enables a series of applications such as photocopying and photodetection. A classic example of photoconductor is polyvinylcarbazole which is used in photocopying.

3.   MECHANISM OF CHARGE TRANSPORTATION IN ORGANIC MATERIALS

While light absorption is the origin of increased charge carrier density, the conductivity also depends on the transportation of the charge carriers. In photoconducting organic compounds including liquid crystals, charge conductivity in the material is increased dramatically as a result of light induced excitons. Excitons are bound electron-hole pairs, and its dissociation will result in free electrons and holes as the charge carriers. In inorganic semiconductors, incident light can directly induce free electrons and holes. In organic semiconductors, this transient state of exciton has a much longer lifetime and can transport itself through the material. The mechanism of exciton transport is still not clear to date. One of the proposed theories is that excitons transport by diffusion, another is that the transport is in the form of energy resonance, while others indicate it is the hopping of excited states at different sites in the material. Photoconductivity of organic semiconductors results from the increased density of charge carriers under illumination.

There are different mechanisms of conductivity in organic compounds including liquid crystals. The most universal mechanism is ionic conduction. No matter how rigorously the material is purified, there always exist some ionic impurities. Under applied electric field these ions will migrate to the corresponding electrodes and contribute to the current. This current is characteristic of exponential increase with increasing temperature. One needs to pay attention to decipher the contribution of ionic conductance if the overall conductance of the sample is small.

A simplified model depicted in Figure 11-1 can qualitatively illustrate the origin of the conductance beyond ionic transportation. Each molecule can be modeled as a box, or a quantum potential well confining motion of electrons. For more conjugated molecules, charge is delocalized within a larger volume and the energy gap is smaller, as depicted in the model as a bigger box. From a simplified quantum

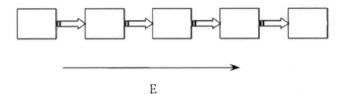

E

Figure 11-1. A simplified model of intermolecular charge transport in electric field. Each box represents a molecule. The size of the boxes is determined by the conjugation of the molecule, and the distances between boxes are the distances between molecules in macroscopic crystalline structure. The bias electric field (E) provides a directed transportation of excitons or charge carriers generated after dissociation of exciton

mechanics reasoning, we could infer that the bigger the box is, the smaller the energy gap is between the ground state and the excited state. As depicted in Figure 11-1 electrons are delocalized within each box, and the relatively bigger size of the box allows for formation of excited states at relatively low energy input. In order to be macroscopically conductive, these excited states need to transfer from one box to another and dissociate. The occurrence of this transportation is mostly termed as a hopping process. If external bias is applied, this hopping leads to directional transport, in another word, a current.

In reality, the boxes can be of complex internal structure determined by the molecular structure. Conjugation within a molecule determines the size of each of these boxes and energy levels within them. For multi-atomic molecules, the energy levels of the molecules adopt a band structure for which the most interesting energy levels are the highest occupied molecular orbital (HOMO) and lowest unoccupied molecular orbital (LUMO). The band gap between the HOMO and LUMO determines the tendency for excitation of the molecule.

Considering intermolecular charge transport, short distances between molecules or closely packed molecules are favorable for charge transport. Hopping occurs more easily in this arrangement, especially for those involving quantum tunneling. For band transport to occur, it is not only the distance between molecules but also the degree of order that makes a difference. The more periodic the positional order is, the more possibly a band transportation exists which will dramatically increase the conductance. In addition to positional order, orientational order is important because organic covalent bonds are directional, so orientationally ordered molecules will facilitate molecular orbital overlap. This cannot be seen in our simplified box model, but if the box is changed to a less symmetric shape, the orientation effect can be represented. It is based on their highly conjugated structure having an aromatic conjugated moiety as a core part that DLCs are more favorable for the purpose as charge generating and transporting materials.

In summary, the dominant factors in determining the magnitude of the charge carrier mobility are (1) the degree of order, (2) the size of the conjugation within each molecule, and (3) the extent of electronic orbital overlap between adjacent molecules. Among these factors, the structure ordering is the most important.

The crystalline order should be ideal in a defect free situation, but unfortunately it is very difficult to achieve defect free single crystals of large volume or single crystalline film of large area. The inevitably existing grain boundaries and defects act as deep traps and reduce the charge mobility dramatically. In addition, polycrystalline materials are intrinsically inhomogeneous. Therefore, for applications requiring uniformity, they are not a good choice even though the low conductivity could be tolerated for some occasions. On the other hand, for amorphous materials in which uniformity can be guaranteed, the lack of order at the molecular level decreases the charge mobility and renders the material poorly conducting. A system with large volume uniformity/homogeneity and microscopic/molecular level order is desired.

Two major categories of 1-D organic conducting materials are conjugated polymers which rely on intra-chain charge transport, and π-stacked small molecules which depend on intra-columnar transport. In the conjugated polymer system the charge transport is based on conjugation and electron delocalization within each molecule, and the charge transport between different polymer chains is based on hopping. In the π-stacked small molecule system the charge transport is governed by the tendency of molecular excitation and the overlap of molecular frontier orbitals between adjacent molecules.

Liquid crystals especially DLCs are a solution for this problem due to their special properties as we will describe below.

4.   UNIQUE STRUCTURES OF DISCOTIC LIQUID CRYSTALS

Liquid crystal systems are unique in their partial ordering. The absence of complete order leaves the freedom to self-repair misorientation, i.e. defect in orientational order. This property endows liquid crystals a self-healing property and enables easy achievement of a large area single domain free of grain boundaries. Generally, DLCs have a discotic core surrounded by some flexible chains (Figure 11-2). The discotic core unit is usually a conjugated structure such as benzene, naphthalene, triphenylene or perylene (Figure 11-3). The conjugated structure is a prerequisite for the DLC photoconductivity since it will result in a relatively small band gap between HOMO and LUMO of the molecule which is a desirable property to excite the molecule with a stimulus of a reasonable magnitude. It should be noted here that porphryins and phthalocyanines among these discogens (Figure 11-3) have superior optic-electro properties due to their low energy band gaps between HOMO and LUMO. The structure of porphyrin or phthalocyanine also enables them to bind with some metal ions. The metal complexes at the center of porphyrins or phthalocyanines can tune the molecular energy level without degradation and sometimes can even improve the mesophase stability.

In general, the partial order increases intermolecular interaction and the partial fluidity makes relaxation possible so as to reduce the defect and suppress grain boundary formation.

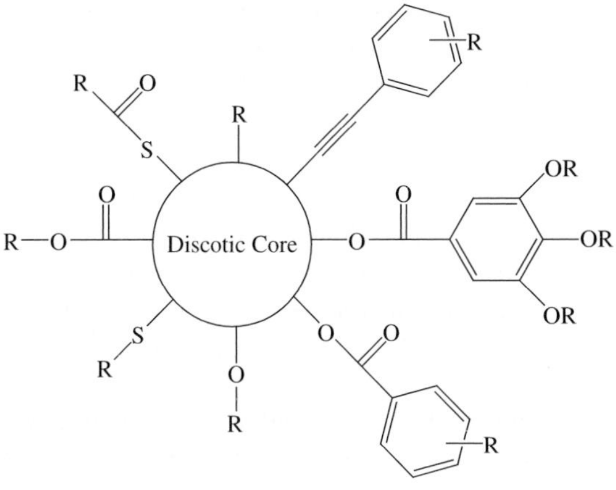

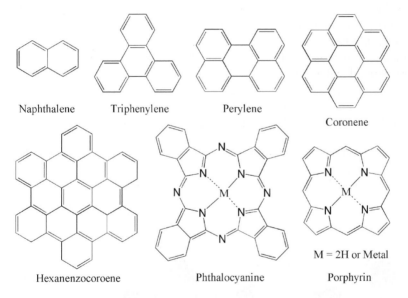

Figure 11-2. A general structural template for DLCs

Naphthalene   Triphenylene   Perylene

Coronene

M = 2H or Metal

Hexanenzocoroene   Phthalocyanine   Porphyrin

Figure 11-3. Structure of some discotic mesogenic cores

For organic electronic materials, the higher degree of conjugation is a preferred molecular property which corresponds to a bigger box as we modeled in Figure 11-1 and will result in a smaller energy band gap. Fortuitously, this is an intrinsic structural property for most of the DLCs since conjugation is also the origin of rigidity which is necessary for many thermotropic mesogens.

Besides molecular structure, higher ordered phase structure is also beneficial to achieve high charge mobility. Based on the charge transport mechanism, closer packing and more chance of overlapping of molecular orbital are favorable. However, this is a dilemma in the sense that higher conjugation will result in a more viscous phase which defects are expected to occur and the resulting material is more difficult to process.

There have been many phase behaviors and classifications on DLCs reported [3]. Most commonly, DLCs exhibit columnar phases and discotic nematic phases (Figure 11-4 and 11-5). From the molecular interaction point of view, discotic mesogens usually incorporate a big conjugated plane (disc) which makes the inter-action between molecules much stronger than in rod-like systems. As a result, the appearance of the nematic phase is rare in discotic materials. Columnar phases (Col) are more commonly observed in DLCs and are more ordered than the discotic nematic phases. In a columnar phase, the molecules aggregate into columns and these columns are arranged into a 2-D periodic structure. Within each column, the molecules are fluid-like, i.e. the distance and molecular orien-tation are fluctuating at a small magnitude. According to the degree of order, the ordered columnar phase (Col_o) and disordered columnar phase (Col_d) are specified. The most typical arrangement of columns is in a hexatic manner. However, in the case of oval shaped molecules and oblique molecular orientation with respect to the columns, other arrangements are possible. According to the symmetry of this 2-D lattice, the columnar phases are classified as hexagonal columnar phase (Col_h), rectangular columnar phase (Col_r), and oblique columnar phase (Col_{ob}), etc. Columnar phases are crystal in 2-D and fluid in 1-D. The higher ordered columnar phases are very solid, yet still feature a much smaller shear elastic modulus than normal crystals. When the correlation between molecules in different columns exists, the columnar liquid crystal becomes a regular 3-D crystal. The discotic nematic phase (N_D) is similar to the nematic phase formed by rod-like molecules. The director is the molecular rotational symmetry axis, i.e. the norm of the molecular plane. N_D is less viscous than other higher ordered discotic columnar phases.

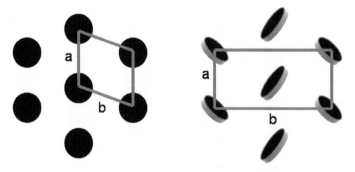

Figure 11-4. Molecular positions within the hexagonal *columnar phase* (left) and rectangular *columnar phase* (right)

Figure 11-5. Molecular arrangements within the discotic nematic phase

Overall, DLCs are better candidates than the calamitic liquid crystals and amorphous organic compounds for electronic application due to their generally larger conjugation of the molecular structure. Also, columnar liquid crystalline phases are highly ordered phases. The preferred charge transport channel is along the columns. A very good property of discotic columnar phases is that due to their high viscosity at low temperature a glassy state with frozen columnar order is usually achieved during transition from the higher temperature less ordered phase without introducing many defects and grain boundaries, avoiding to a great extent the number of deep charge traps.

5. CHARACTERIZATION METHOD

The most widely used experimental method to measure photoconductivity is the time of flight (TOF) measurement. In this method, the electron and hole mobility can be separately measured. The set-up of TOF measurement is composed of a cell which resembles a standard liquid crystal cell with transparent or semi-transparent electrodes sandwiching a thin layer of aligned liquid crystal material. A short pulse laser is shed on the surface of the material to induce charge separation within the material. A bias voltage is applied to provide a field to drive the positive and negative charge carriers to drift to corresponding electrodes. The photocurrent after the pulse is recorded as a function of time. By choosing the polarity of the bias voltage the charge mobility for both sign charge carriers can be measured. If the laser is shed on the positive electrode side, positive charge carrier mobility can be obtained, and vice versa. The schematic diagram of the set-up is shown in Figure 11-6.

The stimulating laser needs to be of a short wavelength and narrow pulse since the quantum transition requires the incident photon energy to match the energy gap between molecular energy levels, which often needs blue photons or even higher energy photons. From the shape of the photo-current vs. time graph, the nature of the charge transportation can be determined.

TOF measurement needs the material to be aligned as shown in Figure 11-7 (left). A schematic TOF transit is shown in Figure 11-7 (right). The total photocurrent comes from a superposition of all the charge transport mechanisms. Ionic conduction contribution is universal but the value is small. Electronic conduction contribution

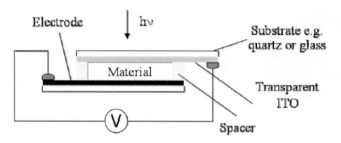

Figure 11-6. Schematic diagram of time-of-flight (TOF) experiment. The incident irradiation is a narrow beam-width short pulse laser.

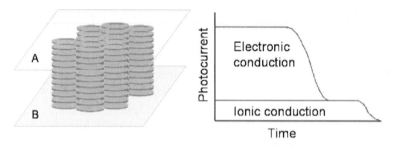

Figure 11-7. The structure of homeotropically aligned columnar phase (left); an ideal transient photo-current in a TOF measurement in a trap-free sample (right). The fast mode is electronic conduction including electron and hole mobilities; the slow mode is ionic charge transport

including hole transport and electron transport is a major component of the overall conduction [4].

It should be noted here that while TOF technique is widely used, there is another experimental method to measure conductivities and mobilities, namely pulse radiolysis-time resolved microwave conductivity (PR-TRMC) measurement as shown in Figure 11-8.

In this method, a known low concentration (micromolar) of electron-hole pairs is created uniformly throughout the material using a nanosecond pulse of ionizing radiation [5, 6]. Microwaves are used to quantitatively probe (without the need of electrode contacts) the radiation induced conductivity within the sample. The mobility is determined from the end of pulse value of the resulting transient conduc-tivity. The material does not need to be aligned.

The PR-TRMC mobility is considered to be the more basic value, however, the electron and hole mobilities cannot be separately measured. The measured mobility is the sum of the one-dimensional (1D) electron and hole carrier mobilities $(\Sigma\mu_{1D} = \mu_+ + \mu_-)$. In contrast, the TOF mobility is considered to be more charac-teristic of bulk sample which is susceptible to defects, impurities and grain bound-aries, etc. and is more relevant to real applications. Generally, the TOF mobility is lower than PR-TRMC value.

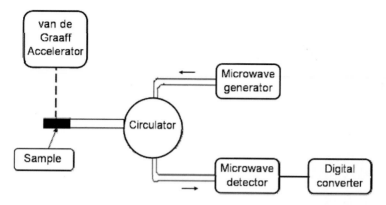

Figure 11-8. Brief schematic diagram of pulse radiolysis-time resolved transform microwave conductivity

6. PROCESSING OF THE PHOTOCONDUCTING DISCOTIC LIQUID CRYSTALS

In the previous sections it was assumed that the photoconducting DLC materials are well aligned at the molecular level. However, the alignment of the materials is not trivial all the time. Surface treatment, thermal treatment including annealing, and mechanical treatment including shearing and structural modification are used to achieve the goal of large area uniform orientation of the disc molecules.

It is known that calamitic (rod-like) liquid crystal alignment can be achieved by well established surface treatment such as surface polymer coating to achieve reliable homeotropic and homogeneous alignment in their nematic phase. Electric field or magnetic field can also be used to achieve their uniform alignment. These will not be detailed here since they are well-developed techniques over a long time. Unfortunately, the alignment of DLCs especially with large conjugation system is completely different from calamitic liquid crystals and is very hard to achieve. The well established alignment technique for calamitic liquid crystals does not usually work for DLCs because the columnar mesophases have much higher viscosity than the calamitic liquid crystals, and accumulation of their ends at the substrate-DLC interface might cost a great deal of entropy since the columns of disc-like molecules are long.

The alignment of photoconducting DLCs is crucial for achieving high conductivity needed for possible applications. There are two typical alignments of DLCs, namely homeotropic and homogeneous alignments in which the columns are perpendicular and parallel to the electrode surface, respectively (Figure 11-9). Homeotropic alignment in columnar phases can provide a most efficient path for electrons and holes along the columnar axis which is most favorable for high conductivity and applications in photovoltaic cells and organic light emitting diodes. Interestingly, most of the homeotropic alignments reported to date contain domain boundaries and disclinations which suppress the carrier transport [7]. It is a challenge to achieve the

Figure 11-9. Schematic illustrations top: un-aligned hexagonal columnar phase; middle: homotropic alignment; bottom: homogeneous alignment

film with perfect homeotropic alignment over a large area. A homogenous alignment of the columnar phase of the photoconducting DLCs has potential applications such as organic thin-film transistors.

In contrast to well studied calamitic liquid crystals, DLCs are relatively less studied, maybe because they are not the major component for the information display industry which boomed during the past decades. But the increasing interest of the photoconducting DLCs as electronic materials calls for material processing method and protocol to propel the development in this field. Application-directed efforts on photoconducting DLCs were particularly fueled by their unique anisotropic optical and electronic properties [8]. Optical and electronic applications, however, require a perfect alignment of the self-organizing material which is usually obtained by utilizing interfacial forces or by applying external fields. Although alignment of calamitic liquid crystals on surfaces or in liquid crystal cells is a well-established technology, few of the alignment methods developed for calamitic liquid crystals can affect the alignment of the DLCs [9]. As a result, there are few reports on applying an external field such as magnetic, electric and mechanical fields as an effective alignment method.

The magnetic field might be most effective on photoconducting DLCs. This is because the large anisotropic molecular magnetic susceptibility and the highly conjugated structure of DLCs usually result in a pronounced diamagnetic effect. The alignment usually takes place at the transition from isotropic or less ordered discotic nematic phase to the columnar phase when the elastic free energy and viscosity are small. The more viscous the DLC material is, the larger the magnetic field is needed. The uniform alignment of the less ordered and viscous discotic nematic phase of a triphenylene derivative sandwiched in a liquid crystal cell was achieved when a 5 T magnetic field was applied parallel to the cell surface during the cooling process. The director was perpendicular to the surface when the cell was coated with polyimide alignment. The quality of homeotropically aligned films was found to be depending on the type of alignment layer [10]. To achieve homogeneous alignment, a rotation method can be used.

Electric field might be also an effective method for alignment of photoconducting DLCs. The anisotropic electronic susceptibility results in the aligning effect of the molecular plane parallel to the applied field. Similar as magnetic field alignment, the aligning effect also takes place at the transition from isotropic to ordered mesophase. However, due to the electric conductivity the electric field applied on the material is restricted to a certain extent to protect the material from short circuit damage.

Based on the anisotropic rheological properties, flow alignment induced by a mechanical perturbation of the system can be used to align the highly viscous columnar phases. The mechanical force aligns the columns parallel to the applied shear field, i.e. parallel to the long axis of the columns. In a liquid crystal cell, shearing the cell in one direction might result in homogeneous alignment with the column axis along the shear direction [11]. The homogeneously aligned sample by a shear is shown in Figure 11-10. In a similar way, carefully prolonged DLC fiber within its columnar phase temperature could align the material with column axis parallel with the fiber axis [12].

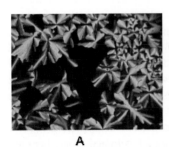 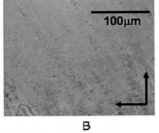

A   **B**

Figure 11-10. Polarizing micrographic textures of a discotic liquid crystalline porphyrin **P6** (A: before shear; B: spontaneously homogeneous alignment achieved by shearing. The sample extinguishes twice as it rotates 180o between crossed polarizers)

The shortcoming of the alignment by external fields is that the alignment will be destroyed and can not restore alignment once heated to isotropic phase. Self-healing alignment needs surface anchoring or a permanent bulk restoring force.

Surface anchoring has been successful in liquid crystal displays in which the calamitic liquid crystals are predominantly aligned by interactions with specially treated or prepared surfaces [13]. However, surface treatments developed for calamitic liquid crystals have surprisingly little effect on the alignment of DLCs. There is little development on surface alignment methods for higher ordered and more viscous discotic columnar mesophases reported. Alignment of DLCs on a surface appears to depend very little on the nature of surface but more likely a property of the DLC material itself. In fact, hexaalkoxy triphenylene aligns homeotropically when cooled down from its isotropic phase, independent of the type of alignment layer [14]. More viscous DLCs such as compounds based on phthalocyanines show small domains of more homogeneously aligned columns, unaffected by the type of surface.

One of the successful methods to obtain monodomains of homogeneously aligned discotic columnar mesophases is the Langmuir-Blodgett (LB) technique [15, 16]. The approach preferentially works in DLC systems with non-symmetrically substituted molecules that have polar and non-polar side-chains at different position of the periphery of the molecule. However, this technique results in planar alignment with in-plane degeneracy of alignment. Flow effect introduced during the dipping process is needed to determine in-plane molecular orientation.

Surface modified by self-assembling monolayers (SAMs) exhibited similar properties as LB films. In a report using gold substrate and functionalized thiol surface layer material, the cores of triphenylene derivatives with 1-2 thiols attached to their side-chains align homogeneously and form domains of columnar stacks with different in-plane orientation [17]. A homeotropic alignment was obtained for derivatives with thioethers attached to the core. The situation with terminal thiols attached to more than two side chains per molecule resulted in non-uniform layers.

It was found that infrared laser irradiation of the hexagonal columnar mesophase (Col_{hp}) shown by a discotic liquid crystaline triphenylene causes alignment change forming a new domain with uniform alignment. This phenomenon is thought to be achieved by the excitation of the selected vibration mode of a chemical bond [18]. Recent studies on this phenomenon have also shown that a relationship was found for the directions of the polarization of incidence and the transition moment of the vibration excitation. The results imply that infrared irradiation is a possible technique for device fabrication using columnar mesophase as a liquid crystalline semiconductor.

Mixing columnar DLCs with a low molar mass non-discotic compound gives rise to rod-shaped phase separated discotic domains having a hexagonal cross-sectional area. The columns are oriented preferentially along the rod-axis. Such areas may be of considerable interest for electro-optic applications. Because an oriented matrix can be used to orient all these domains along a given direction, the surrounding matrix could be later removed by melting away at a proper temperature without jeopardizing the global alignment [19].

Watson and his coworkers reported alignment of discotic liquid crystalline hexa-*peri*-hexabenzocoronene (HBC) by zone-casting technique onto a quartz substrate (Figure 11-11) [20]. They also reported anisotropic optical and electric properties of film aligned by this technique. It is found that the conduction in the direction of casting (columnar alignment direction) is favored by at least a factor of 10 over the perpendicular direction. The anisotropy in its optical absorption at room temperature is not obvious which is attributed to the close to 45° tilt of the HBC molecular plane in the herringbone arrangement of the columns, characteristic of the crystalline solid. [21]

An important effort in Quan Li's group is to develop the novel functional DLCs capable of efficiently absorbing sunlight as well as forming spontaneously homeotropic alignment. A series of well-defined light-harvesting liquid crystalline porphyrins with the same basic structure of the best photoreceptor in nature, chlorophyll, have been designed and synthesized. Spontaneous alignment in porphyrin based DLCs has been successfully achieved by tuning the molecular structure (Figure 11-12) [22]. Further investigation is in progress.

For polymeric discotic liquid crystals, the alignment is different. One approach is to polymerize aligned monomer to achieve aligned polymer. The other is to perform polymerization first and then using mechanical force to align the polymer. In both cases a polymerizable side group needs to be attached chemically to the discotic mesogen. One successful example is the so-called Fuji film by Fuji Photo Film Company in which the discotic liquid crystalline monomer was first aligned in an electric field, and then photo-polymerized under this alignment condition to fix the molecular orientation. This film provides exact compensation that corrects the viewing angle problem by the calamitic liquid crystal display [23].

Another approach of aligning polymeric DLCs is to use the mechanical force after polymerization. Cross-linked discotic liquid crystalline elastomer can be aligned by stressing or elongating the elastomer in one direction [24]. Furthermore, if a small

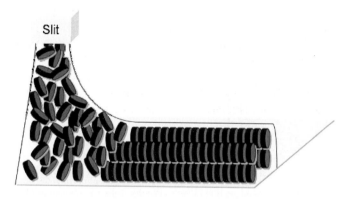

Figure 11-11. Schematic illustration of zone-casting alignment

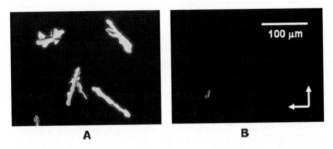

Figure 11-12. Crossed polarizing micrographic textures of spontaneous homeotropically aligned columnar phase of liquid crystalline porphyrin **P8** at room temperature (A and B with different cooling rate. The dark areas are homeotropic alignment where the planes of the disk-shaped molecules are parallel to the substrates. Bright domains represent areas where the molecular planes are tilted with respect to the substrates.)

amount of polymer (typically around 3% to 10%) does not cause detrimental effect, the alignment can be achieved by shearing polymer dispersed discotic liquid crystal (PDDLC) as well. The cross-linked polymer network will act as a bulk restoring force that will provide an alignment force on the discotic inclusions as shown in Figure 11-13.

It is worth noting here that highly viscous discotic columnar mesophases do not easily form monodomain and prefer a homogeneous alignment to homeotropic alignment. Homotropic alignment is more favorable for optic-electro applications such as photovoltaics. As has been discussed in the previous sections, larger conjugated core structure will increase the electronic conducting property, but it will also increase the viscosity of the mesophase of the material which makes the alignment more difficult. The very viscous mesophase exhibited by phthalocyanines and hexabenzocoronenes, for example, remains until decomposition and makes the material very difficult to be homeotropically aligned and capillary-filled into a cell.

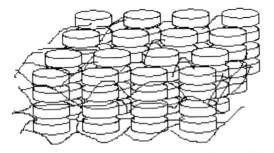

Figure 11-13. Illustration of mechanical stress induced alignment of DLC in elastomer/polymer dispersed discotic liquid crystal. The stressed polymer network will favor aligned molecular arrangement to adjust to the mechanically deformed domains separated by the polymer strands

7. PHOTOCONDUCTING DISCOTIC LIQUID CRYSTALS

Many photoconducting DLC systems have been studied since the pioneer work by Haarer et al. [2] In the category of the DLCs, triphenylenes have been the focus of research because of the facile synthesis of their analogs with various possible structural variation, the easy alignment and their chemical stability up to a reasonably high temperature [25, 26, 27]. According to the structure-mobility relationship discussed previously, degree of conjugation, degree of order and degree of molecular orbital overlapping are the key factors for the magnitude of the charge carrier mobility. Some photoconducting DLCs with different molecular structure are shown in Figure 11-14–11-15

Simple Col_h derivatives such as **1** (HAT_{11}) give low mobilities (**1**, $\mu = 1.0 \times 10^{-4}\,cm^2\,V^{-1}\,s^{-1}$ at 60° C). The mobilities gradually increase to $\mu = 1.9 \times 10^{-3}\,cm^2\,V^{-1}\,s^{-1}$ when the chain length decreases down to pentyl group (HAT_5) [28, 29, 30, 31]. This could be counted by the increase of ordering due to the decrease of side chain length and the resulting less fluid nature. Interestingly, there is

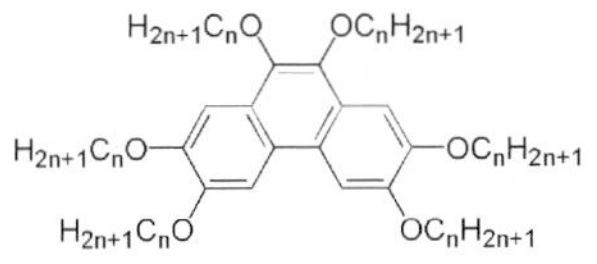

1 (HATn: n=3-11)

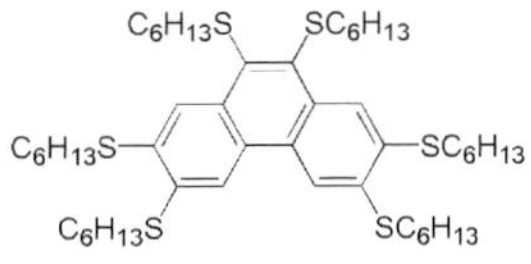

2

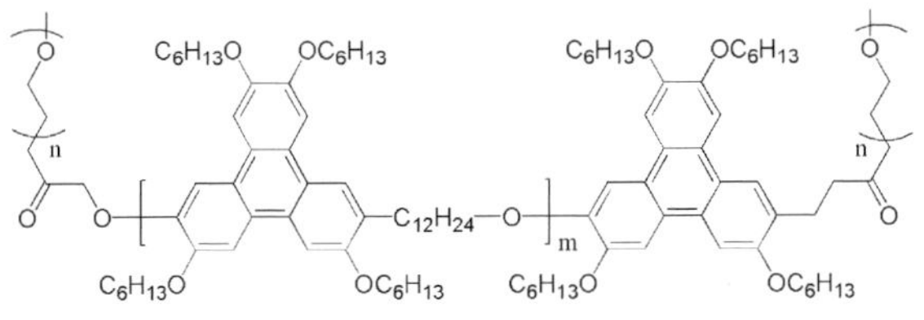

3 (m~12 and n~45)

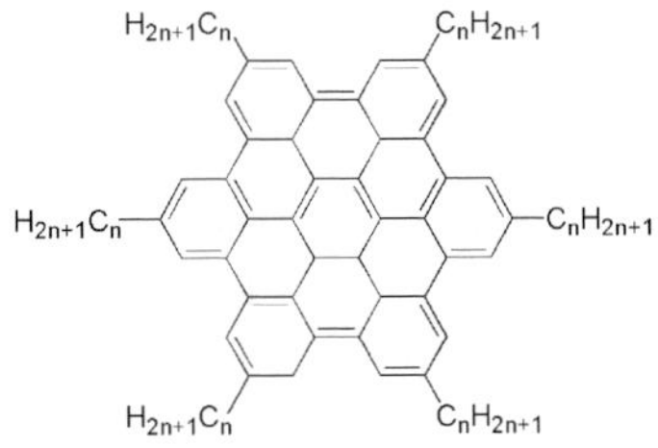

4 (HBCn: n =10, 12 and 14)

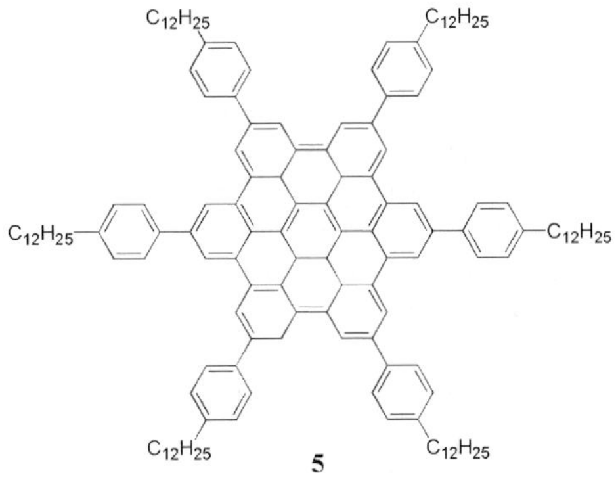

5

Figure 11-14. Examples of photoconducting DLCs

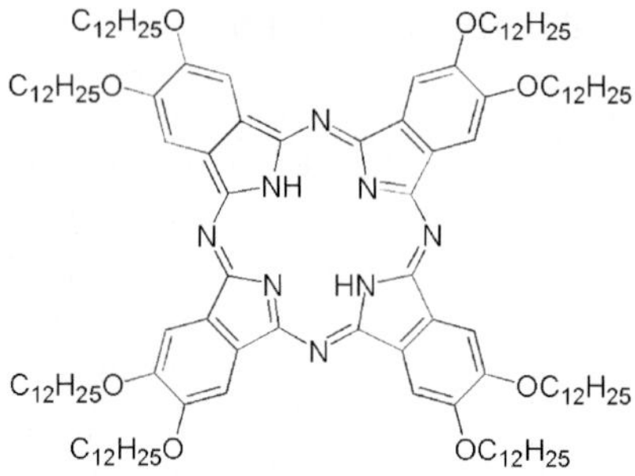

Figure 11-15. Examples of photoconducting DLCs

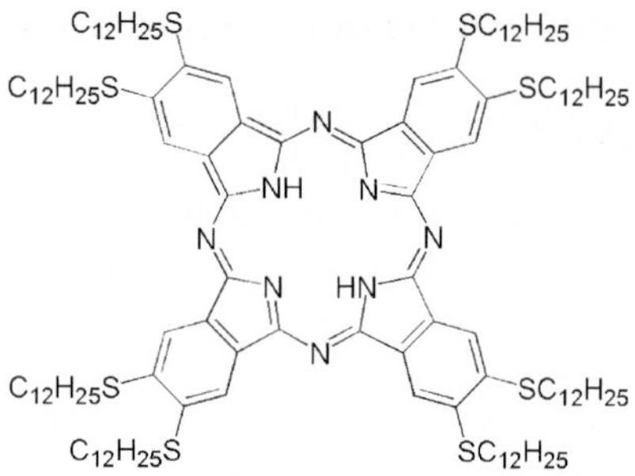

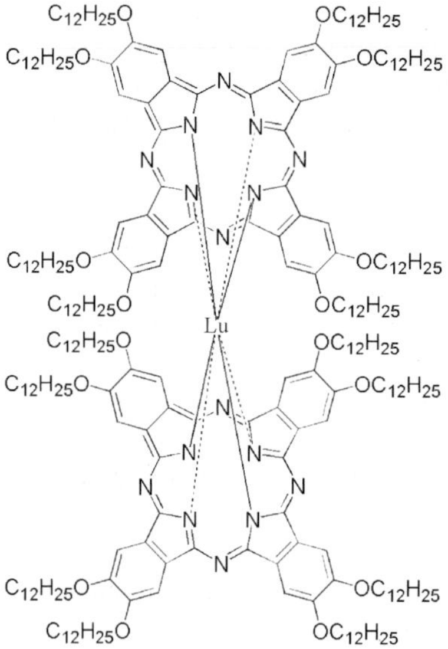

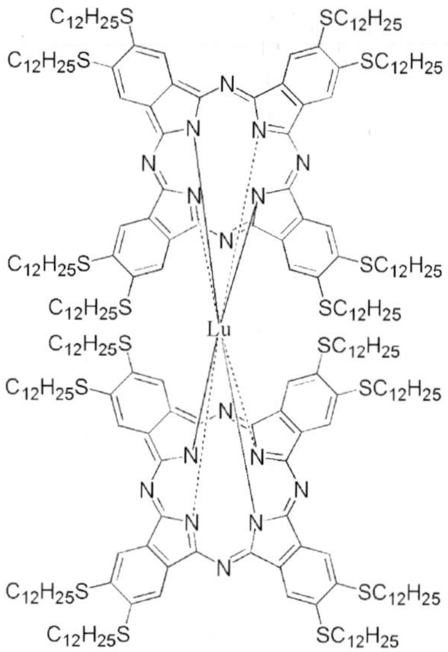

a dramatic increase in the charge mobility when the chain length decreases to butyl (HAT$_4$) or propyl (HAT$_3$). This is because that HAT$_3$ ($\mu = 1.2 \times 10^{-2}$ cm^2 V^{-1} s^{-1} at 130° C) and HAT$_4$ ($\mu = 2.5 \times 10^{-2}$ cm^2 V^{-1} s^{-1} at 80° C) have a much more highly ordered columnar plastic phase.[29, 31] When alkoxy side chain in **1** (HAT$_n$) is replaced with alkylthio, a much faster carrier mobility is observed, e.g. **2** (HTT$_6$, $\mu = 1 \times 10^{-1}$ cm^2 V^{-1} s^{-1} at 65° C) and **1** (HAT$_6$, $\mu = 4 \times 10^{-4}$ cm^2 V^{-1} s^{-1} at 65° C).[31] Recently Hanna et al. investigated the electron carrier transport in the DLC **2** using TOF method and reported that the very fast ambipolar transport was observed in the ordered columnar phase (H phase).[32] Incidentally, both electron and hole mobilities are $\mu = 0.8 \times 10^{-1}$ cm^2 V^{-1} s^{-1} at 45° C.

Donovan et al. reported a triblock polymeric discotic liquid crystalline **3** which has electrons as the major charge carriers. The material is a triblock copolymer consisting of a polymeric main chain discotic liquid crystalline triphenylene capped at either end with blocks of poly(ethylene oxide) [4]. An upper limit of $\mu = 8.4 \times 10^{-3}\,\text{cm}^2\,\text{V}^{-1}\,\text{s}^{-1}$ for the electron mobility at 50° C was obtained in this system.

Larger poly-aromatic cores generally result in higher mobilities (easier molecular orbital overlapping) and more efficient exciton formation and dissociation (smaller band-gap). Warman et al. reported that the charge carrier mobility as high as $0.2 \sim 0.5\,\text{cm}^2\,\text{V}^{-1}\,\text{s}^{-1}$ was measured by PR-TRMC in the hexagonal columnar liquid crystalline phase of the hexabenzocoronenes **4** (HBC$_{10}$, $\Sigma\mu_{1D} = 0.26\,\text{cm}^2\,\text{V}^{-1}\,\text{s}^{-1}$ at 133° C; HBC$_{12}$, $\Sigma\mu_{1D} = 0.38\,\text{cm}^2\,\text{V}^{-1}\,\text{s}^{-1}$ at 110° C; and HBC$_{14}$, $\Sigma\mu_{1D} = 0.31\,\text{cm}^2\,\text{V}^{-1}\,\text{s}^{-1}$ at 116° C) and **5** ($\Sigma\mu_{1D} = 0.46\,\text{cm}^2\,\text{V}^{-1}\,\text{s}^{-}$ at 192° C) [33]. Columnar phases exhibited by phthalocyanines (**6**, $\Sigma\mu_{1D} = 5.1 \times 10^{-2}\,\text{cm}^2\,\text{V}^{-1}\,\text{s}^{-1}$; and **7**, $\Sigma\mu_{1D} = 0.28\,\text{cm}^2\,\text{V}^{-1}\,\text{s}^{-1}$) and their rare-earth metal sandwich complexes (**8**, $\Sigma\mu_{1D} = 0.17\,\text{cm}^2\,\text{V}^{-1}\,\text{s}^{-1}$; and **9**, $\Sigma\mu_{1D} = 0.7\,\text{cm}^2\,\text{V}^{-1}\,\text{s}^{-1}$) have high charge carrier mobilities. Replacement of oxygen in **6** and **8** with sulfur results in an increase in charge mobility [34, 35].

DLCs are not the only system which exhibits photoconductivity or high resistant semiconductivity. Conjugated polymers and other highly ordered liquid crystals such as smectic liquid crystals also show similar properties. By comparing these systems, we could conclude that each system has their advantages and disadvantages. Smectic liquid crystals as another category of ordered liquid crystals has been reported that the photoconductive charge mobility is at the order of $10^{-2} \sim 10^{-5}\,\text{cm}^2\,\text{V}^{-1}\,\text{s}^{-1}$ [36, 37, 38]. For polymeric organic semiconductor, Warman et al. using the PR-TRMC technique compared the mobilities in the columnar discotic liquid crystals and the conductivity of conjugated polymers [39]. The upper limit values on the order of $10\,\text{cm}^2\,\text{V}^{-1}\,\text{s}^{-1}$ were measured for single-crystal polydiacetylenes (PDAs) polymerized either thermally excited or with small amount of irradiation although PDAs have been found to be unsuitable for device applications. However, for the same polymer which is typically amorphous or polycrystalline, much lower values ranging from 0.009 to $0.125\,\text{cm}^2\,\text{V}^{-1}\,\text{s}^{-1}$ were observed. It was understood that the morphology and the disorder in the polymer conjugated backbone structure was responsible for this decrease. Large mobilities are found for discotic materials with maximum values close to $1\,\text{cm}^2\,\text{V}^{-1}\,\text{s}^{-1}$ in both the crystalline solid and liquid crystalline phases owing to their self-organizing nature and high degree of structural order, which compensates for the weaker electronic coupling between monomeric units in the discotic materials compared with covalently bonded conjugated polymers.

The highest PR-TRMC (short-range) mobilities (the sum of holes and electrons) to date reported are the DLC **9** ($\Sigma\mu_{1D} = 0.7\,\text{cm}^2\,\text{V}^{-1}\,\text{s}^{-1}$). The highest TOF (long-range) hole mobility so far reported is the phthalocyanine **10** (8H$_2$Pc, Figure 11-16) in the columnar phase.[40, 41] The mobility is $2 \times 10^{-1}\,\text{cm}^2\,\text{V}^{-1}\,\text{s}^{-1}$ at 100° C on cooling form the Col$_h$ phase to the Col$_r$ although there is no obvious X-ray evidence for high order.

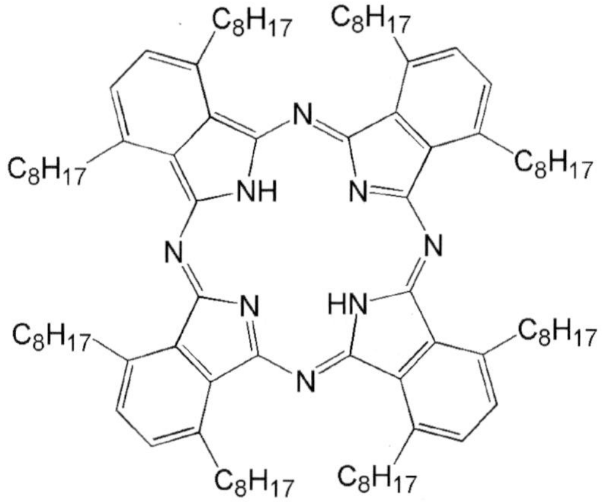

10 (8H₂Pc)

Figure 11-16. Molecular structure of photoconducting DLC 1,4,8,11,15,18,22,25-octaoctyl-phthalocyanine

Similar as inorganic semiconductors, impurities are an important factor that can change the energy level and the charge transport properties in organic semiconductors including photoconducting DLCs. The effect of the impurity, i.e. the other component of the system, depends on the nature of the impurity.

Investigation of the influence of charge trapping impurities on the charge mobility shows that even trace amount of such impurities will ruin the charge transport property of the liquid crystal material. The electronic transportation can be completely destroyed leaving only ionic conductance. The presence of impurities, even of very small amount, will change the charge transportation at orders of magnitude.

While charge trapping impurities are undesirable for improving charge mobility, selectively doping an electron donar with an electron acceptor and vice versa will result in improved properties in a controlled way if the energy band is aligned. Chen et al. investigated the compound system of copper tetra-(octyl-alkoxy-carbonyl)-phthalocyanine **11** (CuPc-C₈, p-type) and 3,4,9,10-tetra-(octyl-alkoxy-carbonyl)-perylene **12** (Pery-C₈, n-type) (Figure 11-17). [42] The novel **11/12** composite was prepared by the solution-blending method. Atomic force microscopy (AFM) demonstrated the formation of the bulk heterojunction structure in their cast-coated films. Enhanced photosensitivity was observed in the photoreceptor made from the **11/12** composite, which was understood to be a result of the large interfacial area between the two components due to the existence of the bulk heterojunction structure.

Besides doping, the hetero-junction formed by semiconductors of different nature can enhance charge transfer. Savenije and his coworkers using the "flash-photolysis time-resolved microwave conductivity" technique (FP-TRMC) studied

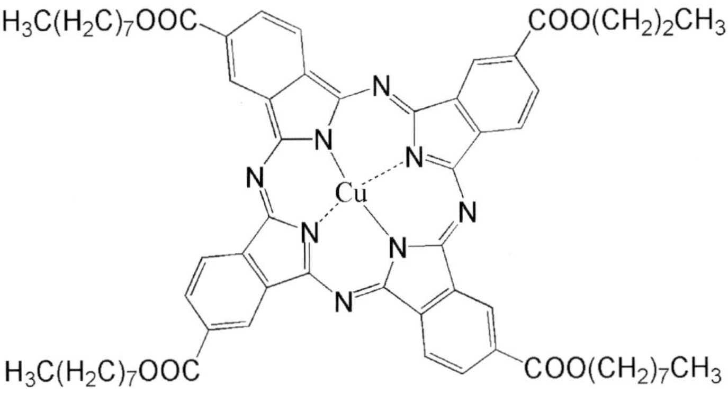

11 (CuPc-C$_8$)

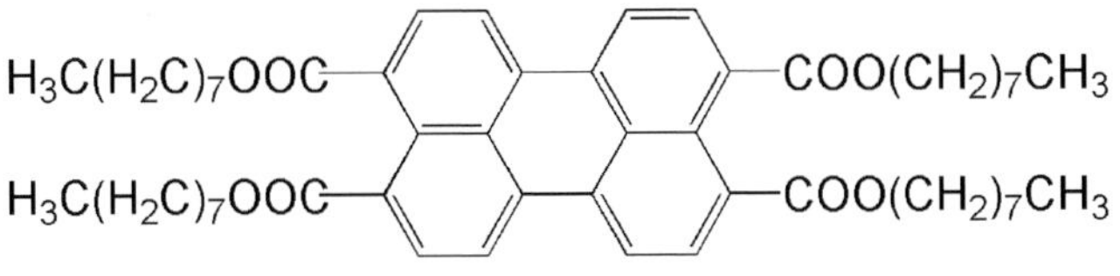

12 (Pery-C$_8$)

Figure 11-17. Examples of photoconducting DLCs

the bilayer system of porphyrin and TiO$_2$ thin film, i.e. free base tetra-*para*-octylphenyl porphyrin (**13**, H$_2$TOPP, Figure 11-18), spin-coated onto a smooth layer of TiO$_2$ [43]. The freshly spin-coated double layer consisting of 50 nm thick shows an incident-photon-to-charge-separation efficiency (IPCSE) of 10% on excitation at 430 nm. They concluded from the amplitude and the temporal shape of the photo-conductivity transients that electron injection occurs both from the singlet and the triplet excited states for the freshly spin-coated double layer. These triplet states can travel by energy transfer over distances of at least 9.6 nm. However, by heating the sample above the crystalline-discotic-lamellar phase-transition temperature of **13**, the alignment of the molecules was changed to homogeneous aligned lamellar phase (with molecules edge-on TiO$_2$ layer) and the IPCSE drops to about 1% as the molecules are dominantly random after spin-coating. This drop of mobility could result from the anisotropic charge mobility.

A mixture of photoconducting materials can act synergistically and provide improved photoconducting properties compared with single component. Rybak et al. reported the mixture system of a DLC (**14**, hole transport material) and perylene derivative (**15**, electron transport material) with photoconducting copolymer **16** using surface potential decay technique (Figure 11-19) [44]. In pure DLC only very weak photogeneration in the visible range was observed, and in pure perylene derivative only negative mobile charge carriers were photogenerated. However, when these two compounds are mixed together, a synergetic effect of photogener-

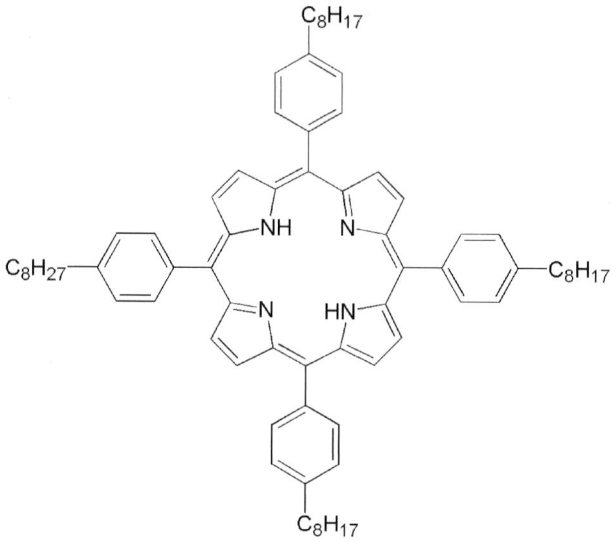

13 (H$_2$TOPP)

Figure 11-18. Molecular structure of photoconducting DLC 5,10,15,20-tetrakis(4-n-octylphenyl) porphyrin

ation of both positive and negative mobile charge carriers was observed. This is because the excitons photogenerated at perylene derivative dissociate into electron–hole pairs, and then electrons are transported via perylene derivative while the holes are transferred onto the DLC molecules.

8.   APPLICATION

Due to many unique properties of photoconducting DLCs, they offer tremendous new possibilities in applications such as organic photovoltaics, organic light emitting diodes, organic photosensors, and organic image receivers. Regardless of the photoresponse in conductivity, the DLCs with reasonable conductivity can also be used for field effect transistor application. Among all the promising application areas, photovoltaic (PV) is of significant importance since it is related to energy generation which is the most important scientific and technological issue facing humanity in the 21st century.

First used in about 1890, the word photovoltaic is a compound word made of two parts: photo, derived from the Greek word for light, and volt, relating to the electricity pioneer Alessandro Volta. Therefore, photovoltaics could literally be translated as light-electricity. A photovoltaic device or solar cell can convert directly the absorbed photons into electrical charges that are used to energize an external circuit. In principle, a PV material is a semiconductor in nature, characterized by its band-gap. Quantum mechanics determined the input energy in the form of a photon

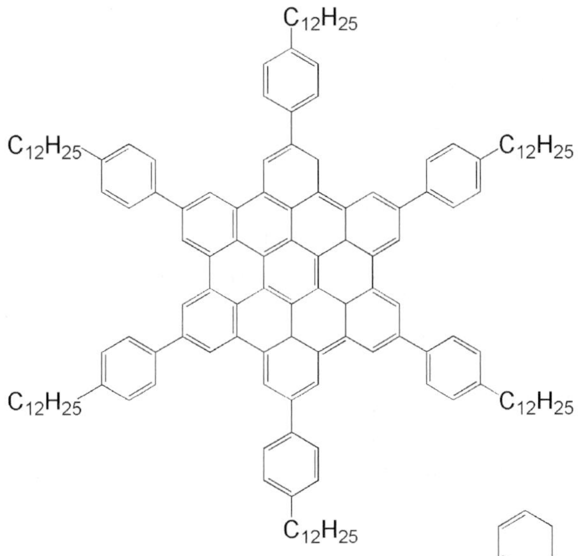

14 (HBC-PhC$_{12}$)

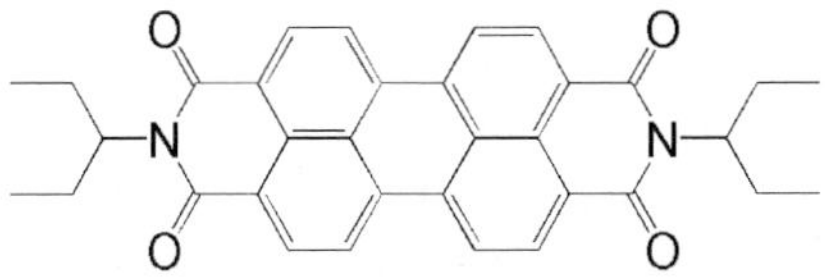

15 (PTCDI)

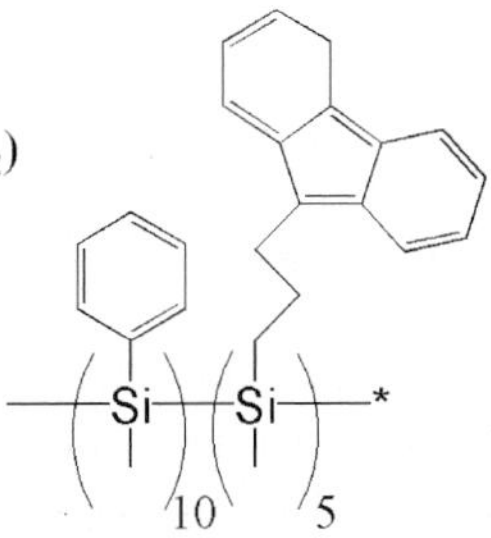

16

Figure 11-19. Molecular structures of **14** (HBC-PhC$_{12}$ as hole transport material), **15** (PTCDI as electron transport material) and Copolymer **16**

needs to have the energy of the band-gap energy to effectively excite the valence band electron. Crystalline silicon has a bandgap energy of 1.1 eV. The bandgap energies of other effective PV semiconductors range from 1.0 to 1.6 eV. The photon energy of light varies according to the different wavelengths of the light. The entire spectrum of sunlight, from infrared to ultraviolet, covers a range of about 0.5 eV to about 2.9 eV. The power conversion efficiency of this process is defined as the ratio of the electric power provided to the external circuit to the solar power incident on the active area of the device and is typically measured under standard simulated solar illumination conditions.

There are many challenges to overcome before organic photovoltaics become technically practical and commercially viable. The excitation efficiency is determined by the intrinsic properties of the material such as HOMO and LUMO. Although theoretical efficiency for organic PV material can be expected to reach a power conversion efficiency as high as crystalline silicon devices, i.e. around $20 \sim 30\%$, the actual efficiency is quite low at this stage with recently reported results around 3% [45, 46]. From the point of view of both material property and device fabrication, band alignment at the interfacial property and microstructures are also an important issue to consider. Moreover, contact resistance needs to be minimized as much as possible. The charge mobility affected by random arrangement of molecules in amorphous semiconducting materials is still far below that in crystalline semiconductors.

The crystalline silicon photovoltaic cells, though efficient, are far too expensive to compete with primary fossil energy. The organic PV technology would hold the promise for the cost reduction since the organic PV materials are potentially cheap, easy to process, and capable of being deposited on flexible substrates and bent where their inorganic competitors, e.g. crystalline silicon, would crack [47]. However, the organic solar cell efficiency based on a currently widely used organic PV material Cu phthalocyanine (CuPc) was reported as about 3% [48]. The major impedance of cell efficiency is that the polycrystalline materials suffer from the scattering of excitons and charge carriers at the grain boundaries of the small crystals. So a challenge for organic PV with the possibility of significant cost reduction is to make them in a desired macroscopic order to improve charge generation and transportation [47, 49, 50]. Thus the photoconducting DLCs capable of efficiently absorbing sunlight as well as being homeotropically aligned will be a desirable candidate to meet the challenge for PV.

Macroscopically ordered microscopic structure can facilitate the hopping process in which the charge carriers are transported in most organic semiconductors, and even make possible a quasi-band transportation in a reduced dimension lattice. This ordered structure can be implemented by columnar mesophases as described earlier. Also, it has already been well established that a liquid crystal, provided with effective alignment method, is capable of forming a large volume single domain (much larger than what could be achieved for a crystal), or large area single domain thin film (the structure of liquid crystal display). Improved charge transport properties should be expected in such systems.

Porphyrins and phthalocyanines are commonly used as organic photoconductors which might be best used for photoconducting devices and related applications due to their intense absorption between 300 and 800 nm and their relatively small HOMO-LUMO gaps (1.5-2.0 eV). Their unique energy levels and big π-conjugated structure make them good candidates as DLCs. These materials can on one hand serve as a high efficiency photovoltaic material determined by the molecular structure, and on the other hand the oriented molecular ordering due to their liquid crystalline property and the ability of self-assembling could retain a large area single domain to suppress charge carrier scattering which has been a major challenge in current OPV devices.

It is worth noting here that the efficiency of the charge-generation process is usually low when a single material forms the organic layer (Figure 11-20). This is because in conjugated materials the binding energy of the lowest singlet exciton (i.e., the strength of the Coulombic attraction between the electron and hole) is significant; this makes excitons (electron and hole) rather stable species. As a result,

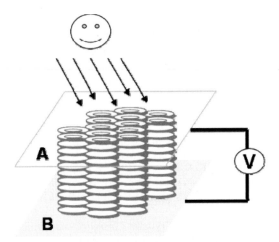

Figure 11-20. Schematic diagram of the prototype homeotropically aligned photovoltaic cell (A: transparent ITO electrode; B: Au electrode or other electrode)

current organic solar cells reply on either blends made from an electron-donor component and an electron-acceptor component or multilayer structures. Porphyrin, which is the basic structure of chlorophyll in nature, is a superior electron donor (p-type semiconducting material). A suitable n-type material for use therewith may be photosentive dye, carbon nanotube, and/or C_{60} (fullerene). For example, C_{60} is an excellent electron acceptor, so semiconducting liquid crystalline porphyrin-C_{60} blends will be a perfect marriage. The discotic liquid crystallineporphyrin absorbs the light and transfers an electron from its excited state to n-type material (e.g. C_{60}, C_{60} derivative or carbon nanotube).

9. OUTLOOK

The past century has seen a boom in electronics and organic chemistry: both of which improved people's lives dramatically. Now the age is coming when electronics and organics will merge and develop a new and better life for people in the near future.

From a molecular point of view, enlargement of conjugated system such as porphyrin systems should be favorable to the charge carrier mobility; however the photoconducting DLCs with large conjugated structure are very difficult to be spontaneously homeotropically aligned on the electrode surface. As a result, a very important research challenge is to focus on the development of the photoconducting DLCs capable of forming homeotropic alignment as well as showing excellent optic-electric properties. New organic PV materials capable of efficient solar energy absorption and new approaches based on engineered nanostructurs can revolutionize the technology used to produce solar electricity [51]; however the availability of such new materials with tailored properties has undoubtedly posed a bottleneck to the OPV technology. A breakthrough of new material development is urgently needed to boost the excellence and prevalence of OPV technology. A single revolutionary application of the fascinating photoconducting DLCs especially on photovoltaics will open a spectacular picture overnight just as the development of cyano biphenyl LCs greatly accelerated the advancement of LC displays over thirty years ago.

10. REFERENCE

1. Chandrasekhar, S.; Sadashiva, B. K.; Suresh, K.A. *Pramana* 1977, **9**, 471–480.
2. Adam, D.; Closs, F.; Frey, T.; Funhoff, D.; Haarer, D.; Ringsdorf, H.; Schuhmacher, P.; Siemensmeyer. Transient photoconductivity in a discotic liquid crystal. *Phys. Rev. Lett.* 1993, **70**, 457–460.
3. Collings, P. J.; Hird, M. In *Introduction to Liquid Crystals: Chemistry and Physics*; Gray, G. W., Goodby, J. W., Fukuda, A., Eds.; Taylor & Francis, Ltd.: London, 1997.
4. Bunning, J. C.; Donovan, K. J.; Bushby, R. J.; Lozman, O. R.; Lu, Z. Electron photo-generation in a triblock co-polymer discotic liquid crystal. *Chem. Phys.* 2005, **312**, 145–150.
5. Warman, J. M.; Schouten, P. G. Charge carrier dynamics in mesomorphic hexakis(pentyloxy)triphenylene. *J. Phys. Chem.* 1995, **99**, 17181–17185.
6. Warman, J. M.; Gelinck, G. H.; de Hass, M. P. The mobility and relaxation kinetics of charge carriers in molecular materials studied by means of pulse-radiolysis time-resolved microwave conductivity: dialkoxy-substituted phenylene-vinylene polymers. *J. Phys.: Condens. Matter* 2002, **14**, 9935–9954.
7. van de Craats, A. M.; Warman, J. M.; Hasebe, H.; Naito, R.; Ohta, K. Charge transport on the mesomorphic free-radical compound bis(octakis(dodecyloxy) phthalo-cyaninato)lutetium(III). *J. Phys. Chem. B* 1997, **101**, 9224-9232.
8. Eichhorn, H. Mesomorphic phthalocyanines, tetraazaporphyrins, porphyrins and triphenylenes as charge-transporting materials. *J. Porph. Phthal.* 2000, **4**, 88–102.
9. Chandrasekhar, S. In *Handbook of Liquid Crystals*, vol. 2B; Demus, D., Goodby, J., Gray, G. W., Spiess, H.-W., Vill, V., Eds.; Wiley-VCH: Weinheim; 1998.
10. Ikeda, S.; Takanishi, Y.; Ishikawa, K.; Takezoe, H. Magnetic field effect on the alignment of a discotic liquid crystal. *Mol. Cryst. Liq. Cryst.* 1999, **329**, 1201-1207.
11. Li, L. *Thesis* 2005.
12. Van Winkle, D. H.; Clark, N. A. Freely suspended strands of tilted columnar liquid crystal phases: One-dimentional nematics with orientational jumps. *Phys. Rev. Lett.* 1982, **48**,1407–1410.
13. Jerome, B. Surface effects and anchoring in liquid crystals. *Rep. Prog. Phys.* 1991, **54**, 391–452.
14. Eichhorn, S. H.; Adavelli, A.; Li, H. S.; Fox, N. Alignment of discotic liquid crystals. *Mol. Cryst. Liq. Cryst.* 2003, **397**, 347–358.
15. Tsukruk, V. V.; Reneker, D. H.; Bengs, H.; Ringsdorf, H. Atomic force microscopy of ordered monolayer films from discotic liquid crystals. *Langmuir* 1993, **9**, 2141–2144.
16. Henderson, P.; Beyer, D.; Jonas, U.; Karthaus, O.; Ringsdorf, H.; Heiney, P. A.; Maliszewskyj, N. C.; Ghosh, S. S.; Mindyuk, O. Y.; Josefowicz, J. Y. Complex

ordering in thin films of di- and trifunctionalized hexaalkoxytriphenylene derivatives. *J. Am. Chem. Soc.* 1997, **119**, 4740–4748.

17. Schönherr, H.; Kremer, F. J. B.; Kumar, S.; Rego, J. A.; Wolf, H.; Ringsdorf, H.; Jaschke, M.; Butt, H.-J.; Bamberg, E. Self-assembled monolayers of discotic liquid crystalline thioethers, discotic disulfides, and thios on gold: Molecular engineering of ordered surfaces. *J. Am. Chem. Soc.* 1996, **118**, 13051–13057.

18. Monobe, H.; Terasawa, N.; Shimizu, Y.; Kiyohara, K.; Heya, M.; Awazu, K. Change of DLC domains by IR irradiation for a plastic columnar mesophase. *Mol. Cryst. Liq. Cryst.* 2005, **437**, 1325–1332.

19. Brandl, B.; Wendorff, J. H. Eutectic mixtures with plastic columnar discotics: molecular structure, phase morphology and kinetics of phase separation. *Liq. Cryst.* 2005, **32**, 553–563.

20. Tracz, A.; Jeszka, J. K.; Watson, M. D.; Pisula, W.; Müllen, K.; Pakula, T. Uniaxial alignment of the columnar super-structure of a hexa (alkyl) hexa-peri-hexabenzocoronene on untreated glass by simple solution processing. *J. Am. Chem. Soc.* 2003, **125**, 1682–1683.

21. Piris, J.; Pisula, W.; Warman, J. M. Anisotropy of the optical absorption and photocon-ductivity of a zone-cast film of a discotic hexabenzocoronene. *Synthetic Metals* 2004, 147, 85–89.

22. Li, Q.; Zhou, X. unpublished results.

23. Kawata, K. Orientation control and fixation of discotic liquid crystal. *Chemical Record* 2002, **2**, 59–80.

24. Disch, S.; Finkelmann, H.; Ringsdorf, H.; Schuhmacher, P. Macroscopically ordered discotic columnar networks. *Macromolecules* 1995, **28**, 2424–2428.

25. Cammidge, A. N.; Bushby, R. J. Synthesis and structural features of discotic liquid crystals. *Handbook of Liquid Crystals* 1998, 2B, 693–748.

26. Cammidge, A. N.; Lozman, O. R. Discotic liquid crystals 25 years on. *Curr. Opin. Coll. Interf. Sci.* 2002, **7**, 343–354.

27. Chandrasekhar, S.; Prasad, S. K. Recent developments in discotic liquid crystals. *Contemp. Phys.* 1999, **40**, 237–245.

28. Kreouzis, T.; Scott, K.; Donovan, K. H.; Boden, N.; Bushby, R. J.; Lozman, O. R.; Liu, Q. Enhanced electronic transport properties in complementary binary discotic liquid crystal systems. *Chem. Phys* 2000, **262**, 489–497.

29. Boden, N.; Bushby, R. J.; Donovan, K. J.; Lozman, O. R.; Lu, Z.; McNeill, A. Movaghar, B.; Donovan, K.; Kreouzis, T. Enhanced conduction in the discotic mesophase. *Mol. Cryst. Liq. Cryst.* 2004, **410**, 541–549.

30. Bengs, H.; Closs, F.; Frey, T.; Funhoff, D.; Ringsdorf, H.; Siemensmeyer, K. Highly photoconductive discotic liquid crystals-structure-property relations in the homologous series of hexa-alkyoxytriphenylenes. *Liq. Cryst.* 1993, **15**, 565–574.

31. Iino, H.; Hanna, J.-I.; Haarer, D. Electronic and ionic carrier transport in discotic liquid crystalline photoconductor. *Phys. Rev.* 2005, **72**, 1932031–1932034.

32. Lino, H.; Takayashiki, Y.; Hanna, J.-I.; Bushby, R. J.; Haarer, D. High electron mobility of 0.1 cm2 V-1 s-1 in the highly ordered columnar phase of hexahexylthiotriphenylene. *App. Phys. Lett.* 2005, **87**, 192105/3.

33. van de Craats, A. M.; Warman, J. M.; Fechtenkötter, A.; Brand, J. D.; Harbison, M. A.; Müllen, K. Record charge carrier mobility in room temperature discotic liquid-crystlline derivative of hexabenzocoronene. *Adv. Mater.* 1999, **11**, 1469–1471.

34. Ohta, K.; Hatshusaka, K.; Sugibayashi, M.; Ariyoshi, M.; Ban, K.; Madeda, F.; Naito, R.; Nishizawa, K.; van de Craats, A. M.; Warman, J. M. Discotic liquid crystalline semiconductors. *Mol. Cryst. Liq. Cryst.* 2003, **397**, 325–345.
35. Ban, K.; Nishizawa, K.; Ohta, K.; van de Craats, A.; Warman, J. M.; Yamamoto, I.; Shirai, H. Discotic liquid crystals of transition metal complexes 29: mesomorphism and charge transport properties of alkylthiosubstituted phthalocyanine rare-earth metal sandwich complexes. *J. Mater. Chem.* 2001, **11**, 321–331.
36. Iino, H.; Hanna, J.-I. Electronic and ionic transports for negative charge carriers in smectic liquid crystalline photoconductor. *J. Phys. Chem. B.* 2005, **109**, 22120–22125.
37. Yoshimoto, N.; Hanna, J.-I. A novel charge transport material fabricated using a liquid crystalline semiconductor and crosslinked polymer. *Adv. Mater.* 2002, **14**, 988-991.
38. Funahashi, M.; Hana, J.-I. Fast hole transport in a new calmitic liquid crystal of 2-(4'-heptyloxyphenyl)-6-dodecylthiobenzothiazole. *Phys. Rev. Lett.* 1997, **78**, 2184-2187.
39. Warman, J. M.; de Haas, M. P.; Dicker, G.; Grozema, F. C. *Chem. Mater.* 2004, **16**, 4600-4609.
40. Inio, H.; Hanna, J.-I.; Bushby, R. J.; Movaghar, B.; Whitaker, B. J.; Cook, M. J. Very high time-of-flight mobility in the columnar phases of a discotic liquid crystal. *Appl. Phys. Lett.* 2005, **87**, 132102-3.
41. Inio, H.; Takayashiki, Y.; Hanna, J.-I.; Bushby, R. J. Fast ambipolar carrier transport and easy homeotropic alignment in a metal-free phthalocyanine derivative. *Jpn. J. Appl. Phys.* 2005, **44**, 1310-1312.
42. Mo, X.; Chen, H.-Z.; Wang, Y.; Shi, M.-M.; Wang, M. Fabrication and photoconductivity study of copper phthalocyanine/perylene composite with bulk heterojunctions obtained by solution blending. *J. Phys. Chem. B* 2005, **109**, 7659–7663
43. Kroeze, J. E.; Koehorst, R. B. M.; Savenije, T. J. Singlet and triplet exciton diffusion in a self-organizing porphyrin antenna layer. *Adv. Funct. Mater.* 2004, **14**, 992–998.
44. Jung, J.; Rybak, A.; Slazak, A.; Bialecki, S.; Miskiewicz, P.; Glowacki, I.; Ulanski, J.; Rosselli, S.; Yasuda, A.; Nelles, G.; Tomovic, Z.; Watson, M. D.; Müllen, K. Photogeneration and photovoltaic effect in blends of derivatives of hexabenzocoronene and perylene. *Synthetic Metals* 2005, **155**, 150–156.
45. Gur, I.; Fromer, N. A.; Geier, M. L.; Alivisatos, A. P. Air-stable all-inorganic nanocrystal solar cells processed from solution. *Science*, 2005, **310**, 462–465.
46. Campos, L.M.; Tontcheva, A; Günes, S.; Sonmez, G.; Neugebauer, H.; Sariciftci, N. S.; Wudlet, F. *Chem. Mater.* 2005, **17**, 4031–4033.
47. Grätzel, M. Photoelectrochemical cells. *Nature* 2001, **414**, 338–344.
48. Peumans:2001 Peumans, P.; Forrest, S. R. Very-high-effiency double-heterostrucutre copper phthalocyanine/C60 photovoltaic cells. *Appl. Phys. Lett.* 2001, **79**, 126–128.
49. Bard, A.J.; Stratmann, M.; Licht S. In *Semiconductor Electrodes and Photoelectrochemistry: Volume of Encyclopaedia of Electrochemistry*; Wiley-VCH: Darmstadt, 2002.
50. Green, M.A. In *Third Generation Photovoltaics: Advance Solar Energy Conversion*; Springer-Verlag: Berlin, 2004.
51. *Basic research needs for solar energy utilization. Report of the basic energy sciences workshop on solar energy utilization.* April 18-21, 2005, Office of Science, US Department of Energy.

INDEX

Anisotropic chemical shift, 86
APAPA, 97–103, 105, 111

Biaxially aligned, 149
Biot number, 292
BLEW, 123
Broadband-PISEMA, 93, 139

Casimir-like force in smectics, 239
^{13}C NMR, 3, 22, 136, 209, 219, 228, 229, 273
Columnar liquid crystal, 76, 117, 119, 121, 125,
    129–131, 132, 134, 136, 302, 303
Conformational chirality, 71
Correlated 2D NMR, 90, 96, 98,
    100, 106
CPMAS, 86, 90, 202
Cross-polarization, CP, 86, 90, 91, 94,
    95, 112, 121
Crossover behaviour, 242

Dark conglomerates, 70
Dielectric anisotropy, 285, 290
Dielectric heating, 278, 282, 283,
    289–291, 293, 294
Dielectric permittivity, 42, 46, 48, 279,
    281–283, 285, 289, 293
Dielectric relaxation, 278, 279, 281–288,
    289–291, 293–295
Dielectric tensor, 71, 278, 286, 290
Dielectric torque, 277, 278, 285, 286,
    288, 293
Director, 103, 118, 120, 124, 127, 143,
    144, 147–149, 151–159, 236–238,
    277–291, 302, 307
Discotic liquid crystal, 310
Dissolved solutes, 239
DQF-COSY, 108
Dual frequency nematic, 285

EBBA, 105–108, 111, 112, 174–177
Electric displacement, 277–280, 293
External fields, 243, 245, 308

Fast MAS, 121
Fast switching, 277, 283
Filaments, 74–78
Flip-flop, 103, 281, 283, 284

Gradient-echo, 174–176

Halperin, Lubensky and Ma mechanism,
    238, 247
Heat conduction, 290, 291
HHTT, 118, 133, 134
HIMSELF, 90, 93, 113
^{2}H NMR, 86
Homochiral, 2, 67, 71, 72

Intercalated structure, 12

Jelly-roll structure, 78

KOBAYASHI-McMillan theory, 237

Laboratory frame, 89, 92, 93, 119, 120
Landau-Peierls instability, 238, 245
Landau Tricritical Point, 241, 243
Larmor, 173, 176
Layer chirality, 2, 12, 13, 68, 71, 73
LG, 123, 124
Liquid crystal mixtures, 244
Local order parameter, 111, 120,
    129, 241

323